Universitext

T0203029

Daniel Perrin

Algebraic Geometry

An Introduction

Translated from the French by
Catriona Maclean

 Springer

EDP
SCIENCES

Professor Daniel Perrin
Département de Mathématiques
Bât. 425
Faculté des Sciences d'Orsay
Université Paris-Sud 11
F-91405 Orsay Cedex
France

Translator:
Catriona Maclean
Institut Fourier
UFR de Mathématiques
UMR 5582 CNRS/Université J. Fourier
100 rue des Maths, B.P. 74
F-38402 St Martin d'Hères Cedex
France

ISBN: 978-1-84800-055-1   e-ISBN: 978-1-84800-056-8
DOI: 10.1007/978-1-84800-056-8

British Library Cataloguing in Publication Data
A catalogue record for this book is available from the British Library

Library of Congress Control Number: 2007935214

Mathematics Subject Classification (2000): 14-01, 14Axx, 14H50, 14M06

EDP Sciences ISBN 978-2-7598-0048-3
Translation from the French language edition:
Géométrie algébrique by Daniel Perrin
Copyright ©1995 EDP Sciences, CNRS Editions, France.
http://www.edpsciences.org/
http://www.cnrseditions.fr/
All Rights Reserved

Printed on acid-free paper

9 8 7 6 5 4 3 2 1

Springer Science+Business Media
springer.com

Contents

Appendices

Preface

This book is built upon a basic second-year masters course given in 1991–1992, 1992–1993 and 1993–1994 at the Université Paris-Sud (Orsay). The course consisted of about 50 hours of classroom time, of which three-quarters were lectures and one-quarter examples classes. It was aimed at students who had no previous experience with algebraic geometry. Of course, in the time available, it was impossible to cover more than a small part of this field. I chose to focus on projective algebraic geometry over an algebraically closed base field, using algebraic methods only.

The basic principles of this course were as follows:

1) Start with easily formulated problems with non-trivial solutions (such as Bézout's theorem on intersections of plane curves and the problem of rational curves). In 1993–1994, the chapter on rational curves was replaced by the chapter on space curves.

2) Use these problems to introduce the fundamental tools of algebraic geometry: dimension, singularities, sheaves, varieties and cohomology. I chose not to explain the scheme-theoretic method other than for finite schemes (which are needed to be able to talk about intersection multiplicities). A short summary is given in an appendix, in which special importance is given to the presence of nilpotent elements.

3) Use as little commutative algebra as possible by quoting without proof (or proving only in special cases) a certain number of theorems whose proof is not necessary in practise. The main theorems used are collected in a summary of results from algebra with references. Some of them are suggested as exercises or problems.

4) Do not hesitate to quote certain algebraic geometry theorems when the proof's absence does not alter the reader's understanding of the result. For example, this is the case for the uniqueness of cohomology or certain technical points in Chapter IX. More generally, in writing this book I tried to privilege understanding of phenomena over technique.

5) Provide a certain number of exercises and problems for every subject
   discussed. The papers of all the exams for this course are given in an
   appendix at the end of the book.

Clearly, a book on this subject cannot pretend to be original. This work
is therefore largely based on existing works, particularly the books by
Hartshorne [H], Fulton [F], Mumford [M] and Shafarevitch [Sh].

I would like to thank Mireille Martin-Deschamps for her careful reading and her remarks. I would also like to thank all those who attended the
course and who pointed out several errors and suggested improvements, notably Abdelkader Belkilani, Nicusor Dan, Leopoldo Kulesz, Vincent Lafforgue
and Thomas Péteul.

I warmly thank Catriona Maclean for her careful translation of this book
into English.

And finally, it is my pleasure to thank Claude Sabbah for having accepted
the French edition of this book in the series *Savoirs Actuels* and for his help
with the editing of the final English edition.

Notation

We denote the set of positive integers (resp. the set of integers, rational numbers, real numbers or complex numbers) by $\mathbf{N}$ (resp. $\mathbf{Z}$, $\mathbf{Q}$, $\mathbf{R}$ or $\mathbf{C}$). We denote by $\mathbf{F}_q$ the finite field with q elements.

We denote the cardinal of a set E by $|E|$. We denote the integral part of a real number by $[x]$. The notation $\binom{n}{p}$ represents the binomial coefficient:

$$\binom{n}{p} = \frac{n!}{p!(n-p)!}.$$

By convention, this coefficient is zero whenever $n < p$.

If $f : G \to H$ is a homomorphism of abelian groups (or modules or vector spaces), we will denote by $\operatorname{Ker} f$ (resp. $\operatorname{Im} f$, resp. $\operatorname{Coker} f$) its kernel (resp. its image, resp. its cokernel). We recall that by definition $\operatorname{Coker} f = H/\operatorname{Im} f$.

An exact sequence of abelian groups (or modules or vector spaces)

$$0 \longrightarrow M' \xrightarrow{\ u\ } M \xrightarrow{\ v\ } M'' \longrightarrow 0$$

is given by the data of two homomorphisms u, v such that

a) u is injective,
b) v is surjective,
c) $\operatorname{Im} u = \operatorname{Ker} v$.

Further definitions and notations are contained in the summary of useful results from algebra.

In the exercises and problems, the symbol ¶ indicates a difficult question.

Introduction

0 Algebraic geometry

Algebraic geometry is the study of algebraic varieties: objects which are the
zero locus of a polynomial or several polynomials. One might argue that
the discipline goes back to Descartes. Many mathematicians—such as Abel,
Riemann, Poincaré, M. Noether, Severi's Italian school, and more recently
Weil, Zariski and Chevalley—have produced brilliant work in this area. The
field was revolutionised in the 1950s and 60s by the work of J.-P. Serre and
especially A. Grothendieck and has since developed considerably. It is now a
fundamental area of study, not just for its own sake but also because of its
links with many other areas of mathematics.

1 Some objects

There are two basic categories of algebraic varieties: affine varieties and pro-
jective varieties. The latter are more interesting but require several definitions.
It it is too early to give such definitions here; we will come back to them in
Chapter II.

To define an affine variety, we take a family of polynomials $P_i \in
k[X_1, \ldots, X_n]$ with coefficients in a field k. The subset V of affine space k^n
defined by the equations $P_1 = \cdots = P_r = 0$ is then an affine algebraic variety.
Here are some examples.

a) If the polynomials P_i are all of degree 1, we get the linear affine subspaces of k^n: lines, planes and so forth.

b) Take $n = 2$, $r = 1$ and $k = \mathbf{R}$, so that k^2 is a real plane and V, defined
by the equation $P(X, Y) = 0$, is a plane "curve". For example, if P is of
degree 2, the curves we get are conics (such as the ellipse $X^2 + Y^2 - 1 = 0$,
the hyperbole $XY - 1 = 0$, or the parabola $Y - X^2 = 0$).

If P is of degree 3, we say that the curve is a cubic. For example, we could consider $Y^2 - X^3 = 0$ (which is a cuspidal curve, *i.e.*, it has a cusp), $X^3 + Y^3 - XY = 0$ (which is a nodal cubic, *i.e.*, it has an ordinary double point or node) or $Y^2 - X(X-1)(X+1) = 0$ (which is a non-singular cubic, also called an elliptic curve, *cf.* below).

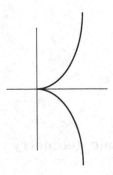

Fig. 1. $X^3 + Y^3 - XY = 0$ **Fig. 2.** $Y^2 - X^3 = 0$

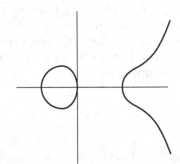

Fig. 3. $Y^2 - X(X-1)(X+1) = 0$

Of course, curves of every degree exist. Let us mention the following two curves in particular: $(X^2+Y^2)^2+3X^2Y-Y^3 = 0$ (the trefoil) and $(X^2+Y^2)^3 - 4X^2Y^2 = 0$ (the quadrifoil).

c) In space k^3 an equation $F(X,Y,Z) = 0$ defines a surface. For example, if F is of degree 2, we get quadric surfaces, such as the sphere $(X^2 + Y^2 + Z^2 - 1 = 0)$ or the one-sheeted hyperboloid $(X^2 + Y^2 - Z^2 - 1 = 0)$.

d) Generally, two equations in k^3 define a space curve. For example, $Y - X^2 = 0$ and $Z - X^3 = 0$ define a space cubic (the set of points of the form (u, u^2, u^3) with $u \in k$).

Fig. 4. $(X^2 + Y^2)^2 + 3X^2Y - Y^3 = 0$ **Fig. 5.** $(X^2 + Y^2)^3 - 4X^2Y^2 = 0$

e) Clearly, the study of algebraic varieties depends heavily on the base field. For example, algebraic geometry over the field of real numbers is sometimes surprising (consider for example the plane "curves" given by the equations $X^2 + Y^2 + 1 = 0$ or $X^2 + Y^2 = 0$). The theory is nicest when k is *algebraically closed* (for example $k = \mathbf{C}$). This is the situation we will deal with in this book. This choice (which is equivalent to giving greater importance to the equations defining varieties than their points) can be partially justified by the fact that any field can be embedded in an algebraically closed field.[1]

Of course, the opposite point of view is just as interesting. It led, for example, to real algebraic geometry ($k = \mathbf{R}$) or arithmetic ($k = \mathbf{Q}$ or even $k = \mathbf{Z}$, or k a finite field). For example, the points of the variety defined over $\mathbf{Z}$ by the equation $X^n + Y^n - Z^n = 0$ are the subject of Fermat's famous theorem. Likewise, the search for rational points on the curve $Y^2 - X(X - 1)(X - \lambda) = 0$ (the arithmetic of elliptic curves) is very much an open question. Two important conjectures concerning these questions have been solved recently, namely Weil's conjecture (solved by Deligne in 1974) and Mordell's conjecture (solved by Faltings in 1982). But this is another story.

f) Furthermore, algebraic varieties appear in many areas of mathematics. One simple example is given by matrices and classical groups. For example, the group

$$\mathrm{SL}_n(k) = \{A \in M_n(k) \mid \det(A) = 1\}$$

is an algebraic variety in the affine space of matrices because the determinant is a polynomial. Likewise, the orthogonal group

$$O_n(k) = \{A \in M_n(k) \mid {}^tAA = I\},$$

or the set of matrices of rank $\leqslant r$, are affine algebraic varieties. The basic concepts of algebraic geometry which we will introduce (such as the dimension

---

[1] When the base field is $\mathbf{R}$ or $\mathbf{C}$, the objects we study also appear in other branches of mathematics (such as topology and differential geometry). It turns out that the difference between these fields depends more on the choice of "good" functions than the objects studied (*cf.* Chapter III).

in Chapter IV and the tangent space in Chapter V) are fundamental tools for the study of these varieties.

g) And finally, let us mention a more complicated example: families of algebraic varieties (for example, the set of lines in k^3 or the set of vector subspaces of dimension d of a space of dimension n) can often themselves be equipped with the structure of an algebraic variety (much as the set of subsets of a set is itself a set) and we can therefore apply algebro-geometric techniques to them.

2 Some problems

One of the principles of this course is to start from easily formulated problems whose solution requires the use of fairly sophisticated algebro-geometric techniques (such as sheaves *cf.* Chapter III and cohomology, *cf.* Chapter VII). Here are two examples: Bézout's problem and the rational curves problem. We will also mention, in Chapter X, the less elementary problem of space curve liaison.

a. Intersection: Bézout's theorem

If we study the intersections of a conic (think of an ellipse) and a line in the plane, we see that they intersect in at most 2 points. A line and a cubic intersect in at most 3 points and two conics intersect in at most 4 points. It is natural to ask whether or not two plane curves C and C', of degrees d and d', always have at most dd' intersection points, and under what circumstances we might get the best possible theorem: that two such curves have exactly dd' intersection points. There are four obvious obstructions to this claim being true.

a) The two curves might have a common component: for example, the curves given by the equations $XY = 0$ and $X(Y - X) = 0$ have the y-axis (*i.e.*, the curve $X = 0$) in common and their intersection is therefore infinite. We will have to assume that the curves C and C' have no common component which means we will have to first specify what we mean by a "component" (*cf.* Chapter I).

b) If k is the real number field, we know that this claim is not always true. For example, the circle $X^2 + Y^2 - 1 = 0$ and the line $X = 2$ do not meet in $\mathbf{R}^2$. On the other hand, they do meet twice in $\mathbf{C}^2$ at the points $(2, \mp i\sqrt{3})$. We will therefore have to assume that the base field is algebraically closed (*cf.* Section 1.e) to get the best possible theorem.

c) Another fundamental counter-example to the best possible theorem is the case of two parallel lines, or a hyperbole and its asymptote, which never meet. Once again, it is clear what we must do to overcome this difficulty: introduce points at infinity. In this context, this means we must work in projective space, not affine space.

d) And finally, returning to the case of a circle and a line, we see there is another case in which the number of intersection points is not two, namely the case where the line is tangent to the circle. For example, the curves $X^2 + Y^2 - 1 = 0$ and $X = 1$ meet at a unique point $(1,0)$. However, if we solve the system formed by these two equations, we obtain the equation $y^2 = 1$, so the solution $y = 0$ is a double root: the point of intersection is a double point and has to be counted twice. Likewise, if we intersect the cubic $Y^2 - X^3 = 0$ with the line $Y = tX$, then we obtain only two points, (t^2, t^3) and $(0,0)$, but the latter is a double point (this happens because the cubic has a singularity at the point in question, *cf.* Chapter V). In short we will have to define carefully what we mean by the intersection multiplicity of two curves at a point. (We suggest the reader try out a trefoil and quadrifoil meeting at $(0,0)$ to convince him or herself that it is not entirely obvious what this definition should be, *cf.* Chapter VI).

Once all these precautions have been taken, we can prove our ideal theorem (*cf.* Chapter VI):

Theorem (Bézout). *Let C and C' be two projective plane curves of degrees d and d', defined over an algebraically closed base field, with no common components. Then the number of intersection points of C and C', counted with multiplicity, is dd'.*

For example, besides the point $(0,0)$, which has multiplicity 14, the trefoil and the quadrifoil meet at four (simple) real points in the plane and two imaginary points at infinity, each of which is counted with multiplicity 3, which does indeed give a total of 24.

b. Parameterisations, rational curves and genus

Let C be a plane curve whose equation is $f(X, Y) = 0$. A rational parameterisation of C is given by two rational fractions $\alpha(T)$ and $\beta(T)$ such that the identity $f(\alpha(T), \beta(T)) = 0$ holds. On calculating the intersection of a cusp cubic $Y^2 - X^3 = 0$ with a line passing through the origin we obtain an example of a rational parameterisation, $x = t^2, y = t^3$, and the basic question is to determine which curves possess such a parameterisation. (These curves are said to be rational.) Here are two reasons (other than the possibility of carrying out effective calculations in the real case) for being interested in such curves.

1) Diophantine equations. These are polynomial equations to which we seek integral solutions. Given a rational parameterisation this problem is easy. As an example, let's solve the equation $x^2 + y^2 - z^2 = 0$ in $\mathbf{Z}$ or, alternatively, $(x/z)^2 + (y/z)^2 - 1 = 0$ in $\mathbf{Q}$, which comes down to the same thing. We have to try to find rational points on the circle $X^2 + Y^2 - 1 = 0$. To do this we parameterise the circle by $\cos u$ and $\sin u$, or, even better, on setting $t = \tan(u/2)$, by

$$x = \frac{1 - t^2}{1 + t^2} \quad , \quad y = \frac{2t}{1 + t^2}.$$

On taking $t \in \mathbf{Q}$, we obtain all the rational points of the circle (since conversely we can calculate t given x and y using the equation $t = y/(1 + x)$). This immediately gives us all the integral solutions of $X^2 + Y^2 = Z^2$, which are given by $x = a^2 - b^2$, $y = 2ab$ and $z = a^2 + b^2$, where $a, b \in \mathbf{Z}$.

2) Calculating integrals. Although this problem is not as important as it once was, it was one of the driving forces behind the development of algebraic geometry in the XIXth century, notably in the work of Abel and Riemann.

Let us consider an integral of the form

$$\int \sqrt{ax^2 + bx + c}\, dx.$$

This integral involves the conic $y = \sqrt{ax^2 + bx + c}$ (*i.e.*, $y^2 = ax^2 + bx + c$). More generally, let $y = \varphi(x)$ be an algebraic function (*i.e.*, a function which, like the above example, contains radicals). We suppose that $y = \varphi(x)$ is the solution to an implicit equation $f(x, y) = 0$, where f is a polynomial. We seek the integral

$$\int g(x, \varphi(x))\, dx,$$

where g is a rational function. (For example, the special case where $\varphi(x) = \sqrt{x(x - 1)(x - \lambda)}$, which arises when calculating the length of the arc of an ellipse, gave rise to the theory of elliptic functions and integrals.)

If the curve $f(x, y) = 0$ has a rational parameterisation $x = \alpha(t)$, $y = \beta(t)$, then this integral can be written in the form

$$\int g(\alpha(t), \beta(t))\, \alpha'(t)\, dt,$$

which we know how to calculate since it is the integral of a rational function.

To be or not to be rational: some examples.

1) Lines are rational curves.

2) So are conics: fix a point m_0, for example $(0, 0)$, on the curve and consider the intersection of the curve with a varying line $y = tx$ passing through this point. The second intersection point can then be rationally parameterised using the parameter t.

3) The same method can be applied to singular cubics. We have seen the case $Y^2 - X^3 = 0$ dealt with above. Likewise, $X^3 + Y^3 - XY = 0$ can be parameterised by

$$x = \frac{t}{1 + t^3} \quad y = \frac{t^2}{1 + t^3}.$$

4) On the other hand, non-singular cubics are not rational (the arithmetic of elliptic curves would otherwise be a much easier problem!). For example, the curves

$$Y^2 = X(X-1)(X+1) \quad \text{or} \quad X^3 + Y^3 - 1 = 0$$

are not rational. More generally, we will show that if the characteristic of the field does not divide n, then the curve $X^n + Y^n - 1 = 0$ with $n \geqslant 3$ is not rational (otherwise solutions of Fermat's equation would exist!).

Assume given a parameterisation $x = p(t)/r(t)$, $y = q(t)/r(t)$ such that $p, q, r \in k[t]$ have no common factor. We would then have $p^n + q^n - r^n = 0$, and hence p, q, r are mutually coprime. On differentiating this equation we get

$$p^{n-1}p' + q^{n-1}q' - r^{n-1}r' = 0.$$

We may suppose that the degree of p is at least as large as the degrees of q and r. After multiplying by r we get

$$p^{n-1}(rp' - pr') = q^{n-1}(qr' - rq'),$$

and since p and q are coprime, p^{n-1} divides $(qr' - rq')$, which, since $n \geqslant 3$, is impossible for degree reasons.

5) Warning: as the example of the curve C whose equation is $y - x^3 = 0$ and which is clearly rational shows, it is not enough to consider singular points in the plane (C has none) but also those at infinity (where C has a cusp).

To solve the rationality of curves problem we will introduce in Chapters VIII and IX an invariant of C, the geometric genus $g(C)$, which is an integer $\geqslant 0$, and we will prove the equivalence

$$C \text{ rational} \iff g(C) = 0.$$

We still need to be able to calculate the genus to check whether or not it vanishes. For a plane curve of degree d which is non-singular (even at infinity) we prove the very simple formula $g = (d-1)(d-2)/2$. We hence prove that if $d \geqslant 3$, such a curve is not rational.

On the other hand, if C has singular points, the genus can be smaller than the value given by the above formula: every ordinary double point (*i.e.*, a double point with distinct tangent vectors) diminishes the genus by 1. More generally, an ordinary multiple point of order r diminishes the genus by $r(r-1)/2$. Hence a curve of degree 4 is rational whenever it has a triple point (for example, the trefoil is rational) or three double points (for example, the curve of equation

$$4(X^2 + Y^2)^2 - 4X(X^2 - 3Y^2) - 27(X^2 + Y^2) + 27 = 0$$

is rational).

If the multiple points are not ordinary (which is the case at turning points, for example), things are more complicated and the genus can be even smaller.

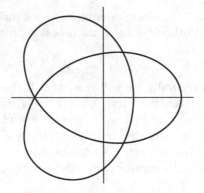

Fig. 6. $4(X^2 + Y^2)^2 - 4X(X^2 - 3Y^2) - 27(X^2 + Y^2) + 27 = 0$

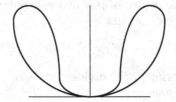

Fig. 7. $(X^2 - Y)^2 + Y^3(Y - 1) = 0$

For example, the curve of degree 4 whose equation is $(X^2 - Y)^2 + Y^3(Y-1) = 0$ is rational even though it has only one double point. In Chapter IX we will give an algorithm allowing us to calculate the genus in all cases.

I

Affine algebraic sets

Throughout this chapter, k is a commutative field.

1 Affine algebraic sets and the Zariski topology

Let n be a positive integer. Consider the space[1] k^n. If $x = (x_1, \ldots, x_n)$ is a point in k^n and $P(X_1, \ldots, X_n)$ is a polynomial, we denote $P(x_1, \ldots, x_n)$ by $P(x)$. The first fundamental objects we encounter are the affine algebraic sets defined below.

Definition 1.1. *Let S be an arbitrary subset of $k[X_1, \ldots, X_n]$. We set*

$$V(S) = \{x \in k^n \mid \forall P \in S, \ P(x) = 0\},$$

i.e., the $x \in V(S)$ are the common zeros of all the polynomials in S. We call $V(S)$ the affine algebraic set defined by S. When the set S is finite, we will often write $V(F_1, \ldots, F_r)$ instead of $V(\{F_1, \ldots, F_r\})$.

Examples 1.2.
1) We have $V(\{1\}) = \varnothing$ and $V(\{0\}) = k^n$, so both the empty set and the whole of k^n are affine algebraic sets.

2) If $n = 1$ and $S \neq \{0\}$, then $V(S)$ is a finite set: the affine algebraic subsets of a line are the line and the finite sets.

3) If $n = 2$, then the affine sets, other than the empty set and the plane, are the "curves" of the form $V(F)$ and the finite sets of points: $V(X, Y) = \{(0,0)\}$, $V(X(X-1), Y) = \{(0,0), (1,0)\}, \ldots$.

---

[1] In fact, what follows is valid in n-dimensional affine k-space, denoted $\mathbf{A}^n(k)$, and is independent of the choice of basis: the affine group action (translations and bijective linear maps) is not a source of any serious problems, *cf.* Example 6.3.2.

Remarks 1.3.

0) The function V is decreasing: if $S \subset S'$, then $V(S') \subset V(S)$.

1) If $S \subset k[X_1, \ldots, X_n]$, then we denote by $\langle S \rangle$ the ideal generated by S: $\langle S \rangle$ is composed of polynomials f of the form $f = \sum_{i=1}^{r} a_i f_i$, where $f_i \in S$ and $a_i \in k[X_1, \ldots, X_n]$. We then have $V(S) = V(\langle S \rangle)$. (The decreasing property implies $V(\langle S \rangle) \subset V(S)$. Conversely, if $x \in V(S)$, then all the $f_i \in S$ vanish on x and hence so do all the $f \in \langle S \rangle$.) As far as affine algebraic sets are concerned we can therefore restrict ourselves to the case where S is an ideal, or, alternatively, to the case where S is the set of generators of an ideal.

2) Since $k[X_1, \ldots, X_n]$ is Noetherian, every ideal is finitely generated: $I = \langle f_1, \ldots, f_r \rangle$, and hence every affine algebraic set is defined by a finite number of equations $V(I) = V(f_1, \ldots, f_r) = V(f_1) \cap \cdots \cap V(f_r)$.

The sets of the form $V(f)$ are called hypersurfaces (strictly speaking, we should reserve this notation for the case where f is non-constant and k is algebraically closed, *cf.* Chapter IV). We have shown above that every affine algebraic set is a finite intersection of hypersurfaces.

3) Note that two polynomials can define the same affine algebraic set. For example, in k^2, $V(X) = V(X^2)$. (Later on, however, we will want to say that $V(X^2)$ is the y-axis counted twice.)

4) A point of k^n is an affine algebraic set:
if $a = (a_1, \ldots, a_n)$, then $\{a\} = V(X_1 - a_1, \ldots, X_n - a_n)$.

5) An arbitrary intersection of affine algebraic sets is an affine algebraic set:

$$\bigcap_j V(S_j) = V \left(\bigcup_j S_j \right).$$

(If we want to restrict ourselves to using ideals, we must replace the union of the sets S_j by their sum.)

6) A finite union of affine algebraic sets is an affine algebraic set. It is enough to prove this for the union of two sets defined by ideals I and J. We now prove that $V(I) \cup V(J) = V(IJ)$. Indeed, $IJ \subset I, J$ (*cf.* Summary 1.2.a) and hence $V(I) \cup V(J) \subset V(IJ)$ by the decreasing property of V. Conversely, suppose that $x \in V(IJ)$ and $x \notin V(I)$. There is then a $P \in I$ such that $P(x) \neq 0$. For all $Q \in J$ we have $PQ \in IJ$, and hence $(PQ)(x) = 0$, so $Q(x) = 0$ and $x \in V(J)$. (The same argument shows that $V(I) \cup V(J) = V(I \cap J)$.)

7) It follows from 4) and 6) that any finite set is an affine algebraic set.

1.4. The Zariski topology. Remarks 5) and 6) above show that the affine algebraic sets are the closed sets of a topology on k^n, which we call the Zariski topology. Of course, any subset X of k^n inherits an induced topology (again called the Zariski topology) whose closed sets are the sets of the form $X \cap V(I)$; in particular, if X is an affine algebraic set, then the closed sets of X are the affine algebraic sets contained in X.

Warning: the Zariski topology is very different from the usual topologies, and the reader will need some time to develop an intuition for its behaviour. Simplifying, the closed sets of this topology are very small: in k^3 the closed sets are all surfaces, curves or points (compare this with the closed balls of usual topologies). On the other hand, the open sets are very large. For example, two non-empty open sets always meet (and hence this topology is not separated). We will encounter another difference when we come to compare the Zariski topology on k^2 and the product of the Zariski topologies on k (*cf.* Problem I).

1.5. Standard open sets. Consider $f \in k[X_1, \ldots, X_n]$ and let $V(f)$ be the hypersurface defined by f. The set $D(f) = k^n - V(f)$ is a Zariski open set of k^n, which we call a *standard* open set. The standard open sets are a basis for this topology (*cf.* Summary 1.8); more precisely, any open set U is a finite union of standard open sets. This is the dual statement to Remark 1.3.2 on intersections of hypersurfaces.

2 Ideal of an affine algebraic set

We introduce an operation I, which is essentially the dual of V, which associates an ideal in the polynomial ring to a set of points.

Definition 2.1. *Let V be a subset of k^n. The set*

$$I(V) = \{f \in k[X_1, \ldots, X_n] \mid \forall x \in V,\ f(x) = 0\}$$

is called the ideal of V.

In other words, $I(V)$ is the set of polynomial functions which vanish on V. To show that it is indeed an ideal, we consider the ring homomorphism

$$r : k[X_1, \ldots, X_n] \longrightarrow \mathcal{F}(V, k),$$

with image in the ring of all k-valued functions on V associating to a polynomial the restriction of the associated polynomial function to V. The kernel of r is $I(V)$ (which is therefore an ideal) and the image of r is the ring $\Gamma(V)$ of polynomial (or regular) functions on V, which is isomorphic to $k[X_1, \ldots, X_n]/I(V)$. This ring, which is a k-algebra of finite type (*cf.* Summary 1.5), is called the *affine algebra of V* and will play a key role throughout the rest of this book.

The guiding philosophy of this book is to associate to the geometric object V an algebraic object $I(V)$ (or $\Gamma(V)$) and set up a dictionary allowing us to translate geometric properties into algebraic properties and vice versa.

Remarks 2.2.

0) The map I is decreasing.

1) If V is an affine algebraic set, then $V(I(V)) = V$. Indeed, it is clear that $V \subset V(I(V))$. Conversely, if $V = V(I)$, then $I \subset I(V)$ and hence $V = V(I) \supset V(I(V))$.

2) It follows that the map $V \mapsto I(V)$ is injective, and hence if $V \subset W$ and $V \neq W$, then there exists a polynomial which vanishes on V and does not vanish on W.

3) Conversely, $I \subset I(V(I))$, but *NB*: equality does not hold in general. There are two basic obstructions to equality.

a) When the field k is not algebraically closed, $V(I)$ can be abnormally small: for example, if $k = \mathbf{R}$ and $I = (X^2 + Y^2 + 1)$, then $V(I) = \varnothing$ (whereas we would have expected to obtain a curve) and hence $I(V(I)) = k[X_1, \ldots, X_n] \neq I$. The same thing happens if $I = (X^2 + Y^2)$.

b) The operation I forgets powers: if $n = 2$ and $I = (X^2)$, then $V(I)$ is the y-axis and $I(V(I)) = (X) \neq I$.

The relationship between I and $I(V(I))$ is fundamental and will be dealt with in Section 4.

2.3. Some examples.

a) We have $I(\varnothing) = k[X_1, \ldots, X_n]$.

b) For $I(k^n)$ we have the following proposition.

Proposition 2.4. *Assume that k is infinite. Then $I(k^n) = 0$.*

In other words, if a polynomial function vanishes everywhere, then the polynomial vanishes.

NB: the proposition is false if k is finite; consider the polynomial $X^q - X$ on $\mathbf{F}_q$.

Proof. We proceed by induction on n. The case $n = 1$ is clear since a non-zero polynomial has only a finite number of roots. At the nth level, if $P \neq 0$ and P is not constant, then we can write, for example,

$$P = a_r(X_1, \ldots, X_{n-1})X_n^r + \cdots,$$

where $r \geqslant 1$ and $a_r \neq 0$. By induction, there exist $(x_1, \ldots, x_{n-1}) \in k^{n-1}$ such that $a_r(x_1, \ldots, x_{n-1}) \neq 0$. Hence the polynomial $P(x_1, \ldots, x_{n-1}, X_n)$ has at most r roots and hence is not zero for every $x_n \in k$.

c) We have $I(\{(a_1, \ldots, a_n)\}) = (X_1 - a_1, \ldots, X_n - a_n)$.

The inclusion $\supset$ is obvious. Conversely, if $P(a_1, \ldots, a_n) = 0$, then we can divide P successively by the terms $X_i - a_i$ (*cf.* Summary 1.1.c) and we write

$$P = (X_1 - a_1)Q_1 + \cdots + (X_n - a_n)Q_n + c,$$

with $c \in k$. But c is simply $P(a_1, \dots, a_n)$, which vanishes, so P is contained in the ideal $(X_1 - a_1, \dots, X_n - a_n)$.

d) Let us assume that k is infinite and calculate the ideal

$$I(V) = I(V(Y^2 - X^3))$$

in $k[X, Y]$. It is clear that $(Y^2 - X^3) \subset I(V)$. Conversely, we know that any point in V can be written as (t^2, t^3), with $t \in k$ (*cf.* Introduction; if $x \neq 0$, we take $t = y/x$ and if $x = 0$, we take $t = 0$). Suppose that $P \in I(V)$. We divide P by $Y^2 - X^3$ with respect to the variable Y (*cf.* Summary 1.1.c):

$$P = (Y^2 - X^3)Q(X, Y) + a(X)Y + b(X).$$

It follows that for any $t \in k$, $P(t^2, t^3) = 0 = a(t^2)t^3 + b(t^2)$. Since k is infinite, we deduce that $a(T^2)T^3 + b(T^2) = 0$ in $k[T]$. Separating the terms of odd and even degrees we get $a = b = 0$ and we have hence proved that $I(V) = (Y^2 - X^3)$.

3 Irreducibility

Let us consider in k^2 the affine algebraic set defined by $XY = 0$. This set is the union of two coordinate axes which are themselves affine algebraic sets and are hence Zariski closed subsets. This is the kind of situation we are now going to deal with. The idea is that in such cases we will essentially be able to restrict ourselves to studying each of the pieces separately.

Proposition-Definition 3.1. *Let X be a* non-empty *topological space. The following are equivalent:*

i) If we can write X in the form $X = F \cup G$, where F and G are closed sets in X, then $X = F$ or $X = G$.

ii) If U, V are two open sets of X and $U \cap V = \varnothing$, then $U = \varnothing$ or $V = \varnothing$.

iii) Any non-empty open set of X is dense in X.

Under these conditions we say that X is irreducible.

These conditions are never satisfied for usual topologies: for example, no separated topological space which is not a point is ever irreducible.

For affine algebraic sets we have a very simple characterisation of irreducible spaces in terms of the ideal $I(V)$. (This is one of the first examples of an algebra-geometry translation.)

Theorem 3.2. *Let V be an affine algebraic set equipped with its Zariski topology. Then, V irreducible $\Leftrightarrow I(V)$ prime $\Leftrightarrow \Gamma(V)$ integral.*

Proof. It will be enough to prove the first equivalence. Assume that V is irreducible and let f, g be such that $fg \in I(V)$. We then have

$$V = V(I(V)) \subset V(f) \cup V(g) \text{ and hence } V = (V(f) \cap V) \cup (V(g) \cap V),$$

and since V is irreducible, we may assume $V(f) \cap V = V$, *i.e.*, $V \subset V(f)$ and $f \in I(V)$.

Conversely, if $I(V)$ is prime and we can write $V = V_1 \cup V_2$ in such a way that V_i is a closed set and $V_i \neq V$, then $I(V) \subset I(V_i)$ and $I(V) \neq I(V_i)$ (*cf.* Remark 2.2.2). We then consider elements $f_i \in I(V_i) - I(V)$, and it follows that $f_1 f_2$ vanishes on V and is hence a member of $I(V)$, which is impossible.

Corollary 3.3. *Assume that k is infinite. Then the affine space k^n is irreducible.*

Proof. We have $I(k^n) = (0)$ (*cf.* Proposition 2.4), and this ideal is prime since $k[X_1, \ldots, X_n]$ is integral.

If k is finite, then the corollary is false because k^n is finite, and therefore k^n is a finite union of points, which are closed sets.

Application 3.4 (Extension of algebraic identities). *Assume that k is infinite. Let V be an affine algebraic set $\neq k^n$ and $P \in k[X_1, \ldots, X_n]$. Assume that P is zero outside of V. Then P is identically zero.*

This is obvious. This theorem enables us to use over an arbitrary field density arguments analogous to those used on $\mathbf{R}$ or $\mathbf{C}$ with the usual topology. A classical application is proving that a certain identity which holds for invertible square matrices holds in fact for all matrices (since the determinant is a polynomial function, the set of non-invertible matrices is a closed set). For example, if we denote the coefficient of X^i in the characteristic polynomial $\det(I - XA)$ by $c_i(A)$, then $c_i(AB) = c_i(BA)$. (We deal first with the case where B is invertible, using the relation $AB = B^{-1}(BA)B$.)

The following proposition will be useful in Chapter IV.

Proposition 3.5. *Let X be a topological space and let Y be a subspace of X. Then if Y is irreducible, so is its closure $\overline{Y}$. If U is an open set of X, then the maps $Y \mapsto \overline{Y}$ and $Z \mapsto Z \cap U$ are mutually inverse bijections between the irreducible closed sets Y in U and the irreducible closed sets Z in X which meet U.*

Proof. If $\overline{Y} = F_1 \cup F_2$, where F_i is a closed set of $\overline{Y}$ and is hence a closed set of X, then $Y = (F_1 \cap Y) \cup (F_2 \cap Y)$, and hence since Y is irreducible, $Y = F_i \cap Y$, or, alternatively, $Y \subset F_i$. But we then have $\overline{Y} \subset F_i$, and hence $\overline{Y} = F_i$.

We now show how to reduce to the irreducible case.

Theorem-Definition 3.6. *Let V be a non-empty affine algebraic set. We can write V uniquely (up to permutation) in the form $V = V_1 \cup \cdots \cup V_r$, where the sets V_i are irreducible affine algebraic sets and $V_i \not\subset V_j$ for $i \neq j$. The sets V_i are called the irreducible components of V.*

Proof.

1) Existence. We proceed by contradiction. We assume there exist non-decomposable affine algebraic sets and we pick one whose ideal is maximal amongst all such sets. (Such a V exists since the ring $k[X_1, \ldots, X_n]$ is Noetherian.) Since V is not irreducible, we can write $V = F \cup G$, where $F, G \neq V$. It follows by injectivity of I that $I(F), I(G) \supset I(V)$ and $I(F), I(G) \neq I(V)$. By maximality of $I(V)$, it follows that F and G are decomposable: $F = F_1 \cup \cdots \cup F_r$, $G = G_1 \cup \cdots \cup G_s$, but V is then decomposable, which gives us a contradiction.

2) Uniqueness. Assume given two expressions: $V = V_1 \cup \cdots \cup V_r = W_1 \cup \cdots \cup W_s$. We set $V_i = V \cap V_i = (W_1 \cap V_i) \cup \cdots \cup (W_s \cap V_i)$. Since V_i is irreducible, there is a j such that $V_i = W_j \cap V_i$, i.e., $V_i \subset W_j$. Likewise, there is a k such that $W_j \subset V_k$, and hence $V_i \subset V_k$, which implies by hypothesis that $i = k$ and hence $V_i = W_j$.

Remark 3.7. If W is an irreducible closed set of V, then W is contained in an irreducible component. It follows that the irreducible components are exactly the maximal closed irreducible subsets of V.

See the exercises for some examples of decompositions.

4 The Nullstellensatz (or Hilbert's zeros theorem)

This is one of the first fundamental theorems of algebraic geometry. It controls the correspondence between affine algebraic sets and ideals; in particular, it enables us to calculate $I(V(I))$. We have already seen (Remark 2.2.3.a) that certain problems arise when k is not algebraically closed. We therefore assume henceforth that:

k **is algebraically closed.**

This hypothesis enables us to avoid the case where the affine algebraic sets are too small. For example, the reader can easily check (without using the Nullstellensatz!) that if $F \in k[X_1, \ldots, X_n]$ is non-constant, the hypersurface $V(F)$ is infinite (if $n \geqslant 2$). The following result is similar.

Theorem 4.1 (Weak Nullstellensatz). *Let $I \subset k[X_1, \ldots, X_n]$ be an ideal different from $k[X_1, \ldots, X_n]$. Then $V(I)$ is non-empty.*

Proof. The proof which follows is valid whenever k is not countable ($k = \mathbf{C}$ for example). For a proof in the general case, *cf.* Problem III, 4.

Embedding I in a maximal ideal if necessary, we may assume that I is maximal. Let $K = k[X_1, \ldots, X_n]/I$ be the residue field. Since $k[X_1, \ldots, X_n]$ is a vector space of at most countable dimension over k, the same is true of K. We then have the following lemma.

Lemma 4.2. *Let k be an uncountable algebraically closed field and let K be an extension of k whose dimension is at most countable. Then $K = k$.*

Proof (of 4.2). It will be enough to show that K is algebraic over k. Otherwise, K would contain a transcendental element, and K would therefore contain a subfield isomorphic to the field of fractions $k(T)$. But this field contains an uncountable family $1/(T-a)$, $a \in k$, and this family is free: given an identity

$$\sum_{i=1}^{n} \frac{\lambda_i}{T - a_i} = 0,$$

we multiply by $T - a_i$ and on setting $T = a_i$ we get $\lambda_i = 0$.

Returning to 4.1, we consider the images $a_1, \ldots, a_n$ of the variables X_i in $K = k$. If $P(X_1, \ldots, X_n) \in I$, then $P(a_1, \ldots, a_n) = 0$, or in other words, the point $(a_1, \ldots, a_n) \in k^n$ is in $V(I)$. QED.

In order to state the Nullstellensatz, we introduce the *radical* of an ideal I in A, which is the ideal

$$\mathrm{rac}(I) = \{x \in A \mid \exists r \in \mathbf{N},\ x^r \in I\}.$$

Theorem 4.3 (Nullstellensatz). *Let I be an ideal of $k[X_1, \ldots, X_n]$. Then $I(V(I)) = \mathrm{rac}(I)$.*

Proof. We set

$$R = k[X_1, \ldots, X_n], \quad I = (P_1, \ldots, P_r) \quad \text{and} \quad V = V(I).$$

It is clear that $\mathrm{rac}(I) \subset I(V(I))$. To prove the converse, consider $F \in I(V)$. We want to show that $F^m \in I$ for large enough m. This can be easily translated in terms of the local ring R_F obtained on inverting F (*cf.* Summary 1.6.b). Indeed, it will be enough to show that the ideal IR_F generated by I in R_F is equal to $(1) = R_F$ since we would then have

$$1 = \sum_i \frac{P_i Q_i}{F^m},$$

and on clearing the denominators we get $F^m \in I$.

But the ring R_F is also isomorphic to $k[X_1, \ldots, X_n, T]/(1 - TF)$ (*cf.* Summary loc. cit.), and hence the condition that $IR_F = (1)$ means that 1 can be written in the form

$$1 = \sum_i P_i Q_i + A(1 - TF),$$

where $A, Q_i \in k[X_1, \ldots, X_n, T]$. Let J be the ideal $(P_1, \ldots, P_r, 1 - TF)$ in $k[X_1, \ldots, X_n, T]$. We have $V(J) = \varnothing$ in k^{n+1}, since if $(x_1, \ldots, x_n, t)$ were in $V(J)$, then the polynomials P_i would vanish at the point $x = (x_1, \ldots, x_n)$ and hence x would be in V, hence F would vanish at x and $1 - TF$ could not vanish there. It follows by the weak Nullstellensatz that $J = (1)$, and the result follows.

Example 4.4. The phenomena involving powers seen in Remark 2.2.3.b arise once more in this context. For example, if $I = (X, Y^2)$, then $I(V(I)) = (X, Y)$.

Remark 4.5. It is clear that the ideal $I(V)$ is equal to its radical (in this case we say that $I(V)$ is *radical*) and hence $I(V(I)) = I$ if and only if I is a radical ideal (in particular, this is the case if I is a prime ideal). The ideal $I(V)$ is radical if and only if the ring $\Gamma(V)$ is reduced (*i.e.*, has no nilpotent elements, *cf.* Summary 1.2.d). We will alter this condition later on when we want to talk about multiple structures.

4.6. Applications of the Nullstellensatz: an algebra-geometry dictionary. Let V be an affine algebraic set. We associate to V its ideal $I(V)$ and its affine algebra $\Gamma(V) \simeq k[X_1, \dots, X_n]/I(V)$, which is a reduced k-algebra of finite type.

a) The case $V = k^n$. The following proposition is an immediate consequence of the Nullstellensatz and 3.2.

Proposition 4.7. *There is a decreasing bijection $W \mapsto I(W)$, whose inverse is $I \mapsto V(I)$, between affine algebraic sets in k^n and radical ideals in $k[X_1, \dots, X_n]$. Moreover, the following properties are equivalent:*
   a) W irreducible $\Leftrightarrow I(W)$ prime $\Leftrightarrow \Gamma(W)$ integral,
   b) W is a point $\Leftrightarrow I(W)$ maximal $\Leftrightarrow \Gamma(W) = k$.

(To prove b) we use the weak Nullstellensatz and the decreasing property of the maps I and V.)

Another example of such a translation is the following:

Proposition 4.8. *The following are equivalent: V is finite $\Leftrightarrow \Gamma(V)$ is a finite-dimensional k-vector space. (We then say that $\Gamma(V)$ is a finite k-algebra, cf. Summary 1.7.)*

Proof.
   1) Assume that V is finite, $V = \{u_1, \dots, u_r\}$, and consider the ring morphism
$$\varphi : k[X_1, \dots, X_n] \longrightarrow k^r,$$
which associates $(F(u_1), \dots, F(u_r))$ to F. (We equip k^r with its product ring structure, *cf.* Summary 1.1.d.) The kernel of f is simply $I(V)$, so $\Gamma(V)$ can be embedded in k^r and is hence finite dimensional.

   2) Conversely, assume that $\Gamma(V)$ is finite dimensional. Let $\overline{X}_i$ be the image of X_i in $\Gamma(V)$. The elements $1, \overline{X}_i, \dots, \overline{X}_i^s, \cdots$ are not independent, and hence in $\Gamma(V)$ there is an identity
$$a_s \overline{X}_i^s + \cdots + a_1 \overline{X}_i + a_0 = 0,$$
such that $a_j \in k$ and $a_s \neq 0$. If $u = (x_1, \dots, x_n)$ is an arbitrary point of V, it follows that we also have
$$a_s x_i^s + \cdots + a_1 x_i + a_0 = 0$$
and hence there are only a finite number of possible values for the ith coordinate of u and hence also for u: V is finite.

b) The general case. Let V be an arbitrary affine algebraic set. If W is an algebraic affine set contained in V, then $I(V) \subset I(W)$ and $I(W)$ determines an ideal $I_V(W)$ of the ring $\Gamma(V)$ (namely its image, *cf.* Summary 1.2.c), which is simply the set of $f \in \Gamma(V)$ which vanish on W. We have an isomorphism $\Gamma(V)/I_V(W) \simeq \Gamma(W)$, from which it follows that this ideal is also radical. We note that if I is an ideal of $\Gamma(V)$, then we can define $V(I)$ either as the set of zeros of functions of I on V:

$$V(I) = \{x \in V \mid \forall f \in I, \ f(x) = 0\}$$

or, which amounts to the same thing, by setting $V(I) = V(r^{-1}(I))$, where r is the canonical projection of $k[X_1, \ldots, X_n]$ onto $\Gamma(V)$. We then have the following proposition.

Proposition 4.9. *There are mutually inverse decreasing bijections $W \mapsto I_V(W)$ and $I \mapsto V(I)$ between affine algebraic subsets contained in V and radical ideals of $\Gamma(V)$. Moreover, we have the following equivalences:*
    a) W irreducible $\Leftrightarrow I_V(W)$ prime $\Leftrightarrow \Gamma(W)$ integral,
    b) W is a point $\Leftrightarrow I_V(W)$ maximal $\Leftrightarrow \Gamma(W) = k$,
    c) W is an irreducible component of $V \Leftrightarrow I_V(W)$ is a minimal prime ideal of $\Gamma(V)$.

Proof. The existence of these bijections is obvious, as is a) (it is enough to note that I is a radical ideal of $\Gamma(V)$ if and only if $r^{-1}(I)$ is a radical ideal of $k[X_1, \ldots, X_n]$). The claim c) is simply the dual of the proposition stating that irreducible components are maximal irreducible subsets (*cf.* 3.6). We note that there are therefore a finite number of minimal prime ideals (*cf.* Summary 4.3).

For b) we note that to any $x \in V$ there corresponds a homomorphism of k-algebras $\chi_x : \Gamma(V) \to k$ which associates the quantity $f(x)$ to f and whose kernel is the maximal ideal

$$m_x = I(\{x\}) = \{f \in \Gamma(V) \mid f(x) = 0\}.$$

The k-algebra homomorphisms $\chi : \Gamma(V) \to k$ are also called the *characters* of $\Gamma(V)$ and they are also in bijective correspondence with the points of V (conversely, we associate to a character χ the point $(\chi(X_1), \ldots, \chi(X_n))$ and we check that this is in V). We have hence proved the following result.

Proposition 4.10. *The points of V are in bijective correspondence with the maximal ideals of $\Gamma(V)$, or, alternatively, with characters of $\Gamma(V)$.*

Examples 4.11.
    a) Consider $V = V(XY) \subset k^2$. Check that the minimal prime ideals of $\Gamma(V)$ are the images of the ideals (X) and (Y) corresponding to the two components of V.

    b) More generally, for any hypersurface the following proposition holds. The proof is left to the reader as an exercise. (*cf.* also Exercise I, 3):

Proposition 4.12. *Consider $F \in k[X_1, \ldots, X_n]$, $F = F_1^{\alpha_1} \cdots F_r^{\alpha_r}$, where the F_i are irreducible and non-associated and $\alpha_i > 0$. We then have:*

1) $I(V(F)) = (F_1 \cdots F_r)$. In particular, if F is irreducible, then $I(V(F)) = (F)$.

2) The decomposition of $V(F)$ into irreducible components is given by $V(F) = V(F_1) \cup \cdots \cup V(F_r)$. In particular, if F is irreducible, then $V(F)$ is as well.

An arbitrary algebraic set V possesses, as in 1.5, standard open sets which form a basis for its Zariski topology.

Proposition-Definition 4.13. *Let V be an affine algebraic set and let $f \in \Gamma(V)$ be non-zero. The set*

$$D_V(f) = V - V(f) = \{x \in V \mid f(x) \neq 0\}$$

(which we denote by $D(f)$ when there is no risk of confusion) is called a standard open set of V. Every open set in V is a finite union of standard open sets.

5 A first step towards Bézout's theorem

We will now show that the intersection of two plane curves without common components is finite. In this section, k is an arbitrary commutative field.

Theorem 5.1. *Let $F, G \in k[X, Y]$ be non-zero polynomials without common factors. Then $V(F) \cap V(G)$ is finite.*

The proof will also give us the following result, which is interesting to compare with 4.8:

Theorem 5.2. *Under the hypotheses of 5.1 the ring $k[X, Y]/(F, G)$ is a finite-dimensional k-vector space.*

Proof. We start by proving the following lemma.

Lemma 5.3. *Under the hypotheses of 5.1 there is a non-zero polynomial $d \in k[X]$ and polynomials $A, B \in k[X, Y]$ such that $d = AF + BG$. (In other words $d \in (F, G)$.)*

Proof (of 5.3). We leave the details of the proof to the reader: we simply apply Bézout's (elementary) theorem to the principal ring $k(X)[Y]$ and clear denominators.

We can now prove 5.1. If $(x, y) \in V(F) \cap V(G)$, then by 5.3 $d(x) = 0$ and hence there are a finite number of possible values x. The same reasoning applied to y shows that the intersection is finite.

To prove 5.2 we use a similar argument, applied to the images of the monomials $X^i Y^j$ in the quotient ring: by 5.3 we see that a finite number of these monomials generate $k[X, Y]/(F, G)$.

Remark 5.4. An example of a polynomial which can be used as the $d(X)$ in Lemma 5.3 is the resultant of F and G, considered as polynomials in Y. If F and G are of degrees p and q, we can show that the degree of the resultant is $\leqslant pq$ and deduce (using a clever trick) that $|V(F) \cap V(G)| \leqslant pq$, which is part of Bézout's theorem.

6 An introduction to morphisms

In this section we will assume that the field k is infinite (which is the case if, for example, k is algebraically closed).

We have defined our objects: affine algebraic sets. However, it is the morphisms rather than the objects which will determine the behaviour of our theory. Indeed, it is by now an established principle of mathematics that the same objects (for example, affine algebraic sets such that $k = \mathbf{C}$) can give rise to radically differing theories depending on the maps allowed between them. We may, for example, consider continuous maps, real differentiable maps, analytic maps or polynomial maps between objects. We then get topology, differential geometry, analytic geometry or algebraic geometry. Of course, in the case at hand we will use polynomial maps. More precisely:

Definition 6.1. *Let $V \subset k^n$ and $W \subset k^m$ be two affine algebraic sets and let $\varphi : V \to W$ be a map which we can write in the form $\varphi = (\varphi_1, \ldots, \varphi_m)$, where $\varphi_i : V \to k$. We say that f is* regular *(or a* morphism*) if its components f_i are polynomial (i.e., are elements of $\Gamma(V)$). We denote the set of regular maps from V to W by* $\mathrm{Reg}(V, W)$.

Remark 6.2. It is clear that we obtain in this way a *category*: the identity is a morphism, as is the composition of two morphisms. All the usual notions—isomorphisms, automorphisms, and so forth—therefore apply. We note that morphisms are continuous maps for the Zariski topology (which is to say that the preimage of an algebraic set under a morphism is again an algebraic set), but the converse is false (for example, any bijective map from k to k is continuous for the Zariski topology but is not necessarily polynomial).

Examples 6.3.

1) The elements of $\Gamma(V)$, particularly the coordinate functions, are morphisms from V to k.

2) The bijective affine maps from k^n to itself are isomorphisms: they correspond to polynomials of degree 1.

3) Consider $V \subset k^n$. The projection f from V to k^p, $p \leqslant n$, given by $\varphi(x_1, \ldots, x_n) = (x_{i_1}, \ldots, x_{i_p})$, is a morphism.

4) Let V be the parabola $V(Y - X^2)$ and let f be the projection $\varphi : V \to k$ given by $\varphi(x, y) = x$. Then f is an isomorphism, whose inverse is given by $x \mapsto (x, x^2)$.

5) The map $\varphi : k \to V(X^3 + Y^2 - X^2)$, given by the parameterisation $x = t^2 - 1$, $y = t(t^2 - 1)$ (obtained by intersecting with the line $Y = tX$), is a morphism but not an isomorphism (φ is not injective).

6) The map $\varphi : k \to V(Y^2 - X^3)$ given by the parameterisation $t \mapsto (t^2, t^3)$ is a bijective morphism, but we will see further on that it is not an isomorphism.

We have associated to an affine algebraic set V its affine algebra $\Gamma(V)$ and started to set up a dictionary allowing us to pass from one to the other. Of course, we will have to extend this correspondence to morphisms: in other words, we must show that it is *functorial*. This is done in the following trivial proposition.

Proposition-Definition 6.4. *Let* $\varphi : V \to W$ *be a morphism. For any* $f \in \Gamma(W)$, *we set* $\varphi^*(f) = f \circ \varphi$. *Then* φ^* *is a morphism of k-algebras,* $\varphi^* : \Gamma(W) \to \Gamma(V)$.

Remarks 6.5.

1) We now have a *contravariant functor*, which we denote by Γ, from the category of affine algebraic sets with regular maps to the category of k-algebras with k-algebra morphisms which associates $(\Gamma(V), \varphi^*)$ to (V, φ). (The word contravariant means that the direction of arrows is reversed, and functoriality means that $(g \circ f)^* = f^* \circ g^*$ and the identity is sent to the identity.)

2) We can calculate φ^* in the following way: let $V \subset k^n$ and $W \subset k^m$ be two affine algebraic sets and let $\varphi : V \to W$ be a morphism, written in the form $\varphi = (\varphi_1, \ldots, \varphi_m)$, where $\varphi_i \in \Gamma(V)$. We denote by η_i the ith coordinate function on W, which is the image of the variable Y_i in $\Gamma(W)$. Then $\varphi^*(\eta_i) = \varphi_i$. If the functions φ_i are restriction to V of polynomials $P_i(X_1, \ldots, X_n)$, then the homomorphism

$$\varphi^* : k[Y_1, \ldots, Y_m]/I(W) \longrightarrow k[X_1, \ldots, X_n]/I(V)$$

is given by $Y_i \mapsto \overline{P}_i(X_1, \ldots, X_n)$.

3) If $\varphi(x) = y$, then it is easily checked that with the notation of 4.9 $(\varphi^*)^{-1}(m_x) = m_y$.

Examples 6.6.

1) If φ is the projection $\varphi : V(F) \subset k^2 \to k$, where $\varphi(x, y) = x$, then φ^* is the map from $\Gamma(k) = k[X]$ to $k[X, Y]/(F)$ which associates X to X.

2) Consider the parameterisation of $V(Y^2 - X^3)$ by t^2, t^3. We have

$$\varphi^* : k[X, Y]/(Y^2 - X^3) \longrightarrow k[T],$$

which is given by $\varphi^*(\overline{X}) = T^2$ and $\varphi^*(\overline{Y}) = T^3$.

We will now study the properties of the functor Γ. Its behaviour on morphisms is as good as we could have hoped.

Proposition 6.7. *The functor* Γ *is fully faithful. In other words the map* $\gamma : \varphi \mapsto \varphi^*$ *from* $\mathrm{Reg}(V, W)$ *to* $\mathrm{Hom}_{k\text{-alg}}(\Gamma(W), \Gamma(V))$ *is bijective.*

Proof. We assume $V \subset k^n$ and $W \subset k^m$. We denote the coordinate functions on W by η_i (*cf.* 6.5.2).

1) Γ is faithful, *i.e.*, if φ and ψ are two morphisms from V to W such that $\varphi^* = \psi^*$, then $\varphi = \psi$ (injectivity of γ). Indeed, this follows from the formula which gives the components of φ: $\varphi_i = \varphi^*(\eta_i)$ (*cf.* 6.5.2) and the analogous formula for ψ.

2) Now let $\theta : \Gamma(W) \to \Gamma(V)$ be a homomorphism of k-algebras. We set $\varphi_i = \theta(\eta_i) \in \Gamma(V)$. We consider the map $\varphi : V \to k^m$ whose coordinates are the elements φ_i. If we can show that the image of φ is contained in W, then we will have (*cf.* 6.5.2) that $\theta = \varphi^*$, which would establish the surjectivity of γ. Consider $F(Y_1, \ldots, Y_m) \in I(W)$ and $x \in V$. We calculate $F(\varphi(x)) = F(\theta(\eta_1), \ldots, \theta(\eta_m))(x)$. Since θ is a morphism of algebras, $F(\theta(\eta_1), \ldots, \theta(\eta_m)) = \theta(F(\eta_1, \ldots, \eta_m))$, and since $F(\eta_1, \ldots, \eta_m)$ is the image in $\Gamma(W)$ of $F(Y_1, \ldots, Y_m) \in I(W)$, it vanishes and we are done.

Corollary 6.8. *Let $\varphi : V \to W$ be a morphism. Then φ is an isomorphism if and only if φ^* is an isomorphism. It follows that V and W are isomorphic if and only if their algebras $\Gamma(V)$ and $\Gamma(W)$ are isomorphic.*

Application 6.9. *The morphism $\varphi : k \to V = V(Y^2 - X^3)$ given by $\varphi(t) = (t^2, t^3)$ is not an isomorphism.*

Indeed, if f were an isomorphism, then the image of $\Gamma(V)$ under φ^* would be the whole ring $\Gamma(k) = k[T]$. But the image of φ^* is the subring $k[T^2, T^3]$ of $k[T]$, which is strictly smaller.

In fact, the two curves are not isomorphic because their rings are not isomorphic. The homomorphism φ^* is injective (*cf.* 2.4.d or 6.11 below), so $\Gamma(V)$ is isomorphic to $k[T^2, T^3]$. The element T is contained in the field of fractions of $\Gamma(V)$ (it is T^3/T^2), and it is integral over $\Gamma(V)$ (since it satisfies the equation $X^2 - T^2 = 0$), but it is not in $\Gamma(V)$, and hence this ring is not integrally closed. On the other hand, the ring $k[T]$ is principal, and hence is integrally closed. This phenomenon is caused by the singular point of the curve V (*cf.* Chapter V).

There is also a dictionary for morphisms; here is an example. We start with a definition.

Definition 6.10. *Let $\varphi : V \to W$ be a morphism. We say that φ is dominant if the closure of its image (in the Zariski topology) is equal to the whole of W, $\overline{\varphi(V)} = W$.*

We then have the following proposition.

Proposition 6.11. *Let $\varphi : V \to W$ be a morphism.*
1) φ dominant $\Leftrightarrow$ φ^* injective.
2) Assume that φ is dominant and V is irreducible. Then W is irreducible.

Proof.

1) If φ is dominant and $f \in \operatorname{Ker} \varphi^*$, then $f\varphi = 0$ and hence f vanishes on $\varphi(V)$ and, since f is continuous, f vanishes everywhere. Conversely, set $X = \overline{\varphi(V)}$. This is an affine algebraic set contained in W. Assume $X \neq W$. Then (*cf.* 2.2.2) there exists a non-zero $f \in \Gamma(W)$ which vanishes on X. But then $f\varphi = \varphi^*(f) = 0$, which is a contradiction.

2) This follows from 1) and 3.2. (We can also argue directly assuming W to be of the form $F \cup G$.)

Remark 6.12. We note that the conditions are dual to each other because of the contravariance of Γ. Be careful, however: φ^* injective does not imply φ surjective (consider the projection of the hyperbole $XY = 1$ on the x-axis).

We finish this section by showing that when the field k is algebraically closed, the situation is as good as it could be.

Theorem 6.13. *Assume that k is algebraically closed. The functor Γ is then an equivalence of categories between the category of affine algebraic sets with regular maps and the category of reduced k-algebras of finite type with homomorphisms of k-algebras.*

(This means that the functor is fully faithful (*cf.* Proposition 6.7) and, moreover, it is essentially surjective: if A is a reduced k-algebra of finite type, then there exists a V such that A is isomorphic to $\Gamma(V)$.)

Proof. We have already proved full faithfulness in Proposition 6.7. To prove surjectivity, consider A a reduced k-algebra of finite type. Since A is of finite type, we can write $A \simeq k[X_1, \ldots, X_n]/I$ (*cf.* Summary 1.5), and since A is reduced, the ideal I is radical. We set $V = V(I)$. We have $I(V) = \operatorname{rac}(I) = I$ by the Nullstellensatz, and hence $A \simeq \Gamma(V)$.

Remark 6.14. This theorem is the culmination of the programme of translation between geometry and algebra undertaken in this chapter. In the affine setting this translation is more or less optimal, but in projective geometry we will have to use functions defined only on open sets.

In the following chapters we will need the notion of a rational function on V, which we will re-examine in detail in Chapters VIII and IX.

Definition 6.15. *Let V be an irreducible affine algebraic set, so the ring $\Gamma(V)$ is integral. The field of fractions of $\Gamma(V)$ is called the field of rational functions on V and is denoted $K(V)$.*

Remark 6.16. If $f \in K(V)$, then f can be written in the form $f = g/h$, where $g, h \in \Gamma(V)$ and $h \neq 0$. We can therefore consider f to be a function defined on the standard open set $D(h)$ defined by $h(x) \neq 0$.

Exercises

1) Is the set $\{(t, \sin t) \mid t \in \mathbf{R}\}$ algebraic?

2) Let V be an affine algebraic set, $V \subset k^n$, and consider $x \notin V$. Show that there is an $F \in k[X_1, \ldots, X_n]$ such that $F(x) = 1$ and $F|_V = 0$.

3) Let $F \in k[X, Y]$ be an irreducible polynomial. Assume that $V(F)$ is infinite. Prove that $I(V(F)) = (F)$.
   Application: let F be of the form $F_1^{\alpha_1} \cdots F_r^{\alpha_r}$, where the polynomials F_i are irreducible and the sets $V(F_i)$ are infinite. Find the irreducible components of $V(F)$.

4) **Some results on irreducibility.** (All spaces in this question are arbitrary topological spaces.)
   a) If X is irreducible and U is an open subset of X, show that U is irreducible.
   b) If X is of the form $U_1 \cup U_2$, where the sets U_i are open and irreducible, and $U_1 \cap U_2 \neq \varnothing$, show that X is irreducible.
   c) If $Y \subset X$ and Y is irreducible, show that $\overline{Y}$ is irreducible.

5) A ring A is said to be *connected* if every idempotent in A is trivial (*i.e.*, if every element e in A such that $e^2 = e$ is equal to 0 or 1).
   a) Prove that every integral domain is connected.
   b) If A is the direct product of two non-trivial rings, prove that A is not connected.
   c) Conversely, if A possesses a non-trivial idempotent e, prove that $A \simeq A/(e) \times A/(1 - e)$.
   d) Let V be an affine algebraic set over an algebraically closed field k. Prove that V is connected (in the Zariski topology) if and only if $\Gamma(V)$ is connected. (If V has two connected components, start by finding a function which is 0 on one and 1 on the other.) Is this still the case if k is not algebraically closed?

6) Assume that k is infinite. Determine the function rings A_i ($i = 1, 2, 3$) of the plane curves whose equations are $F_1 = Y - X^2$, $F_2 = XY - 1$, $F_3 = X^2 + Y^2 - 1$. Show that A_1 is isomorphic to the ring of polynomials $k[T]$ and that A_2 is isomorphic to its localised ring $k[T, T^{-1}]$. Show that A_1 and A_2 are not isomorphic (consider their invertible elements). What can we say about A_3 relative the two other rings? (Treat separately the cases where -1 is or is not a square in k, and pay special attention to the characteristic 2 case.)

7) Let $f : k \to k^3$ be the map which associates (t, t^2, t^3) to t and let C be the image of f (the space cubic). Show that C is an affine algebraic set and calculate $I(C)$. Show that $\Gamma(C)$ is isomorphic to the ring of polynomials $k[T]$.

8) Assume that k is algebraically closed. Determine the ideals $I(V)$ of the following algebraic sets.

$$V(XY^3 + X^3Y - X^2 + Y), \quad V(X^2Y, (X-1)(Y+1)^2),$$
$$V(Z - XY, Y^2 + XZ - X^2).$$

II

Projective algebraic sets

Throughout this chapter, k will be a commutative field.

0 Motivation

We have already seen the main reason for introducing projective space in the Introduction when discussing Bézout's theorem. In affine space, results on intersections always contain a certain number of special cases due to parallel lines or asymptotes. For example, in the plane two distinct lines meet at a unique point except when they are parallel. In projective space, there are no such exceptions.

Historically, projective space was introduced in the XVIIth century by G. Desargues, but was mostly developed in the XIXth century (by Monge, Poncelet, Klein and others). We have known that it is the natural setting for most geometries since Klein's Erlangen programme (1872).

In algebraic geometry, it is also the setting which gives the most satisfying results. However, affine space remains important since it is a local model of projective space.

1 Projective space

a. Definition

Let n be an integer $\geqslant 0$ and let E be a k-vector space of dimension $n+1$. We introduce the equivalence relation $\mathcal{R}$ on $E - \{0\}$:

$$x \mathcal{R} y \iff \exists \lambda \in k^*, \, y = \lambda x.$$

The relation $\mathcal{R}$ is simply collinearity and the equivalence classes for $\mathcal{R}$ are the lines in E passing through 0 with 0 removed.

Definition 1.1. *The projective space associated to E, denoted by $\mathbf{P}(E)$, is the quotient of $E - \{0\}$ by the relation $\mathcal{R}$. When $E = k^{n+1}$ (i.e., given a basis), we write $\mathbf{P}(E) = \mathbf{P}^n(k)$ and we call this space standard n-dimensional projective space.*

We denote by p the canonical projection $k^{n+1} - \{0\} \to \mathbf{P}^n(k)$. If $x = (x_0, x_1, \ldots, x_n)$ is $\neq 0$ and $\bar{x} = p(x)$, then we say that $\bar{x}$ is a *point* of $\mathbf{P}^n(k)$, whose *homogeneous coordinates* are $(x_0, x_1, \ldots, x_n)$. We note that the elements x_i are not all 0 and if $\lambda \in k$ is $\neq 0$, then $(\lambda x_0, \lambda x_1, \ldots, \lambda x_n)$ is another system of homogeneous coordinates for $\bar{x}$, which justifies our terminology.

Remarks 1.2.
1) When $k = \mathbf{R}$ or $\mathbf{C}$, projective space has a natural topology, namely the quotient of the topology on $k^{n+1} - \{0\}$. Projective space is then easily checked to be compact and connected.

2) The fact that the projective space associated to k^{n+1} is of dimension n corresponds to the fact that lines passing through the origin are contracted to points.

b. Projective subspaces

Using the above notation, we consider a subspace F in E of dimension $m+1$, where m is an integer satisfying $0 \leqslant m \leqslant n$.

Definition 1.3. *The image of $F - \{0\}$ in $\mathbf{P}(E)$ is called a projective subspace of dimension m, denoted by $\overline{F}$.*

(This is justified, amongst other things, by the fact that the restriction to F of the collinearity relation on E is simply the collinearity relation on F.)
When $m = 0$, we call $\overline{F}$ a point: when $m = 1, 2, \ldots, n-1$, we call it a line, plane,$\ldots$, projective hyperplane, and we can set up a theory of projective geometry, analogous to affine geometry, whose intersection theorems have no special cases.

Proposition 1.4. *Let V, W be two projective subspaces of $\mathbf{P}(E)$ of dimensions r and s such that $r + s - n \geqslant 0$. Then $V \cap W$ is a projective subspace of dimension $\geqslant r + s - n$. (In particular, $V \cap W$ is non-empty.)*

Proof. This follows immediately from theorems on intersections of vector subspaces.

Example 1.5. If $n = 2$, then two distinct lines in the projective plane meet at a unique point. If $n = 3$, a plane and a line meet at least at one point and this point is unique if the line is not contained in the plane; two distinct planes meet at a line, and so forth.

2 Homographies

If E is a vector space, then the linear group $GL(E)$ acts on E. We consider $u \in GL(E)$; since u is injective and preserves collinearity, u induces a bijection $\overline{u}$ of $\mathbf{P}(E)$.

Definition 2.1. *A bijection of* $\mathbf{P}(E)$ *induced by an element* u *in* $GL(E)$ *is called a* homography.

Remarks 2.2.

a) If $\overline{F}$ is a projective subspace of dimension d in $\mathbf{P}(E)$ and $\overline{u}$ is a homography, then $\overline{u}(\overline{F}) = \overline{u(F)}$: the image of $\overline{F}$ is a projective subspace of dimension d. When $d = 1$, we see that homographies preserve alignment.

b) For an explanation of the word homography, *cf.* 3.1.1.

c) It is clear from the definition that the group k^* of homotheties acts trivially on $\mathbf{P}(E)$ and it is easy to check that these are the only elements which act trivially on $\mathbf{P}(E)$. The group of homographies on $\mathbf{P}(E)$ (or the projective group of E) is therefore the quotient $PGL(E) = GL(E)/k^*$.

d) Homographies are automorphisms of projective space, in the sense given in Chapter III.

3 Relation between affine and projective space

What follows is an intuitive introduction to the affine-projective link. We will deal in more detail with this link in Chapter III.

Let the $n + 1$-dimensional space E be equipped with a basis, so that $\mathbf{P}(E) = \mathbf{P}^n(k)$, with coordinates $(x_0, x_1, \ldots, x_n)$. Let H be the hyperplane of equation $x_0 = 0$ and let $\overline{H}$ be the associated projective hyperplane. Set $U = \mathbf{P}^n(k) - \overline{H}$. There is then a bijection $\varphi : U \to k^n$ which associates to $\overline{x}$ (where $x = (x_0, x_1, \ldots, x_n)$) the point $(x_1/x_0, \ldots, x_n/x_0)$. This map is well defined, since x_0 does not vanish on U, and its image is independent of the system of coordinates chosen for $\overline{x}$. It is also a bijection whose inverse is given by $(x_1, \ldots, x_n) \mapsto (1, x_1, \ldots, x_n)$.

Moreover, since the hyperplane $\overline{H}$ is a projective space of dimension $n-1$, the foregoing gives a description of projective space $\mathbf{P}^n(k)$ of dimension n as being a disjoint union of an affine space k^n of dimension n and a projective space $\overline{H}$ of dimension $n-1$. Alternatively, we have embedded a copy of affine space k^n in a projective space of the same dimension. The points of k^n are said to be "at finite distance" and the points of $\overline{H}$ are said to be "at infinity".

Of course, the notion of infinity depends on the choice of hyperplane $\overline{H}$ and it is entirely possible to change it by taking another hyperplane of the form $x_i = 0$, or indeed a more general hyperplane. In fact, there is no infinity in projective space: infinity is an affine concept!

Examples 3.1.

1. The projective line. We take $n = 1$: we denote the coordinates of k^2 by (x, t) and we take $t = 0$ to be the "hyperplane" at infinity $\overline{H}$. In fact, since all the points $(x, 0)$ in H are collinear, $\overline{H}$ contains a unique point $\infty = (1, 0)$ and we identify k and $\mathbf{P}^1(k) - \{\infty\}$ via the map $x \mapsto (x, 1)$. It follows that the projective line is an affine line to which we add a unique point at infinity. This description enables us to calculate the cardinal of the projective line when k is finite. When $k = \mathbf{R}$ or $\mathbf{C}$, we get topological information: the projective line is the Alexandrov compactification of the affine line and is hence a circle for $k = \mathbf{R}$ and a sphere for $k = \mathbf{C}$.

We can check that under this identification the homographies of $\mathbf{P}^1(k)$ are the maps $x \mapsto ax + b/cx + d$, extended to infinity in the usual way: these maps are indeed homographies in the usual sense.

2. The projective plane. We use coordinates (x, y, t) and take $t = 0$ to be the hyperplane at infinity. In this case $\overline{H}$ consists of points with *homogeneous* coordinates $(x, y, 0)$ and is hence a projective line denoted by D_∞. The complement of $\overline{H}$ is formed of points $(x, y, 1)$ and is isomorphic to the affine plane k^2 via the map which forgets the third coordinate.

a. The projective lines in the plane. Let us determine the projective lines in $\mathbf{P}^2(k)$. Such a line $\overline{D}$ is the image of a subspace of dimension 2 in k^3, and is therefore given by a unique non-trivial linear equation $ux + vy + wt = 0$, where u, v and w are not all 0. There are two cases to consider.

1) If $u = v = 0$, then we may assume $w = 1$ and that $\overline{D}$ is simply D_∞.

2) Otherwise, we consider the restriction D of $\overline{D}$ to the affine plane k^2: we obtain the points (x, y) such that $ux + vy + w = 0$, *i.e.*, an affine line. On the other hand, we consider the restriction of $\overline{D}$ to D_∞: we obtain the points $(x, y, 0)$ such that $ux + vy = 0$. There is a unique such point, for which we may choose homogeneous coordinates $(v, -u, 0)$. This point at infinity on the line D corresponds to the *direction* of D: moreover, if D' is an affine line parallel to D, then its equation is $ux + vy + w' = 0$ and D' has the same point at infinity as D.

To summarise, the projective lines other than D_∞ are in bijective correspondence with affine lines: each projective line contains an extra point at infinity corresponding to its direction.

b. Conics. Let $\overline{C}$ be the subset of $\mathbf{P}^2$ consisting of points whose homogeneous coordinates (x, y, t) satisfy the equation $xy - t^2 = 0$. (See 4.1 for an explanation of why we must use homogeneous polynomials in projective space.) The intersection of $\overline{C}$ with the affine plane k^2 is the hyperbole $xy = 1$. At infinity, $\overline{C}$ has two points, $(1, 0, 0)$ and $(0, 1, 0)$, corresponding to the asymptotes of C. Furthermore, if we take the intersection of $\overline{C}$ and the projective line $x - t = 0$ corresponding to the affine line $x = 1$, which is parallel to the asymptote $x = 0$, we get one point $(1, 1, 1)$ at finite distance and another point $(1, 0, 0)$ at infinity, corresponding to the direction of the asymptote. If we take the

intersection with the asymptote itself, we get the point at infinity counted
double: the asymptote is tangent to $\overline{C}$ at infinity.

NB: we have seen that the line at infinity in projective space is not intrin-
sically fixed. We now choose $x = 0$ to be the line at infinity. In the (y, t) affine
plane we obtain a curve C', whose equation is $y = t^2$, a parabola with only
one point at infinity, $(0, 1, 0)$. (This parabola is tangent to the line at infinity.)

Continuing, we now take $x + y = 0$ to be the line at infinity and assume
that k is the field of real numbers. After a change of variables (or homography)
$t' = x + y$, $x' = x$, $y' = t$ we get a new equation for $\overline{C}$, $x'^2 + y'^2 - x't' = 0$. In
affine coordinates, relative to the new line at infinity $t' = 0$ this curve is an
ellipse (actually a circle) $x'^2 + y'^2 - x' = 0$, and this time there is no point at
infinity (because we have taken $k = \mathbf{R}$).

The conclusion of this little game is that the familiar conic types (hyper-
bolae, parabolas, ellipses, etc) are affine properties. In projective space, these
properties can be expressed by simply saying that the conic cuts the line chosen to be the line at infinity in two, one or no points. Moreover, we will show
that (*cf.* Exercise V, 3) in projective space there is a unique non-degenerate
conic (up to homography).

4 Projective algebraic sets

We will now repeat in the projective setting the material contained in Chapter I, passing quickly over similar points and emphasising differences.

We work with an *infinite* commutative field k: n is an integer > 0 and
we denote by $\mathbf{P}^n(k)$ or simply $\mathbf{P}^n$ the projective space of dimension n. The
coordinates on $\mathbf{P}^n$ are denoted by $(x_0, x_1, \ldots, x_n)$. We denote by R the ring of
polynomials $k[X_0, \ldots, X_n]$. In small dimensions we will mostly use variables
x, y, z, t and take the hyperplane $t = 0$ to be the hyperplane at infinity.

The first difference with affine sets is that the polynomials F in the ring
$k[X_0, \ldots, X_n]$ no longer define functions on projective space since their value
at a point $\overline{x}$ depends on the chosen system of homogeneous coordinates. For
example, if F is homogeneous of degree d, then

$$F(\lambda x_0, \lambda x_1, \ldots, \lambda x_n) = \lambda^d F(x_0, x_1, \ldots, x_n).$$

However, we can define zeros of polynomials in the following way.

Proposition-Definition 4.1. *Consider $F \in k[X_0, \ldots, X_n]$ and $\overline{x} \in \mathbf{P}^n$. We
say that $\overline{x}$ is a zero of F if $F(x) = 0$ for any system of homogeneous coordinates x for $\overline{x}$. We then write either $F(x) = 0$ or $F(\overline{x}) = 0$. If F is homogeneous, it is enough to check that $F(x) = 0$ for any system of homogeneous
coordinates. If $F = F_0 + F_1 + \cdots + F_r$, where F_i is homogeneous of degree i,
then it is necessary and sufficient that $F_i(x) = 0$ for all i.*

Proof. Only the last statement needs to be proved. If $F(\lambda x) = \lambda^r F_r(x) + \cdots + \lambda F_1(x) + F_0(x) = 0$ for any λ, then since k is infinite all the values $F_i(x)$ vanish. The converse is obvious.

This is the first appearance of homogeneous polynomials, which will play an essential role in the study of projective space.

Definition 4.2. *Let S be a subset of $k[X_0, \dots, X_n]$. We set*

$$V_p(S) = \{x \in \mathbf{P}^n \mid \forall F \in S, \ F(x) = 0\}$$

(in the sense of 4.1, of course). We say that $V_p(S)$ is the projective algebraic set defined by S. When there is no risk of confusion, we denote this set by $V(S)$.

Remark 4.3. It is clear that if I is the ideal generated by S, then $V_p(I) = V_p(S)$. Since $k[X_0, \dots, X_n]$ is Noetherian, we can therefore assume that S is finite and by 4.1 we can even assume S is a finite set of homogeneous polynomials.

Examples 4.4.

a) We have $V_p((0)) = \mathbf{P}^n$.

b) Let $m = R^+ = (X_0, \dots, X_n)$ be the ideal of polynomials with constant term 0. We have $V_p(m) = \varnothing$. (The homogeneous coordinates of a point in $\mathbf{P}^n$ are not all 0.) We call this ideal the "irrelevant" ideal. *NB*: this is the case even if k is algebraically closed, which is an important difference with affine geometry (*cf.* Chapter I, 4.1).

c) Points are projective algebraic sets: consider $x = (x_0, x_1, \dots, x_n) \in \mathbf{P}^n$. One of the component x_i—for example, x_0—is not 0, so we can assume $x_0 = 1$. We then have $\{x\} = V_p(X_1 - x_1 X_0, \dots, X_n - x_n X_0)$.

d) If $n = 2$, projective plane curves are defined by homogeneous equations: $Y^2 T - X^3 = 0$, $X^2 + Y^2 - T^2 = 0, \dots$

Remarks 4.5. As in the affine case, the following hold.

a) The map V_p is decreasing.

b) An arbitrary intersection or finite union of projective algebraic sets is a projective algebraic set, so there is a (Zariski) topology on $\mathbf{P}^n$ whose closed sets are the projective algebraic sets. Of course, the Zariski topology on subsets of $\mathbf{P}^n$ is simply the restriction of the Zariski topology on $\mathbf{P}^n$. We will see in Chapter III that if we embed affine space k^n in projective space, then we recover by this method the Zariski topology on k^n.

c) Let $V \subset \mathbf{P}^n$ be a projective algebraic set; we associate to V its cone $C(V)$, which is the inverse image of V under the projection $p : k^{n+1} - \{0\} \to \mathbf{P}^n(k)$, plus the origin of k^{n+1}. If I is a homogeneous ideal (*cf.* 7.2 below) different from R and $V = V_p(I)$, then $C(V) = V(I) \subset k^{n+1}$ (in the affine category). If $I = R$, then $C(V) = V(R^+) = \{0\}$. This type of argument sometimes enables us to reduce a projective problem to a similar affine problem (*cf.* 5.4).

5 Ideal of a projective algebraic set

Definition 5.1. *Let V be a subset of $\mathbf{P}^n$. We define the ideal of V by the formula*

$$I_p(V) = \{F \in k[X_0, \ldots, X_n] \mid \forall x \in V, \ F(x) = 0 \ \textit{(in the sense of 4.1)}\}.$$

Remarks 5.2.
   a) $I_p(V)$ is a homogeneous radical ideal by 4.1 (*cf.* 7.2)
   b) The operation I_p is decreasing.
   c) If V is a projective algebraic set, then $V_p(I_p(V)) = V$. If I is an ideal, then $I \subset I_p(V_p(I))$.
   d) We have $I_p(\mathbf{P}^n) = (0)$ and $I(\varnothing) = k[X_0, \ldots, X_n]$.

5.3. Irreducibility. The definitions and results of Chapter I can be easily translated *mutatis mutandis* into projective geometry.

Assume now that the field k is algebraically closed. There is then a projective version of the affine Nullstellensatz. The main difference is the existence of the irrelevant ideal $R^+ = (X_0, \ldots, X_n)$.

Theorem 5.4 (Projective Nullstellensatz). *Assume that k is algebraically closed. Let I be a homogeneous ideal of $k[X_0, \ldots, X_n]$ and set $V = V_p(I)$.*

1)
$$V_p(I) = \varnothing \iff \exists\, N \ \textit{such that} \ (X_0, \ldots, X_n)^N \subset I$$
$$\iff (X_0, \ldots, X_n) = R^+ \subset \mathrm{rac}(I).$$

2) If $V_p(I) \neq \varnothing$, then $I_p(V_p(I)) = \mathrm{rac}(I)$.

Proof. If $I = R$, then $V = V_p(I) = \varnothing$ and 1) is trivially true. Assume therefore that $I \neq R$. We apply the affine Nullstellensatz to the cone of V: $C(V) = V(I) \subset k^{n+1}$ (*cf.* 4.5.c). The statement that $V = V_p(I)$ is empty means exactly that $C(V)$ contains only the origin in k^{n+1} and hence that $\mathrm{rac}(I)$ is equal to R^+, which proves 1). We now prove 2). As $V = V_p(I)$ is non-empty, $I_p(V) = I(C(V))$, so this ideal is equal to $\mathrm{rac}(I)$ by the affine Nullstellensatz.

Remark 5.5. It can be checked that if I is a homogeneous ideal, then so is its radical. We therefore obtain a bijection between non-empty projective algebraic sets in $\mathbf{P}^n$ and homogeneous reduced ideals of R which do not contain the irrelevant ideal R^+. The prime ideals still correspond to irreducible projective algebraic sets; but be careful—points no longer correspond to maximal ideals.

6 A graded ring associated to a projective algebraic set

Let $V \subset \mathbf{P}^n$ be a projective algebraic set and let $I_p(V)$ be its ideal. Since $I_p(V)$ is homogeneous, the quotient ring

$$\Gamma_h(V) = k[X_0, \ldots, X_n]/I_p(V)$$

is graded (*cf.* 7.1 below). This is one of the graded rings naturally associated to V. We leave it to the reader to check that there is still a dictionary translating homogeneous radical ideals in $\Gamma_h(V)$ into projective algebraic sets contained in V.

Remarks 6.1.
   a) NB: the elements of $\Gamma_h(V)$, unlike their affine analogues, do not define functions on V. However, if $f \in \Gamma_h(V)$ and $x \in \mathbf{P}^n$, then the statement that x is a zero of f is meaningful and independent of the choice of representative of x. The ring $\Gamma_h(V)$ differs in two other fundamental ways from affine rings: we will see in Chapter III, 11.6 that this ring depends fundamentally on the embedding of V in $\mathbf{P}^n$, and moreover, even for a fixed embedding there are several graded rings naturally associated to V (*cf.* Chapter III, 9.8).
   b) The following is a consequence of the Nullstellensatz. If $f \in \Gamma_h(V)$ is homogeneous of degree > 0 and vanishes on the closed set W in V, then f is contained in the radical of the ideal $I_V(W)$.

The open sets $D^+(f)$. These are the projective analogues of the open affine sets $D(f)$.

Definition 6.2. *Let V be a projective algebraic set and consider a homogeneous element $f \in \Gamma_h(V)$ of degree > 0. We set*

$$D^+(f) = \{x \in V \mid f(x) \neq 0\}.$$

   It is clear that the sets $D(f)$ are open sets of V. Moreover:

Proposition 6.3. *With the notations of 6.2, every non-empty open set of V is a finite union of open sets of the form $D^+(f)$.*

Proof. Let U be a non-empty open set of V, so that $V - U = V_p(I)$, where I is a homogeneous ideal of R. It follows that $I = (F_1, \ldots, F_r)$, where the $F_i \in k[X_0, \ldots, X_n]$ are homogeneous polynomials of degree > 0. If f_i is the image in $\Gamma_h(V)$ of F_i, then $U = D^+(f_1) \cup \cdots \cup D^+(f_r)$.

Example 6.4. Considering the open sets $D^+(X_i)$ for $i = 0, \ldots, n$ in $\mathbf{P}^n$ we obtain an open cover of $\mathbf{P}^n$ by open sets which were shown in Section 3 to be in natural bijection with affine spaces. This remark will enable us to define the structure of a variety on $\mathbf{P}^n$ in the next chapter.

Remark 6.5. For reasons which will become apparent in the next chapter, we will only consider the open sets $D^+(f)$ for elements f of degree > 0. If f is of degree 0, hence constant, the corresponding open set is trivial—empty if $f \neq 0$ and equal to V if $f = 0$. If f is of degree > 0, then $D^+(f)$ is non-empty and can only be equal to V if V is finite (*cf.* IV, 2.9).

To conclude, the main differences between affine and projective geometry are the following:

1) We have to use homogeneous polynomials and replace rings by graded rings and ideals by homogeneous ideals in the projective setting.

2) Polynomials (and more generally elements of all the graded rings involved) are no longer functions.

3) The irrelevant ideal $(X_0, \ldots, X_n)$ plays a very special role.

4) We can recover affine geometry from projective geometry using the open sets $D^+(X_i)$ or, more generally, open sets of the form $D^+(f)$.

7 Appendix: graded rings

Definition 7.1. *A k-algebra R is said to be* graded *if it can be written as a direct sum*

$$R = \bigoplus_{n \in \mathbf{N}} R_n,$$

where the subspaces R_n of R satisfy $R_p R_q \subset R_{p+q}$. The elements of R_p are said to be homogeneous of degree p *and this condition is the usual rule for the degree of a product.*

We note that R_0 is a subalgebra of R and $m = R^+ = \bigoplus_{n>0} R_n$ is an ideal of R whose quotient is isomorphic to R_0. The canonical example of a graded ring is the ring of polynomials, graded by the degree function in the usual way.

Proposition-Definition 7.2. *Let R be a graded k-algebra and let I be an ideal of R. The following are equivalent.*

1) I is generated by homogeneous elements.

2) If $f \in I$ and $f = \sum_0^r f_i$ and f_i is homogeneous of degree i, then $f_i \in I$ for every i.

Such an ideal is said to be homogeneous.

Proof. It is clear that 2) implies 1). Conversely, assume that I is generated by homogeneous elements G_i of degrees α_i. Consider $F = F_0 + \cdots + F_r \in I$, where F_i is homogeneous of degree i. By induction, it will be enough to show that $F_r \in I$. But we can write $F = \sum U_i G_i$, and on identifying terms of highest degree, we get $F_r = \sum U_{i,r-\alpha_i} G_i$, so F_r is contained in I.

Proposition 7.3. *Let R be a graded k-algebra and let I be a homogeneous ideal of R. Let S be the quotient k-algebra $S = R/I$ and p the canonical projection. Then S has a natural grading given by $S_i = p(R_i)$.*

Proof. It will be enough to show that S is the direct sum of the spaces S_i, but this is exactly condition 2) of Proposition 7.2.

Remark 7.4. If $R_0 = k$ and I is homogeneous and different from R, then I is contained in R^+.

Definition 7.5. *Let R be a graded k-algebra. An R-module M is said to be graded if it can be written as a direct sum*

$$M = \bigoplus_{n \in \mathbf{Z}} M_n,$$

where the k-subspaces M_n of M satisfy $R_p M_q \subset M_{p+q}$ for all $p \in \mathbf{N}$ and $q \in \mathbf{Z}$. A homomorphism $\varphi : M \to N$ between two graded R-modules is said to be homogeneous of degree d if, for all n, $\varphi(M_n) \subset N_{d+n}$.

We note that if φ is homogeneous of degree d, then the kernel of φ is a graded submodule of M, i.e., if $x = \sum x_n \in \mathrm{Ker}\, \varphi$, then $x_n \in \mathrm{Ker}\, \varphi$ for all n.

Exercises

1 Homographies

Let E be a k-vector space of dimension $n + 1$ and let $\mathbf{P}(E)$ be the associated projective space. If $u \in \mathrm{GL}(E)$, u induces a bijection $\bar{u}$ from $\mathbf{P}(E)$ to itself which we call a homography.

a) What can we say about u when $\bar{u} = \mathrm{Id}$?
b) Show that the image of a projective subspace of dimension d under a homography is again a projective subspace of dimension d.
c) Conversely, show that if V and W are two projective subspaces of dimension d, then there is a homography $\bar{u}$ such that $\bar{u}(V) = W$.
d) Assume $E = k^2$ and

$$u = \begin{pmatrix} a & b \\ c & d \end{pmatrix}$$

such that $ad - bc \neq 0$. Take the point $(1, 0)$ in $\mathbf{P}^1(k) = \mathbf{P}(E)$ to be the point at infinity, so points x in k can be identified with points $(x, 1)$ in $\mathbf{P}^1(k) - \{\infty\}$. Determine $\bar{u}$ explicitly and explain the origins of the word homography.

2 Markings

Using the same notation as in 1, we denote the canonical projection from $E - \{0\}$ to $\mathbf{P}(E)$ by p. A marking of $\mathbf{P}(E)$ consists of $n+2$ points $x_0, \ldots, x_{n+1}$ of $\mathbf{P}(E)$ such that there is a basis $e_1, \ldots, e_{n+1}$ of E such that $p(e_i) = x_i$ for $i = 1, \ldots, n + 1$ and $p(e_1 + \cdots + e_{n+1}) = x_0$.

a) Assume $n = 1$. Prove that a marking of $\mathbf{P}(E)$ (i.e., the projective line) is exactly the data of three distinct points. (For example, in $\mathbf{P}^1(k)$ we can take $0 = (0, 1)$, $\infty = (1, 0)$ and $1 = (1, 1)$.)

b) Prove that $n + 2$ points $x_0, \ldots, x_{n+1} \in \mathbf{P}(E)$ form a marking if and only if no $n + 1$ of them are contained in a hyperplane.

c) Prove that if $x_0, \ldots, x_{n+1}$ and $y_0, \ldots, y_{n+1}$ are two markings of $\mathbf{P}(E)$, then there is a unique homography which sends each x_i to y_i. Study the case $n = 1$ in detail.

3 Quadrics

Let k be an algebraically closed field. A quadric in $\mathbf{P}^3(k)$ is a projective algebraic set of the form $Q = V(F)$, where F is an irreducible polynomial of degree 2 in X, Y, Z, T and hence gives rise to a quadratic form on k^4 which we assume to be non-degenerate.

a) Prove that if $Q = V(F)$ is a quadric, then there is a homography h such that $h(Q) = V(XT - YZ)$. We assume from now on that Q is of this form.

b) Prove that Q contains two families of lines both of which are indexed by $\mathbf{P}^1$. Prove that a unique line from each family passes through any point of Q, that two lines in the same family are disjoint and that two lines in different families meet at a unique point.

c) Prove that if D_1, D_2, D_3 are three lines in $\mathbf{P}^3$ which are pairwise disjoint, then there is a unique quadric Q containing the lines D_i. (Start by showing that for any 9 points in $\mathbf{P}^3$ we can find a set of the form $V(F)$ with F of degree 2 containing them all, and then show that if a set of this form meets a line in three points, then it contains it.)

4 The space cubic

We assume that k is infinite.

We consider the map $\varphi : \mathbf{P}^1 \to \mathbf{P}^3$ defined by

$$\varphi(u, v) = (u^3, u^2 v, u v^2, v^3).$$

We set $C = \operatorname{Im} \varphi$.

a) Show that $C = V(I)$, where I is the ideal

$$(XT - YZ, \; Y^2 - XZ, \; Z^2 - YT).$$

(For the non-obvious inclusion, consider the affine open sets $X \neq 0$, $T \neq 0$.)

b) Prove that $I(C)$ is equal to I. (Start by proving (by induction on the degree of F relative to Y and Z, for example) that any homogeneous polynomial $F \in k[X, Y, Z, T]$ is equal modulo I to a polynomial of the form $a(X, T) + b(X, T)Y + c(X, T)Z$.)

c) Prove that $I(C)$ cannot generated by two elements. (Consider the terms of degree 2.) We say that C is not a scheme-theoretic complete intersection.

d) ¶ Prove that on the other hand $C = V(Z^2 - YT, F)$, where F is a homogeneous polynomial to be determined. We say that C is a set-theoretic complete intersection, which means that C can be defined by two equations, or, alternatively, that C is the intersection of two surfaces. NB: these surfaces are tangent to each other and C should be thought of as being of multiplicity 2 in this intersection.

e) ¶ Prove there is a "resolution" of $I(C)$ *i.e.*, an exact sequence

$$0 \longrightarrow R(-3)^2 \xrightarrow{\ u\ } R(-2)^3 \xrightarrow{\ v\ } I(C) \longrightarrow 0,$$

where R is the ring $k[X,Y,Z,T]$ and $R(-i)$ is the graded R-module which is simply R with a shifted grading: $R(-i)_n = R_{n-i}$. The homomorphisms u and v are of degree 0 (*i.e.*, they send elements of degree n to elements of degree n). Here, this means that v is given by three homogeneous polynomials of degree 2 which generate $I(C)$ (what could they be!) and u by a 3×2 matrix whose coefficients are homogeneous polynomials of degree 1. Our aim is to calculate u, *i.e.*, the relations (or, to use a nicer word, the syzygies) linking the generators of I.

5 The union of two distinct lines

Let D_1 and D_2 be two distinct lines in $\mathbf{P}^3$.

a) Show that up to homography we can assume $D_1 = V(X,Y)$ and $D_2 = V(Z,T)$.
b) Consider $C = D_1 \cup D_2$. Calculate $I(C)$.
c) ¶ Prove that there is a resolution of $I(C)$ of the following form (*cf.* 4):

$$0 \longrightarrow R(-4) \longrightarrow R(-3)^4 \longrightarrow R(-2)^4 \longrightarrow I(C) \longrightarrow 0.$$

III

Sheaves and varieties

0 Motivation

If we compare the study of affine algebraic sets and projective algebraic sets, we find many similarities and a few fundamental differences, such as the role played by homogeneous polynomials and graded rings in projective geometry. The most important difference, however, is the functions. If V is an affine algebraic set, we have a lovely function algebra $\Gamma(V)$ and an almost perfect dictionary translating properties of V into properties of $\Gamma(V)$. One of the problems of projective geometry is that elements of $\Gamma_h(V)$ do not define functions on V, even in the simplest case, namely a homogeneous polynomial, since if $x \in \mathbf{P}^n$ and F is homogeneous of degree d, then the quantity $F(x)$ depends on the choice of representative: $F(\lambda x) = \lambda^d F(x)$.

To solve this problem we will exploit the idea that projective space $\mathbf{P}^n$ contains open sets $U_i = D^+(X_i)$ which are isomorphic to affine spaces. On these open sets there is a set of well-behaved functions—the polynomials. We might imagine that we could get good functions on $\mathbf{P}^n$ by gluing together good functions on the open sets U_i. It turns out that this method will not give us many functions defined globally on $\mathbf{P}^n$, as the example below shows.

Consider the projective line $\mathbf{P}^1$, with homogeneous coordinates x and t and open sets U_0 ($x \neq 0$) and U_1 ($t \neq 0$). These are isomorphic to k via maps j_0 and j_1 given, respectively, by $\tau \mapsto (1, \tau)$ and $\xi \mapsto (\xi, 1)$ with inverses $(x, t) \mapsto t/x$ and $(x, t) \mapsto x/t$. On U_0 (resp. U_1) a function is good if it is polynomial function $f(\tau) = f(t/x)$ (resp. $g(\xi) = g(x/t)$). To obtain a good function on the whole of $\mathbf{P}^1$ these two functions must coincide on $U_0 \cap U_1$. In other words, for all $x, t \neq 0$ we must have $f(t/x) = g(x/t)$, that is to say,

$$a_n \frac{t^n}{x^n} + \cdots + a_1 \frac{t}{x} + a_0 = b_m \frac{x^m}{t^m} + \cdots + b_1 \frac{x}{t} + b_0,$$

and clearing denominators we get

$$a_n t^{n+m} + a_{n-1} t^{n+m-1} x + \cdots$$
$$+ a_0 t^m x^n - b_0 t^m x^n - b_1 t^{m-1} x^{n+1} - \cdots - b_m x^{m+n} = 0.$$

The above polynomial, which vanishes on the Zariski open set $x \neq 0, t \neq 0$ in k^2, is therefore identically zero (at least if k is infinite). But this implies that all the coefficients a_i, b_i should be zero for $i > 0$ and a_0 and b_0 should be equal. In other words, f and g must be two equal constants and the global function constructed on $\mathbf{P}^1$ is a constant.

The moral of the story is that in projective geometry global functions (*i.e.*, functions defined on the whole of $\mathbf{P}^n$, or on the whole of V, where V is a projective algebraic set) are not enough (most of the time the only such functions are constants), and if we want to deal with functions, we must settle for *locally* defined functions, *i.e.*, functions defined on open sets.

It is this observation which leads us to the notion of *sheaves*.

1 The sheaf concept

a. Sheaves of functions: definition

Definition 1.1. *Let X be a topological space and $\mathbf{K}$ a set. A $\mathbf{K}$-valued sheaf of functions on X is given, for every open set U in X, by a set $\mathcal{F}(U)$ of functions from U to $\mathbf{K}$, which satisfy the following two axioms.*

1) Restriction. If V is an open set contained in U and $f \in \mathcal{F}(U)$, then $f|_V \in \mathcal{F}(V)$.

2) Gluing. If U is covered by open sets U_i ($i \in I$), then for any choice of elements $f_i \in \mathcal{F}(U_i)$ such that $f_i|_{U_i \cap U_j} = f_j|_{U_i \cap U_j}$ there is a unique function $f \in \mathcal{F}(U)$ such that $f|_{U_i} = f_i$.

Remarks 1.2.

a) In the gluing axiom, the existence of a function $f : U \to \mathbf{K}$ such that $f|_{U_i} = f_i$ is clear. The condition simply says that this function is in $\mathcal{F}(U)$.

b) The above axioms are natural. They say that a certain class of functions on X (which we want to use as our class of good functions) has enough good properties to be manageable. More precisely, we ask that the class in question should be stable under restriction (Axiom 1) and local, that is to say, to check that f is a good function it should be enough to check it is locally a good function (Axiom 2).

c) These conditions are satisfied by some highly important classes of functions. The set of all functions is of course a sheaf, but we also have a sheaf of all real or complex valued continuous functions. Alternatively, if X is an open set in $\mathbf{R}^n$, we have a sheaf of all differentiable or analytic functions, etc.

d) We note that restriction defines a map $r_{V,U} : \mathcal{F}(U) \to \mathcal{F}(V)$ such that for any U, $r_{U,U} = \mathrm{Id}_{\mathcal{F}(U)}$, and for all $W \subset V \subset U$, $r_{W,U} = r_{W,V} \, r_{V,U}$.

Notation 1.3. We set $\mathcal{F}(U) = \Gamma(U,\mathcal{F})$. The elements of $\Gamma(U,\mathcal{F})$ are called *sections* of $\mathcal{F}$ over U. When U is equal to X, we call the corresponding sections global sections.

b. General sheaves

We will need more general sheaves than function sheaves. The restriction operation, which is no longer obvious, is now given axiomatically by 1.2.d.

Definition 1.3. *Let X be a topological space. A presheaf on X is given by the following data:*

- *For every open set U in X, a set $\mathcal{F}(U)$;*
- *For every pair of open sets U and V with $V \subset U$, a map $r_{V,U} : \mathcal{F}(U) \to \mathcal{F}(V)$ called the restriction map,*

such that the two following conditions are satisfied:
  i) If $W \subset V \subset U$, then $r_{W,U} = r_{W,V}\, r_{V,U}$,
  ii) We have $r_{U,U} = \mathrm{Id}_{\mathcal{F}(U)}$;
  We set $r_{V,U}(f) = f|_V$.
  We say that $\mathcal{F}$ is a sheaf if in addition it satisfies the gluing Condition 2) of 1.1.

Remarks 1.5.
  a) The presheaf whose sections over U are constant **K**-valued functions is not generally a sheaf. The gluing condition is not satisfied for a non-connected open set.

  b) It is always possible to consider a given sheaf to be a sheaf of functions (in a non-natural way) (*cf.* Exercise III, A.1). This remark enables us to restrict ourselves to sheaves of functions if necessary.

  c) If $\mathcal{F}$ is a presheaf on X (for the sake of simplicity we assume it is a presheaf of functions from X to **K**), we can embed it in a canonical way in a sheaf $\mathcal{F}^+$ called the *associated sheaf* of $\mathcal{F}$. To do this we localise the condition for being an element of $\mathcal{F}(U)$ in the following way. For any open set U in X we set

$$\mathcal{F}^+(U) = \{f : U \longrightarrow \mathbf{K} \mid \forall x \in U,\ \exists V \text{ open},$$
$$\text{such that } x \in V \subset U, \text{ and } g \in \mathcal{F}(V) \text{ such that } f|_V = g\}.$$

The sheaf thus defined is the best possible solution to the problem (*i.e.*, it is the smallest sheaf containing $\mathcal{F}$, *cf.* [H] Chapter II, 1.2).

  d) Given a sheaf $\mathcal{F}$ on X and an open set U in X, the sheaf $\mathcal{F}|_U$ is defined in the obvious way; if V is an open set in U, then we set $\mathcal{F}|_U(V) = \mathcal{F}(V)$.

c. Sheaves of rings

The most important sheaves we will be working with are sheaves of rings (or, more precisely, sheaves of k-algebras). The statement that $\mathcal{F}$ is a sheaf of rings means that the spaces $\mathcal{F}(U)$ are commutative rings and the restriction functions are homomorphisms of rings. This is true of the sheaf of (arbitrary) functions into a ring, or for sheaves of continuous/differentiable functions into $\mathbf{R}$ or $\mathbf{C}$, with the usual addition and multiplication. We can of course define sheaves of other structures (groups, modules or k-algebras) in a similar way.

Definition 1.5. *A ringed space is a topological space X equipped with a sheaf of rings. This sheaf is called the* structural sheaf *of X and is traditionally denoted by $\mathcal{O}_X$.*

"Morally" this sheaf is the sheaf of "good" functions on X and we have therefore assumed that a sum or product of good functions is still a good function.

Warning 1.7. From now on we fix an algebraically closed field k. Unless otherwise specified, the structural sheaf of all ringed spaces considered will be a sheaf of k-valued functions and we will assume it is a sheaf of k-algebras containing the constant functions.

With these precautions taken, there is a natural notion of morphism (and hence isomorphism) of ringed spaces:

Definition 1.6. *Let $(X, \mathcal{O}_X)$ and $(Y, \mathcal{O}_Y)$ be two ringed spaces. A morphism of ringed spaces is given by a continuous map $\varphi : X \to Y$, which transforms good functions into good functions by composition. In other words, for any function $g : U \to k$ such that $g \in \Gamma(U, \mathcal{O}_Y)$ we should have $g\varphi \in \Gamma(\varphi^{-1}U, \mathcal{O}_X)$.*

Remarks 1.9.

a) We note that when, for example, the sheaf in question is the sheaf of differentiable functions, the composition condition is equivalent to the requirement that f be a differentiable map.

b) For any open set U of Y we define a homomorphism of rings

$$\varphi_U^* : \Gamma(U, \mathcal{O}_Y) \longrightarrow \Gamma(\varphi^{-1}U, \mathcal{O}_X)$$

by setting $\varphi_U^*(g) = g\varphi$. These homomorphisms are compatible with restrictions. In other words they satisfy the condition $r_{\varphi^{-1}V, \varphi^{-1}U}\, \varphi_U^* = \varphi_V^*\, r_{V,U}$. For ringed spaces whose sheaves are not simply sheaves of functions a morphism consists not only of the data of a continuous map φ but also a collection of homomorphisms φ_U^* satisfying the above compatibility conditions (*cf.* [EGA] I, 0.4.1).

2 The structural sheaf of an affine algebraic set

Let $V \subset k^n$ be an affine algebraic set. We want to define good functions on the open sets of V. We will be guided by the following two remarks.

1) The good functions on V should be the polynomial functions $\Gamma(V)$.

2) V has a very simple basis of open sets, the sets $D(f)$.

In fact, the following lemma shows that it is enough to define the structural sheaf on a basis of open sets.

Lemma 2.1. *Let X be a topological space, $\mathcal{U}$ a basis of open sets in X and $\mathbf{K}$ a set. We suppose given for every open set $U \in \mathcal{U}$ a set $\mathcal{F}(U)$ of functions from U to $\mathbf{K}$ satisfying the following conditions:*

i) (Restriction) If $V, U \in \mathcal{U}$, $V \subset U$ and $s \in \mathcal{F}(U)$, then $s|_V \in \mathcal{F}(V)$.

ii) (Gluing) If an open set $U \in \mathcal{U}$ is covered by sets U_i indexed by $i \in I$, such that $U_i \in \mathcal{U}$ and if s is a function from U to $\mathbf{K}$ such that $\forall i \in I$, $s|_{U_i} \in \mathcal{F}(U_i)$, then $s \in \mathcal{F}(U)$.

Then there is a unique sheaf $\overline{\mathcal{F}}$ of functions on X such that, for every $U \in \mathcal{U}$, $\overline{\mathcal{F}}(U) = \mathcal{F}(U)$.

Proof. Any open set U is covered by open sets U_i contained in the basis $\mathcal{U}$. We set

$$\overline{\mathcal{F}}(U) = \{s : U \to \mathbf{K} \mid \forall i, \; s|_{U_i} \in \mathcal{F}(U_i)\}.$$

We leave it as an exercise for the reader to check that this definition is independent of the choice of open cover U_i and that $\overline{\mathcal{F}}$ is indeed the required sheaf.

Provided we check Conditions i) and ii) from 2.1, it is enough to define sheaves on bases of open sets.

We also note the following trivial lemma.

Lemma 2.2. *Let X be a topological space equipped with a basis of open sets $\mathcal{U}$, let $\mathcal{F}$ be a sheaf and let $\mathcal{G}$ be a presheaf on X. We assume $\mathcal{F}(U) = \mathcal{G}(U)$ for every $U \in \mathcal{U}$. Then $\mathcal{F} = \mathcal{G}^+$ (cf. 1.5.c).*

In the case in hand we therefore seek to define $\Gamma(D(f), \mathcal{O}_V)$. Since $D(f)$ is the set of points where the function f does not vanish, it is natural to include the inverse function f^{-1} along with the polynomial functions on V in the set of sections $\Gamma(D(f), \mathcal{O}_V)$. More precisely, we consider the restriction homomorphism $r : \Gamma(V) \to \mathcal{F}(D(f), k)$, where $\mathcal{F}(D(f), k)$ denotes the ring of all functions from $D(f)$ to k. Since $r(f)$ is invertible, r can be factorised through the localisation $\Gamma(V)_f$, $r = \rho j$ (cf. Summary 1.6.b) and the homomorphism $\rho : \Gamma(V)_f \to \mathcal{F}(D(f), k)$ is injective. Indeed, if $\rho(g/f^n) = 0$, then $g(x) = 0$ on $D(f)$ and hence $fg = 0$ on V, which implies that g/f^n is zero in the localised ring (cf. Summary loc. cit.). We then have the following definition.

Definition 2.3. *Let V be an affine algebraic set and consider a non-zero $f \in \Gamma(V)$. We set*
$$\Gamma(D(f), \mathcal{O}_V) = \Gamma(V)_f$$
(identified with a subring of the ring of k-valued functions on $D(f)$ via ρ). By this method we define a sheaf of rings on V called the sheaf of regular functions.

In the special case where $\Gamma(V)$ is an integral domain, the ring $\Gamma(V)_f$ is a subring of the field of fractions $K(V)$ of $\Gamma(V)$ (the field of rational functions on V, *cf.* Chapter I, 6.15) and the natural homomorphism $j : \Gamma(V) \to \Gamma(V)_f$ is injective. If $\Gamma(V)$ is not an integral domain (*i.e.*, if V is not irreducible) there is no fraction field and j is no longer necessarily injective.

Check 2.4. We must now check that the conditions given in 2.1 allowing us to construct a sheaf on V are satisfied.

a) *Restriction.* If $D(f) \subset D(g)$, then $V(g) \subset V(f)$, and since f vanishes on $V(g)$, the Nullstellensatz implies $f^n = gh$. Given $u/g^i \in \Gamma(V)_g$, its restriction to $D(f)$ can be written in the form $uh^i/g^ih^i = uh^i/f^{ni}$, and this restriction is indeed contained in $\Gamma(V)_f$.

b) *Gluing.* We assume that V is irreducible (and hence $\Gamma(V)$ is an integral domain): the general case is left to the reader as an exercise (*cf.* 2.5.b).

Let $D(f)$ be a standard open set covered by the sets D_{f_i}, where $f_i \neq 0$. This means that $V(f)$ is the intersection of the sets $V(f_i)$, or, alternatively, that $V(f) = V(I)$, the ideal generated by the functions f_i. Since the ring $\Gamma(V)$ is Noetherian, we can assume that there are only a finite number of functions f_i.

Let s_i be sections of $D(f_i)$ which we write as $s_i = a_i/f_i^n$ (we can use the same n for all the sections s_i since there is a finite number of them). We assume these sections to be coincident on the intersections $D(f_i) \cap D(f_j)$. We have $a_i f_j^n = a_j f_i^n$ on $D(f_i) \cap D(f_j)$, and hence this relation holds on V by density (since V is irreducible).

Since f vanishes on $V(f_1, \ldots, f_r) = V(f_1^n, \ldots, f_r^n)$, the Nullstellensatz tells us that $f \in \mathrm{rac}(f_1^n, \ldots, f_r^n)$. In other words, there is an integer m and functions $b_j \in \Gamma(V)$ such that $f^m = \sum_{j=1}^r b_j f_j^n$.

(We note that when dealing with an open cover of the whole of V there is an identity of the form $1 = \sum_{j=1}^r b_j f_j^n$, which is an algebraic analogue of the partitions of unity used in analysis.)

We now look for a section s on $D(f)$ of the form $s = a/f^m$ (with m as defined above) such that $s|_{D(f_i)} = s_i$, *i.e.*, $a/f^m = a_i/f_i^n$, or, alternatively, $f_i^n a = a_i f^m = a_i \sum_{j=1}^r b_j f_j^n = \sum_{j=1}^r b_j a_i f_j^n = \sum_{j=1}^r b_j a_j f_i^n = f_i^n \sum_{j=1}^r b_j a_j$. This clearly holds if we set $a = \sum_{j=1}^r b_j a_j$.

Remarks 2.5.

a) The careful reader will check that if an open set can be written as a standard open set in two different ways $D(f_1) = D(f_2)$, then the corresponding rings are the same. This follows from the argument used in 2.4.a above.

b) When proving 2.4 in the general case, a little care is needed. The equality of the sections s_i on the intersection then only means that there is a natural number N such that $f_i^N f_j^N (a_i f_j^n - a_j f_i^n) = 0$. We then write $f^m = \sum_j b_j f_j^{n+N}$ and $a = \sum_j a_j b_j f_j^N$.

c) Calculating $\Gamma(U, \mathcal{O}_V)$ for a non-standard open set is harder. Consider for example $U = k^2 - \{(0,0)\}$ (*cf.* Exercises III, A.2).

3 Affine varieties

We have equipped any affine algebraic set V with the structure of a ringed space $(V, \mathcal{O}_V)$ by taking the sheaf of regular functions $\mathcal{O}_V$ defined above. An affine algebraic variety is essentially the same thing.

Definition 3.1. *An affine algebraic variety is a ringed space which is isomorphic as a ringed space to a pair $(V, \mathcal{O}_V)$, where V is an affine algebraic set and $\mathcal{O}_V$ is the sheaf of regular functions on V. A morphism of affine algebraic varieties is simply a morphism of ringed spaces.*

Remarks 3.2.

a) Many authors (especially Americans) reserve the word variety for an irreducible variety.

b) The only advantage affine algebraic varieties have over affine algebraic sets is that they are intrinsic, *i.e.*, their structure does not depend on a choice of embedding in k^n. A typical example of this is a standard open set $D(f)$:

Proposition 3.3. *Let V be an affine algebraic set and consider $f \in \Gamma(V)$. The open set $D(f)$ equipped with the restriction of the sheaf $\mathcal{O}_V$ to $D(f)$ is an affine algebraic variety.*

Proof. Assume that V is embedded in k^n: set $I = I(V)$ and let F be a polynomial whose restriction to V is f. Our aim is to show that $D(f)$ is isomorphic to an affine algebraic set. The trick is to look for this set in k^{n+1}: we consider the map

$$\varphi : (x_1, \ldots, x_n) \longmapsto (x_1, \ldots, x_n, 1/f(x_1, \ldots, x_n))$$

sending $D(f)$ into k^{n+1}. The image W of φ is equal to the set $V(J)$, where $J = I + (X_{n+1}F - 1)$ (*cf.* the proof of the Nullstellensatz). It is clear that φ is a homeomorphism from $D(f)$ onto W whose inverse is given by the projection $p(x_1, \ldots, x_n, x_{n+1}) = (x_1, \ldots, x_n)$. It is easy to check that this is an isomorphism. (It is enough to check this fact on standard open sets, *cf.* 11.1 below.)

Example 3.4. The group of invertible matrices with complex coefficients $\mathrm{GL}(n, \mathbf{C})$ is an affine algebraic variety. In fact, it is an open set of the form $D(f)$ in the affine space of matrices $M(n, \mathbf{C}) = \mathbf{C}^{n^2}$, f being the determinant function, which is a polynomial.

In the following proposition we show that the affine variety morphisms between two affine algebraic sets are exactly the maps defined in Chapter I. We denote by $\mathrm{Hom}_{\mathrm{Var}}(X, Y)$ the set of affine variety morphisms from X to Y.

Proposition 3.5. *Let $(X, \mathcal{O}_X)$ and $(Y, \mathcal{O}_Y)$ be two affine algebraic sets equipped with the affine variety structures given by the structural sheaves $\mathcal{O}_X$ and $\mathcal{O}_Y$. There are natural bijections*

$$\mathrm{Hom}_{Var}(X, Y) \simeq \mathrm{Reg}(X, Y) \simeq \mathrm{Hom}_{k\text{-alg}}(\Gamma(Y, \mathcal{O}_Y), \Gamma(X, \mathcal{O}_X)).$$

Proof. The existence of the second bijection was established in Chapter I, 6.7. We use the notation from Chapter I, 6.7. We establish the existence of the first bijection as follows. If $\varphi : X \to Y$ is a variety morphism, then on considering the coordinate functions η_i on Y it is immediate that $\varphi = (\eta_1\varphi, \dots, \eta_m\varphi)$ is regular (*cf.* 1.8). Conversely, if φ is regular, $D(g)$ is a standard open set of Y and $f = h/g^r \in \Gamma(D(g), \mathcal{O}_Y)$, then $f\varphi = \varphi^*(h)/\varphi^*(g)^r \in \Gamma(D(\varphi^*(g)), \mathcal{O}_X)$, which shows that φ is also a morphism of varieties.

Remark 3.6. The above proposition shows that the category of affine algebraic varieties is equivalent to the category of affine algebraic sets. (However, we have seen that if we do not identify isomorphic objects, the former category contains more objects.)

4 Algebraic varieties

The main use of ringed spaces is that they enable us to define classes of objects which are locally isomorphic to some particular kind of space. For example, a differentiable manifold is a ringed space which is locally isomorphic to an open set of $\mathbf{R}^n$ with the sheaf of differentiable functions. In our case, of course, the local models are affine algebraic varieties.

Definition 4.1. *An algebraic variety is a quasi-compact ringed space (cf. Summary 1.8) which is locally isomorphic to an affine algebraic variety. A morphism of algebraic varieties is simply a morphism of ringed spaces.*

To say that $(X, \mathcal{O}_X)$ is locally isomorphic to an affine algebraic variety means that for any $x \in X$ there is an open set U containing x such that $(U, \mathcal{O}_X|_U)$ is isomorphic to an affine algebraic variety. We leave it to the reader to check that affine varieties are quasi-compact and hence are algebraic varieties.

Definition 4.2. *Let[1] X be an algebraic variety. The open sets of X which are isomorphic to affine algebraic varieties are called the* affine open sets *of X.*

Proposition 4.3. *Let X be an algebraic variety. The affine open sets form a basis of open sets of X. More precisely, any open set in X is a finite union of affine open sets (and is hence quasi-compact).*

Proof. We have $X = \bigcup_{i=1}^{r} U_i$, where the sets U_i are open and affine (this is the definition of a variety: the quasi-compactness of X allows us to take a union of only a finite number of open sets). Let U be an open set of X. We have $U = \bigcup_{i=1}^{r} U \cap U_i$ and so it will be enough to prove the result for $U \cap U_i$. But since U_i is affine, this follows from Chapter I, 4.13 and 3.3.

If X is an algebraic variety and U is an open set in X, then it is easy to check using 4.3 that the sections of $\Gamma(U, \mathcal{O}_X)$ are continuous k-valued functions (where k is equipped with the Zariski topology).

Corollary 4.4. *A non-empty algebraic variety can be uniquely written as a finite union of irreducible closed sets which do not contain each other. These are its* irreducible components.

Proof. By quasi-compactness, we can write $X = U_1 \cup \cdots \cup U_n$, where the sets U_i are open affine sets. We then write each of the sets U_i as a finite union of closed sets $U_{i,j}$ which are closed in U_i. We then have $X = \bigcup_{i,j} \overline{U}_{i,j}$ and the $\overline{U}_{i,j}$ are irreducible by Chapter I, 3.5.

Examples 4.5.
   a) If X is an algebraic variety and U is an open set of X, then U equipped with the sheaf $\mathcal{O}_X|_U$ is an algebraic variety which is called an *open subvariety* of X.
   In particular, any open subset of an affine algebraic variety is an algebraic variety (called a *quasi-affine variety*); but *be careful*: it is not necessarily affine (for example, $k^2 - \{(0,0)\}$ is not affine *cf.* Exercise III, A.4).
   b) *Closed subvarieties.* Let X be an algebraic variety and let Y be a closed set in X. Our aim is to define a sheaf $\mathcal{O}_Y$ on Y. The most natural idea, given that the inclusion map has to be a morphism, is to take the sheaf of functions on open sets of X restricted to Y, *i.e.*, to define the sections over an open set V on Y by

$$\{f : V \to k \mid \exists\, U \subset X,\ \text{open},$$
$$\text{such that } U \cap Y = V \text{ and } \exists\, g \in \mathcal{O}_X(U) \text{ such that } g|_V = f\}.$$

Unfortunately, this formula only defines a presheaf $\mathcal{O}_{0,Y}$ in general. For the gluing condition to be satisfied, we have to "localise" the definition, *i.e.*, we have to consider the sheafification $\mathcal{O}_Y = \mathcal{O}_{0,Y}^{+}$ of this presheaf (*cf.* 1.5.c).

---

[1] By abuse of notation we will often write X when we mean the variety $(X, \mathcal{O}_X)$.

Proposition-Definition 4.6. *Let X be an algebraic variety and let Y be a closed set in X. We define a sheaf of rings $\mathcal{O}_Y$ on Y by setting*

$$\mathcal{O}_Y(V) = \{f : V \to k \mid \forall x \in V, \exists U \subset X, \text{ open,}$$
$$\text{with } x \in U \text{ and } g \in \mathcal{O}_X(U) \text{ such that } g|_{U \cap V} = f|_{U \cap V}\}$$

for any open V in Y. If X is an algebraic variety (resp. an affine algebraic variety), then the same is true of Y with the sheaf $\mathcal{O}_Y$ and the inclusion of Y in X is a morphism.

Proof. It will be enough to prove the result for affine X since the general case then immediately follows. We assume that X is affine: it will then be enough to show that the sheaf $\mathcal{O}_Y$ is equal to the sheaf $\mathcal{R}_Y$ of regular functions on Y. Consider $f \in \Gamma(X)$ and its image in $\Gamma(Y)$, $\overline{f}$. By Lemma 2.2, it will be enough to show that $\mathcal{R}(D(\overline{f})) = \mathcal{O}_{0,Y}(D(\overline{f}))$. Since we know that $D(\overline{f}) = D(f) \cap Y$ and the restriction homomorphism $\Gamma(X)_f \to \Gamma(Y)_{\overline{f}}$ is surjective, we know that $\mathcal{R}(D(\overline{f})) = \Gamma(Y)_{\overline{f}} \subset \mathcal{O}_{0,Y}(D(\overline{f}))$. Conversely, if $\overline{s} \in \mathcal{O}_{0,Y}(D(\overline{f}))$, then $\overline{s}$ is the restriction of a section $s \in \mathcal{O}_X(U)$ defined on an open set U in X such that $U \cap Y = D(\overline{f})$. We cover U with open sets $U_i = D(g_i)$. Their restrictions $D(\overline{g}_i)$ cover $D(\overline{f})$, and since $s|_{U_i} \in \mathcal{O}_X(U_i) = \Gamma(X)_{g_i}$, $\overline{s}|_{D(\overline{g}_i)} \in \Gamma(Y)_{\overline{g}_i} = \mathcal{R}_Y(D(\overline{g}_i))$. But then $\overline{s} \in \mathcal{R}_Y(D(\overline{f}))$ since $\mathcal{R}_Y$ is a sheaf. QED.

Remark 4.7. In fact, what we are trying to prove is that there is a surjective map of sheaves from $\mathcal{O}_X$ to $\mathcal{O}_Y$ (*cf.* §6 below). We note that if V is an open affine set of Y which is the restriction to Y of an open affine set U in X, then the proof above shows that any regular function on V is the image of a regular function on U. In this case, the problem mentioned above does not arise. On the other hand, we have to be more careful when the open sets are not affine, as the example below shows (*cf.* also Exercise III, A.3):

Example 4.8. Let us take $V = \mathbf{P}^2$ (*cf.* §8), let W be the closed set $V(X(X - T), XY)$, a union of the y-axis and the point $(1, 0, 1)$, and consider W as an open set of itself. We seek open sets $\Omega = \mathbf{P}^2 - Z$ in $\mathbf{P}^2$ containing W: this is equivalent to saying that the closed set Z does not meet W and in particular does not meet $V(X)$. By Bézout's theorem, this implies that Z is finite (*cf.* Chapter VI). But then $\Gamma(\Omega, \mathcal{O}_{\mathbf{P}^2}) = \Gamma(\mathbf{P}^2, \mathcal{O}_{\mathbf{P}^2}) = k$ (*cf.* 8.8 and Exercise III, A.2.b). The only functions on W arising from open sets on V are therefore constant functions. However, if $\mathcal{O}_W$ is a sheaf, then there are obviously other functions on W, such as the functions which are constant on each of the two connected components of W.

c) *Subvarieties.* We merge the above two examples.

Definition 4.9. *Let X be an algebraic variety and let Y be a locally closed subset in X (i.e., the intersection of an open set and a closed set). Then Y, equipped with the variety structure defined in a) and b), is called an algebraic subvariety of X.*

5 Local rings

Definition 5.1. *Let X be an algebraic variety and consider $x \in X$. We consider the pairs (U, f), where U is an open set in X containing x and $f \in \Gamma(U, \mathcal{O}_X)$. Two such pairs, (U, f) and (V, g), are said to be equivalent if there is an open set W such that $x \in W \subset U \cap V$ and $f|_W = g|_W$. The equivalence classes for this relation are called the* germs *of functions at x. The germ of (U, f) at x is denoted by f_x. The set of germs at x is denoted $\mathcal{O}_{X,x}$.*

Proposition-Definition 5.2. *With the above notations, the set $\mathcal{O}_{X,x}$ is canonically equipped with a ring structure. This ring is a local k-algebra with maximal ideal $m_{X,x} = \{f \in \mathcal{O}_{X,x} \mid f(x) = 0\}$. We call it the* local ring *of X at x. We have $\mathcal{O}_{X,x}/m_{X,x} \simeq k$.*

Proof. The ring structure is defined in the following way. Given two germs (U, f) and (V, g), we add and multiply them by first restricting to $U \cap V$ and then using the ring structure on $\Gamma(U \cap V, \mathcal{O}_X)$. We check that these operations are well-defined (*i.e.*, do not depend on the choice of representatives). And finally, if $f \in \mathcal{O}_{X,x}$, then we denote by $f(x)$ the value of f at x, which does not depend on the choice of representative either. There is therefore a ring homomorphism $\pi : \mathcal{O}_{X,x} \to k$ which associates $f(x)$ to f. This homomorphism is obviously surjective and its kernel is $m_{X,x}$, which is therefore a maximal ideal. On the other hand, if $f \in \mathcal{O}_{X,x} - m_{X,x}$, then f is invertible in $\mathcal{O}_{X,x}$ (which proves that this ring is local). Indeed, lift f to (U, f). After possibly shrinking U we can assume that U is affine. We then have $x \in D_U(f)$, but f is then invertible in $\Gamma(D_U(f), \mathcal{O}_U) = \Gamma(D_U(f), \mathcal{O}_X)$ and hence f is invertible in $\mathcal{O}_{X,x}$.

Remark 5.3. Of course, the terminology "local ring" comes from this kind of example. The advantage of the local ring over the rings of functions on affine open sets containing x is that it is intrinsic.

Proposition 5.4. *Assume that k is algebraically closed. Let X be an algebraic variety, take a point $x \in X$ and let U be an affine open set containing x. We set $A = \Gamma(U, \mathcal{O}_X)$. Let m be the maximal ideal in A corresponding to the point x (cf. Chapter I, 4.9). We then have $\mathcal{O}_{X,x} \simeq A_m$. In particular, the prime ideals of $\mathcal{O}_{X,x}$ correspond bijectively to closed irreducible subsets of U (and hence of X) containing x.*

Proof. There is a homomorphism r from A to $\mathcal{O}_{X,x}$ which associates to a its germ at x, a_x. This homomorphism factorises through A_m since if $a \notin m$, then $a(x) \neq 0$, and hence a_x is invertible. It is easy to check (using the open sets $D(f)$) that this gives us the required isomorphism. The last claim follows from Chapter I, 4.9 (to obtain the result on the whole of X we simply use closure, *cf.* Chapter I, 3.5).

Remark 5.5. Let $\varphi : X \to Y$ be a morphism of varieties and consider points $x \in X$ and $y = \varphi(x)$. This induces a homomorphism $\varphi^* : \mathcal{O}_{Y,y} \to \mathcal{O}_{X,x}$. Indeed, if $f_y \in \mathcal{O}_{Y,y}$ is represented by a pair (U, f), where U is an open set in Y containing y, then by definition of a morphism there is a homomorphism $\varphi^* : \Gamma(U, \mathcal{O}_Y) \to \Gamma(\varphi^{-1}(U), \mathcal{O}_X)$ which associates the composition $f\varphi$ to f. We then associate to f_y the germ $(f\varphi)_x$, and this gives us the required map.

We note that this homomorphism is *local*, *i.e.*, it sends $m_{Y,y}$ into $m_{X,x}$. If φ is an isomorphism, then so is φ^*.

6 Sheaves of modules

Let $(X, \mathcal{O}_X)$ be a ringed space (for example, an algebraic variety). In this section we will study sheaves of modules. This is a key notion whose usefulness will become apparent as we go on.

Definition 6.1. *A $\mathcal{O}_X$-module is a sheaf $\mathcal{F}$ such that for any open set U in X, $\mathcal{F}(U)$ is a $\mathcal{O}_X(U)$-module and the restriction maps are linear maps.*

Remark 6.2. Warning: the use of the word "linear" in this context may not be what you expect. If $V \subset U$, then there are restriction maps $r : \mathcal{O}_X(U) \to \mathcal{O}_X(V)$ and $\rho : \mathcal{F}(U) \to \mathcal{F}(V)$; $\mathcal{F}(U)$ and $\mathcal{F}(V)$ are therefore both $\mathcal{O}_X(U)$-modules; $\mathcal{F}(U)$ is a $\mathcal{O}_X(U)$-module by definition and $\mathcal{F}(V)$ becomes one via r. We then ask that ρ should be $\mathcal{O}_X(U)$ linear, *i.e.*, $\rho(af) = r(a)\rho(f)$. Since $\mathcal{O}_X(U)$ is a k-algebra, all the spaces $\mathcal{F}(U)$ are k-vector spaces.

Example 6.3. The null sheaf is clearly a $\mathcal{O}_X$-module. A finite direct sum (or finite direct product, which is the same thing) of $\mathcal{O}_X$-modules is a $\mathcal{O}_X$-module. For example, $\mathcal{O}_X^n$, the direct sum of n copies of the structural sheaf, is a $\mathcal{O}_X$-module.

Essentially all the usual A-module constructions, such as homomorphisms, kernels, images, exact sequences, etc., are also possible with $\mathcal{O}_X$-modules.

Definition 6.4. *Let $\mathcal{F}, \mathcal{G}$ be two $\mathcal{O}_X$-modules. A homomorphism $f : \mathcal{F} \to \mathcal{G}$ is given by the data of $\mathcal{O}_X(U)$-linear maps for every U, $f(U) : \mathcal{F}(U) \to \mathcal{G}(U)$ which are compatible in the obvious way with restrictions.*

We can then define the kernel sheaf of f by the formula

$$(\operatorname{Ker} f)(U) = \operatorname{Ker}(f(U)),$$

and we say that f is injective if $f(U)$ is injective for all U, or, alternatively, if $\operatorname{Ker} f = 0$.

On the other hand, we have to be more careful in defining image sheaves and surjective maps. This is one of the fundamental difficulties in sheaf theory. Consider a homomorphism $f : \mathcal{F} \to \mathcal{G}$. It is tempting to define its image by the formula $(\operatorname{Im} f)(U) = \operatorname{Im}(f(U))$. Unfortunately, in general this is not a sheaf, as the following example (which was historically the first such example) shows. (The objects here are in fact abelian groups, not modules, but the principle is the same.)

Example 6.5. We set $X = \mathbf{C}$ and let $\mathcal{F} = \mathcal{O}_{\mathbf{C}}$ be the sheaf of holomorphic functions on the open sets of $\mathbf{C}$. We let $\mathcal{G} = \mathcal{O}_{\mathbf{C}}^*$ be the sheaf of non-vanishing holomorphic functions. There is a homomorphism $f : \mathcal{F} \to \mathcal{G}$ given by the exponential map. If we define $\mathrm{Im}\, f$ as above, this is not a sheaf: the gluing condition is not satisfied. We set $U = \mathbf{C} - \{0\}$. This open set is covered by open sets V and W which are $\mathbf{C}$ without the positive and negative real half-lines. On V and W the identity function z is in the image of the exponential map (these open sets are simply connected and hence the function $\mathrm{Log}(z)$ exists), but, over U, z is not in the image of f since U is not simply connected.

To get around this problem we "localise" and define $\mathrm{Im}\, f$ to be the sheaf associated to this presheaf (*cf.* 1.5.c):

Definition 6.6. *Let* $f : \mathcal{F} \to \mathcal{G}$ *be a homomorphism of* $\mathcal{O}_X$-*modules. We define the sheaf* $\mathrm{Im}\, f$ *as follows. Consider* $s \in \mathcal{G}(U)$*: we say that* $s \in (\mathrm{Im}\, f)(U)$ *if for all* $x \in U$ *there is an open set* $V \subset U$ *such that* $x \in V$ *and* $s|_V \in \mathrm{Im}(f(V))$. *We say that* f *is surjective if* $\mathrm{Im}\, f = \mathcal{G}$.

Remark 6.7. The map f is said to be surjective if it is *locally* surjective. This is true of the exponential map in Example 6.5.

Definition 6.8. *An exact sequence of* $\mathcal{O}_X$-*modules* $\mathcal{F} \xrightarrow{u} \mathcal{G} \xrightarrow{v} \mathcal{H}$ *is given by the data of two homomorphisms of sheaves* u *and* v *such that* $\mathrm{Ker}(v) = \mathrm{Im}(u)$.

We now return to closed subvarieties. We will need the following definition.

Definition 6.9. *Let* $\varphi : Y \to X$ *be a continuous map and let* $\mathcal{F}$ *be a sheaf on* Y. *The direct image of* $\mathcal{F}$, *denoted by* $\varphi_* \mathcal{F}$, *is the sheaf defined on* X *by* $\varphi_* \mathcal{F}(U) = \mathcal{F}(\varphi^{-1}(U))$ *for any open set* U *in* X.

Example 6.10. Let X be an algebraic variety, let Y be a closed subset of X and let $j : Y \to X$ be the canonical injection. We denote by $\mathcal{F}_Y$ the sheaf of all k-valued functions on Y and consider its direct image $j_* \mathcal{F}_Y$. There is a morphism of $\mathcal{O}_X$-modules $r : \mathcal{O}_X \to j_* \mathcal{F}_Y$ which associates to $s \in \mathcal{O}_X(U)$ its restriction $s|_{U \cap Y} \in j_* \mathcal{F}_Y(U) = \mathcal{F}_Y(U \cap Y)$. Using the definition of the structural sheaf on Y given in 4.6 the sheaf $j_* \mathcal{O}_Y$ is then simply the image sheaf of r.

In what follows we will often identify $\mathcal{O}_Y$ and $j_* \mathcal{O}_Y$, which allows us to consider $\mathcal{O}_Y$ as a $\mathcal{O}_X$-module. (See Exercise II, A.9 for a justification of this identification.)

We also consider the kernel of r which we denote by $\mathcal{J}_Y$ (or, alternatively, $\mathcal{J}_{Y/X}$ when we want to keep track of the original space). This is a $\mathcal{O}_X$-module: even better, it is an *ideal* (or sheaf of ideals) in $\mathcal{O}_X$ (in other words, $\mathcal{J}_Y(U)$ is an ideal in $\mathcal{O}_X(U)$ for all U).

After identifying $\mathcal{O}_Y$ and its direct image we get the following exact sequence of $\mathcal{O}_X$-modules (called the fundamental exact sequence associated to the closed space Y):

$$0 \longrightarrow \mathcal{J}_Y \longrightarrow \mathcal{O}_X \longrightarrow \mathcal{O}_Y \longrightarrow 0.$$

Definition 6.11. *Let $\mathcal{F}$ and $\mathcal{G}$ be two $\mathcal{O}_X$-modules. We define the tensor product $\mathcal{F} \otimes_{\mathcal{O}_X} \mathcal{G}$ to be the sheaf associated to the presheaf $U \mapsto \mathcal{F}(U) \otimes_{\mathcal{O}_X(U)} \mathcal{G}(U)$. This is once again a $\mathcal{O}_X$-module.*

7 Sheaves of modules on an affine algebraic variety

Let V be an affine algebraic variety and consider $A = \Gamma(V, \mathcal{O}_V)$. If $\mathcal{F}$ is a $\mathcal{O}_V$-module, then $\Gamma(V, \mathcal{F})$ is an A-module and this correspondence is functorial by 6.4: it is the global sections functor. The aim of this section is to find an "inverse" functor.

Definition 7.1. *Let M be an A-module. We define a $\mathcal{O}_V$-module $\widetilde{M}$ on the standard open sets of V in the following way. If $f \in A$, then we set $\widetilde{M}(D(f)) = M_f = M \otimes_A A_f$. In particular, $\widetilde{M}(V) = \Gamma(V, \widetilde{M}) = M$.*

Remarks 7.2.

1) We can describe the localised module M_f as being the set of pairs (x, s) such that $x \in M$ and $s = f^n$, quotiented by the equivalence relation

$$(x, s) \sim (y, t) \iff \exists u = f^r \quad u(xt - ys) = 0.$$

(NB: the u term may be necessary if A is not an integral domain or M is not torsion free, *i.e.*, if it is possible to have $ax = 0$ satisfied with $a \in A$ and $x \in M$ both non-zero.) We denote the image of (x, s) by x/s. This image is also denoted $x \otimes (1/s)$.

2) We check as in 2.3 that this does indeed define a sheaf.

3) In particular, $\widetilde{A} = \mathcal{O}_V$.

Proposition 7.3. *The correspondence $M \mapsto \widetilde{M}$ is functorial and exact and commutes with direct sums and tensor products.*

Proof.

1) Functoriality. Let $\varphi : M \to N$ be a homomorphism of A-modules. By functoriality of the tensor product there is a map $\varphi_f : M_f = M \otimes_A A_f \to N_f = N \otimes_A A_f$. The functoriality of the correspondence follows.

2) Exactness. Let $0 \to M' \to M \to M'' \to 0$ be an exact sequence of A-modules. Then the sequence $0 \to M'_f \to M_f \to M''_f \to 0$ obtained by localisation is exact. It will be enough to prove the injectivity of $i : M'_f \to M_f$ since the rest is immediate by basic properties of the tensor product (*cf.* Summary 2.2). Assume $i(x'/f^n) = 0$ in M_f. This implies $f^r x' = 0$ in M and hence $f^r x' = 0$ in M', so $x/f^n = 0$ in M'_f. (A more sophisticated way of saying the same thing is that the A-module A_f is flat, *cf.* [Bbki] AC II.)

3) Sums and products. We have $(M \oplus M')_f = M_f \oplus M'_f$ and $(M \otimes_A M')_f = M_f \otimes_{A_f} M'_f$. These formulae are left to the reader as an exercise.

Example 7.4: The exact sequence associated to a closed subset. Let W be a closed subset of V defined by an ideal $I = I_V(W)$ in A. Then we have an exact sequence $0 \to I \to A \to A/I \to 0$ which, on passing to sheaves, gives us $0 \to \widetilde{I} \to \widetilde{A} \to \widetilde{A/I} \to 0$, which is simply the fundamental exact sequence $0 \to \mathcal{J}_W \to \mathcal{O}_V \to \mathcal{O}_W \to 0$. (This can be proved by noting that W is an affine variety and $\Gamma(W) = A/I$, or, alternatively, by a direct calculation in a standard open set $D(f)$.)

Definition 7.5. *A $\mathcal{O}_V$-module isomorphic to a $\mathcal{O}_V$-module of type $\widetilde{M}$ is said to be quasi-coherent. If M is of finitely generated over A, we say that $\widetilde{M}$ is coherent. We will sometimes simply say a quasi-coherent sheaf rather than a quasi-coherent $\mathcal{O}_V$-module.*

Remark 7.6. The functor $M \mapsto \widetilde{M}$ is an equivalence of categories between the category of A-modules and the category of quasi-coherent $\mathcal{O}_V$-modules (this functor has a "quasi-inverse", namely the functor of global sections Γ). Once again, in the affine case, everything essentially depends on the ring $A = \Gamma(V)$. We note that there are non quasi-coherent sheaves on affine algebraic varieties (*cf.* [H] II 5.2.3 or Exercise III, B.5 below) but we will not deal with them in this course.

When dealing with quasi-coherent sheaves on affine algebraic varieties, there is no fundamental difficulty with surjectivity of homomorphisms of sheaves.

Proposition 7.7. *Let $0 \to \mathcal{F}' \to \mathcal{F} \to \mathcal{F}'' \to 0$ be an exact sequence of quasi-coherent sheaves on an affine algebraic variety X. We have the following exact sequence:*

$$0 \longrightarrow \Gamma(X, \mathcal{F}') \longrightarrow \Gamma(X, \mathcal{F}) \xrightarrow{\ \pi\ } \Gamma(X, \mathcal{F}'') \longrightarrow 0.$$

Proof. The only problem is to prove the surjectivity of π. This will be done in Chapter VII when we prove that the cohomology group $H^1(X, \mathcal{F}')$ is zero.

The following proposition, whose proof can be found in [H] II §5, shows that the property of being quasi-coherent is local.

Proposition 7.8. *Let X be an affine algebraic variety and let $\mathcal{F}$ be a $\mathcal{O}_X$-module. Then $\mathcal{F}$ is quasi-coherent (resp. coherent) if and only if there is an affine open covering U_i of X such that $\Gamma(U_i, \mathcal{O}_X) = A_i$ and A_i-modules M_i (resp. of finite type) such that, for every i, $\mathcal{F}|_{U_i} \simeq \widetilde{M_i}$.*

This proposition justifies the following definition.

Definition 7.9. *Let X be an algebraic variety and let $\mathcal{F}$ be a $\mathcal{O}_X$-module. We say that $\mathcal{F}$ is quasi-coherent (resp. coherent) if there is an open affine covering of X, U_i, such that $\mathcal{F}|_{U_i}$ is quasi-coherent (resp. coherent) on each U_i.*

We have the following proposition on tensor products of quasi-coherent sheaves.

Proposition 7.10. *Let X be a variety and let $\mathcal{F}$ and $\mathcal{G}$ be two quasi-coherent sheaves on X. Then*

a) The sheaf $\mathcal{F} \otimes_{\mathcal{O}_X} \mathcal{G}$ is quasi-coherent;

b) For any affine open set U in X

$$(\mathcal{F} \otimes_{\mathcal{O}_X} \mathcal{G})(U) = \mathcal{F}(U) \otimes_{\mathcal{O}_X(U)} \mathcal{G}(U).$$

Proof. If U is an open set, then $(\mathcal{F} \otimes_{\mathcal{O}_X} \mathcal{G})|_U = \mathcal{F}|_U \otimes_{\mathcal{O}_U} \mathcal{G}|_U$ (since these two sheaves are associated to the same presheaf $W \mapsto \mathcal{F}(W) \otimes_{\mathcal{O}_X(W)} \mathcal{G}(W)$ on U). If U is affine and we set $A = \Gamma(U, \mathcal{O}_X)$, $F = \Gamma(U, \mathcal{F})$, $G = \Gamma(U, \mathcal{G})$, then $(\mathcal{F} \otimes_{\mathcal{O}_X} \mathcal{G})|_U = \widetilde{F} \otimes_{\widetilde{A}} \widetilde{G} = \widetilde{F \otimes_A G}$, which proves the above two claims.

8 Projective varieties

a. Definition of the structure sheaf

We use the same notations as in Chapter II.

Let $V \subset \mathbf{P}^n$ be a projective algebraic set with its Zariski topology. We will put an algebraic variety structure on V. Of course, it will be enough to define the structure sheaf on the basis of open sets $D^+(f)$, where $f \in \Gamma_h(V)$ is a homogeneous polynomial of positive degree (*cf.* Chapter II, 6.2). To get an idea of how to define this sheaf, let us look at the open set $U_0 = D^+(X_0)$. We know that there is a bijection $j : k^n \to U_0$ defined by $(\xi_1, \ldots, \xi_n) \mapsto (1, \xi_1, \ldots, \xi_n)$ whose inverse is $(x_0, \ldots, x_n) \mapsto (x_1/x_0, \ldots, x_n/x_0)$. On U_0 the good functions correspond to the polynomial functions on k^n, namely the polynomials in $x_1/x_0, \ldots, x_n/x_0$. In other words, we set $\Gamma(U_0, \mathcal{O}_{\mathbf{P}^n}) = k[X_1/X_0, \ldots, X_n/X_0]$. This ring is contained in the localisation of $k[X_0, \ldots, X_n]$ relative to X_0: however, considering only polynomials in variables X_i/X_0 is equivalent to taking only those members of the localised ring of the form F/X_0^r, with F homogeneous of degree r. This leads us to the following definition.

Definition 8.1. *Let R be a graded ring and let $f \in R$ be a homogeneous element of degree d. We define a grading on the localised ring R_f by setting $\deg(g/f^r) = e - rd$ whenever g is a homogeneous element of R of degree e. The set of elements of degree 0 of R_f is then a subring which we denote by $R_{(f)}$.*

Remark 8.2. If we consider $R = \Gamma_h(V)$ and a homogeneous $f \in R$ of degree > 0, then the elements of $R_{(f)}$ have the following great advantage: they define functions on the open set $D^+(f)$. A change of homogeneous coordinates multiplies the numerator and denominator by the same scalar.

Definition 8.3. *Let V be a projective algebraic set. We define a sheaf of k-valued functions on V by setting*

$$\Gamma(D^+(f), \mathcal{O}_V) = \Gamma_h(V)_{(f)}$$

for any homogeneous $f \in \Gamma_h(V)$ of degree > 0.

The reader will check using Remark II.6.1.b that, as in 2.4, the conditions of 2.1 are satisfied and so this definition does indeed give us a sheaf on V. NB: it is important that we only consider elements f of degree > 0. If it were possible to apply 8.3 to $f = 1$, we would have $D^+(f) = V$ and the global sections of $\mathcal{O}_V$ would be the degree 0 elements of $\Gamma_h(V)$, *i.e.*, the constant functions. But if V is not connected, there are other possible sections on V— namely the functions which are constant on each component of V.

Proposition-Definition 8.4. *Let V be a projective algebraic set. The ringed space $(V, \mathcal{O}_V)$ (with the sheaf defined above) is then an algebraic variety. An algebraic variety which is isomorphic to a projective algebraic set (resp. an open set in a projective algebraic set) with the sheaf defined above will be called a projective variety (resp. a quasi-projective variety).*

Proof. We start by reducing to the case $V = \mathbf{P}^n$. Assume $V \subset \mathbf{P}^n$. Consider $f \in \Gamma_h(V)$, the image of a homogeneous polynomial $F \in k[X_0, \ldots, X_n]$. We have $D^+(f) = V \cap D^+(F)$. Moreover, the restriction homomorphism

$$r : \Gamma(D^+(F), \mathcal{O}_{\mathbf{P}^n}) \longrightarrow \Gamma(D^+(f), \mathcal{O}_V)$$

is clearly surjective. But $\mathcal{O}_V$ is then the image of the sheaf $\mathcal{O}_{\mathbf{P}^n}$ in the sheaf of functions in V (*cf.* 2.2), and if $\mathbf{P}^n$ is a variety, then V is simply the closed subvariety supported on V (*cf.* Example 6.10).

For $\mathbf{P}^n$ it will be enough to show that the open sets $D^+(X_i)$ are affine varieties and by homography it will in fact be enough to prove this for $D^+(X_0)$ (it is clear that homographies are automorphisms of $\mathbf{P}^n$ with its ringed space structure). This is essentially a formal translation of the affine-projective link seen in Chapter II and will be dealt with in the next paragraph.

b. The affine-projective link

We set $U_0 = D^+(X_0)$, the set of points of $\mathbf{P}^n$ whose x_0 coordinate is $\neq 0$. There is a bijection $j : k^n \to U_0$, defined by

$$(\xi_1, \ldots, \xi_n) \longmapsto (1, \xi_1, \ldots, \xi_n),$$

whose converse is given by

$$(x_0, \ldots, x_n) \longmapsto (x_1/x_0, \ldots, x_n/x_0).$$

We consider the Zariski topology on k^n and the Zariski topology induced on U_0 by $\mathbf{P}^n$. The following proposition completes the proof of 8.4.

Proposition 8.5.

1) j is a homeomorphism.

2) j is an isomorphism of ringed spaces between the affine variety $(k^n, \mathcal{O}_{k^n})$ and $(U_0, \mathcal{O}_{\mathbf{P}^n}|_{U_0})$.

To prove this proposition we study the homogenisation and dehomogenisation operators in detail. If P is a homogeneous polynomial of degree d in $T_1, \ldots, T_m$, then the formula

$$(0) \qquad P\left(\frac{T_1}{T_i}, \ldots, \frac{T_m}{T_i}\right) = \frac{P(T_1, \ldots, T_m)}{T_i^d}.$$

holds in the field of fractions. We will use the following convention: polynomials in $k[X_0, \ldots, X_n]$ will be written using capital letters and those in $k[X_1, \ldots, X_n]$ will be written using small letters.

i) *The $\flat$ operator.* This is a surjective ring homomorphism $k[X_0, \ldots, X_n] \to k[X_1, \ldots, X_n]$ given by

$$P(X_0, \ldots, X_n) \longmapsto P_\flat(X_1, \ldots, X_n) = P(1, X_1, \ldots, X_n).$$

Its kernel is the ideal $(X_0 - 1)$. We are particularly interested in the case where P is homogeneous of degree d. In this case we have (in the field of fractions $k(X_0, \ldots, X_n)$)

$$(1) \qquad P_\flat\left(\frac{X_1}{X_0}, \ldots, \frac{X_n}{X_0}\right) = P\left(\frac{X_0}{X_0}, \frac{X_1}{X_0}, \ldots, \frac{X_n}{X_0}\right) = \frac{P(X_0, \ldots, X_n)}{X_0^d}.$$

We note that if P is homogeneous of degree d, then $P_\flat$ is of degree d if and only if X_0 does not divide P.

ii) *The operation $\sharp$.* NB: this operation is not a homomorphism. Contrary to the operation $\flat$, it goes from $k[X_1, \ldots, X_n]$ into $k[X_0, \ldots, X_n]$: if $p \in k[X_1, \ldots, X_n]$, $p^\sharp$ is the homogeneous polynomial of smallest degree of $k[X_0, \ldots, X_n]$ such that $p = (p^\sharp)_\flat$. We can describe it as follows: if $p = p_0 + p_1 + \cdots + p_d$, where p_i is homogeneous of degree i and $p_d \neq 0$, then we set $p^\sharp(X_0, \ldots, X_n) = X_0^d p_0 + X_0^{d-1} p_1 + \cdots + p_d$, or, alternatively,

$$(2) \qquad p^\sharp(X_0, \ldots, X_n) = X_0^d\, p\left(\frac{X_1}{X_0}, \ldots, \frac{X_n}{X_0}\right).$$

iii) *Some remarks.* If $p, q \in k[X_1, \ldots, X_n]$, then $(pq)^\sharp = p^\sharp q^\sharp$ (this follows from Formula (2)).

If $p \in k[X_1, \ldots, X_n]$, then $(p^\sharp)_\flat = p$.

If $P \in k[X_0, \ldots, X_n]$ is homogeneous, then $P = (X_0)^r (P_\flat)^\sharp$, where X_0^r is the largest power of X_0 dividing P. (We reduce to the case where X_0 does not divide P and use (1) and (2), noting that the degree does not decrease on passing from P to $P_\flat$.)

It follows that if P is homogeneous and $P_\flat = 0$, then $P = 0$.

iv) *j is a homeomorphism.* Since the sets $D^+(F)$ and $D(f)$ are bases of open sets on $\mathbf{P}^n$ and k^n respectively, this follows from the following two formulae whose proof is immediate.

a) If $F \in k[X_0, \ldots, X_n]$ is homogeneous, then

$$j^{-1}(D^+(F)) = j^{-1}(D^+(F) \cap U_0) = D(F_\flat),$$

b) If $f \in k[X_1, \ldots, X_n]$, then $j(D(f)) = D^+(f^\sharp) \cap U_0$.

v) *The sheaf isomorphism.* Consider a homogeneous $F \in k[X_0, \ldots, X_n]$ of degree d. It will be enough to prove there is an isomorphism

$$\Gamma(D^+(F) \cap U_0, \mathcal{O}_{\mathbf{P}^n}) \simeq \Gamma(D(F_\flat), \mathcal{O}_{k^n}).$$

We note that $D^+(F) \cap U_0 = D^+(FX_0)$.
We have $\Gamma(D^+(F) \cap U_0, \mathcal{O}_{\mathbf{P}^n}) = k[X_0, \ldots, X_n]_{(FX_0)}$ and the elements of this ring are functions of the form $P/(FX_0)^r$, where P is homogeneous of degree $r(d+1)$. They are also exactly the elements of the form $P/F^r X_0^s$, where P is of degree $rd + s$ (it is enough to multiply the numerator by a power of F or X_0 to recover the above form). On the other hand, $\Gamma(D(F_\flat), \mathcal{O}_{k^n}) = k[X_1, \ldots, X_n]_{F_\flat}$.
To define the required map φ we start with the morphism $\flat$: $k[X_0, \ldots, X_n] \to k[X_1, \ldots, X_n]$. Since $(FX_0)_\flat = F_\flat$, this homomorphism induces a homomorphism ψ on local rings

$$\psi : k[X_0, \ldots, X_n]_{FX_0} \longrightarrow k[X_1, \ldots, X_n]_{F_\flat},$$

and we construct φ by composing ψ with the natural injection

$$i : k[X_0, \ldots, X_n]_{(FX_0)} \longrightarrow k[X_0, \ldots, X_n]_{FX_0}.$$

It remains to prove that φ is indeed an isomorphism. Note that $\varphi(P/F^r X_0^s) = P_\flat/F_\flat^r$.

1) φ is injective.
If $P_\flat/F_\flat^r = 0$, then $P_\flat = 0$ and hence $P = 0$ (*cf.* iii).

2) φ is surjective.
Consider $p/F_\flat^r \in k[X_1, \ldots, X_n]_{F_\flat}$; we have $p/F_\flat^r = \varphi(X_0^s p^\sharp / F^r)$, where $s = rd - \deg p \in \mathbf{Z}$.

This completes the proof of 8.5 (and hence also of 8.4).

Remarks 8.6.
1) Projective space $\mathbf{P}^n$ is irreducible. This follows from 8.5 and Exercise I, 4.b.

2) We will prove in 11.8 that all open sets $D^+(F)$ in $\mathbf{P}^n$ are open affine sets. It follows that the open sets $D^+(f)$ in $V \subset \mathbf{P}^n$ are also open affine sets.

In the following proposition we describe the local ring of $\mathbf{P}^n$ at a point. For a proof see Exercise III, B.4.

Proposition 8.7. *Consider* $x = (x_0, \ldots, x_n) \in \mathbf{P}^n$ *and* $I_x = I_p(\{x\})$ *the homogeneous prime ideal of polynomials vanishing at* x. *Then the local ring* $\mathcal{O}_{\mathbf{P}^n,x}$ *is* $k[X_0, \ldots, X_n]_{(I_x)}$, *the subring of the local ring* $k[X_0, \ldots, X_n]_{I_x}$ *consisting of elements of degree* 0. *If* $x_0 = 1$ *and we set* $\xi = (x_1, \ldots, x_n)$, *then the homomorphism* $\flat$ *induces an isomorphism from* $\mathcal{O}_{\mathbf{P}^n,x}$ *to* $\mathcal{O}_{k^n,\xi}$.

The following proposition confirms that there are no non-trivial global functions on a projective algebraic variety. It will be proved in Chapter VIII (*cf.* also Problem II, 3).

Proposition 8.8. *Let* X *be an irreducible projective algebraic variety. Then* $\Gamma(X, \mathcal{O}_X) = k$ *(the only global sections of* $\mathcal{O}_X$ *are the constant functions).*

9 Sheaves of modules on projective algebraic varieties

Let X be a projective algebraic variety equipped with an embedding into $\mathbf{P}^n$. We will give a definition of sheaves of modules on X similar to the definition of § 7 in the affine case but which acts on *graded* modules (*cf.* Chapter II, 7.5). We set $R = \Gamma_h(X)$. It is important to note that this ring, and particularly its grading, depends on the choice of embedding (*cf.* 11.6 below).

Definition 9.1. *Let* M *be a graded* R-*module. We define a* $\mathcal{O}_X$-*module* $\widetilde{M}$ *on the standard open sets of* X *as follows: if* $f \in R$ *is homogeneous of degree* > 0, *then* $\widetilde{M}(D^+(f)) = M_{(f)}$.

Of course, $M_{(f)}$ denotes the submodule of M_f (*cf.* § 7) consisting of elements of degree 0, *i.e.*, elements of the form x/f^n, where x is homogeneous of degree $n \deg f$.

Remarks 9.2.
  1) We have $\widetilde{R} = \mathcal{O}_X$.

  2) Restricting to the affine open set $D^+(f)$ whose ring is $R_{(f)}$ we check that the sheaf $\widetilde{M}$ is simply the sheaf $\widetilde{M_{(f)}}$ associated to the $R_{(f)}$-module $M_{(f)}$ defined in § 7 in the affine case. In particular, the $\mathcal{O}_X$-module $\widetilde{M}$ is quasi-coherent (and is coherent if M is an R-module of finite type). Conversely, all quasi-coherent $\mathcal{O}_X$-modules are obtained in this way (*cf.* 9.8.2).

We consider the category of graded R-modules whose morphisms are the homomorphisms of degree 0 (*i.e.*, degree-preserving morphisms *cf.* Chapter II, 7.5). We then have the following analogue of Proposition 7.3.

Proposition 9.3. *The correspondence* $M \mapsto \widetilde{M}$ *is functorial, exact and commutes with direct sums and tensor products.*

Proof. The proof is essentially identical to the proof of 7.3, except for the part concerning the tensor product. To do this we start by checking that we can define a graded R-module structure on $P = M \otimes_R N$ by setting $P_n = \sum_{p+q=n} M_p \otimes_k N_q$. We then construct a morphism $\varphi : \widetilde{M} \otimes_{\widetilde{R}} \widetilde{N} \to \widetilde{M \otimes_R N}$. It is enough to do this on the open sets $D^+(f)$, and on these sets we set $\varphi(x/f^r \otimes y/f^s) = (x \otimes y)/f^{r+s}$. We check that φ is an isomorphism of $\mathcal{O}_X$-modules. It is enough to do this on open sets $D^+(f)$ such that f of degree 1, since these open sets cover X. The inverse isomorphism is obviously given over such sets by $(x \otimes y)/f^r \mapsto x/f^p \otimes y/f^q$, where $p = \deg x$ and $q = \deg y$. (If f is not of degree 1, we still have an isomorphism which is slightly less trivial but which will not escape the attentive reader.)

For the exact sequence, the following result, which is more precise, holds.

Proposition 9.4. *Let* $\varphi : M \to N$ *be a homogeneous degree* 0 *morphism between graded R-modules. Assume that, for large n, $\varphi_n : M_n \to N_n$ is surjective. Then* $\widetilde{\varphi} : \widetilde{M} \to \widetilde{N}$ *is a surjective map of sheaves.*

Proof. We choose an open set $D^+(f)$, where f is homogeneous of degree > 0. It will be enough to show that $\varphi_{(f)} : M_{(f)} \to N_{(f)}$ is surjective. Consider $y/f^r \in N_{(f)}$. For large enough s, $f^s y$ is in the image of φ and hence so is $y/f^r = f^s y/f^{r+s}$. QED.

Example 9.5: The exact sequence associated to a closed set.

Let X be a projective algebraic variety, set $R = \Gamma_h(X)$ and let Y be a closed subset of X defined by a homogeneous ideal in R, $I = I_X(Y)$. There is then an exact sequence of graded R-modules $0 \to I \to R \to R/I \to 0$, which on sheafifying gives an exact sequence $0 \to \widetilde{I} \to \widetilde{R} \to \widetilde{R/I} \to 0$. Once again, this sequence is simply the fundamental exact sequence $0 \to \mathcal{J}_Y \to \mathcal{O}_X \to \mathcal{O}_Y \to 0$, as can be checked directly over the open sets $D^+(f)$.

We now define sheaves $\mathcal{O}_{\mathbf{P}^n}(d)$ and $\mathcal{O}_X(d)$ on a projective variety X. Their advantage over the structure sheaf is that whenever $d > 0$ they have global sections, namely the homogeneous polynomials of degree d. We start by defining shifted modules.

Definition 9.6. *Let R be a graded ring and let $M = \bigoplus_{n \in \mathbf{Z}} M_n$ be a graded R-module. The module $M(d)$ is the graded module which is equal to M except that the grading is shifted:* $M(d)_n = M_{d+n}$.

We note that a shift operation does not alter the exactness of exact sequences of modules.

Definition 9.7. *Let X be a projective algebraic variety embedded in $\mathbf{P}^n$ and consider $R = \Gamma_h(X)$. The sheaf $\mathcal{O}_X(d)$ is the sheaf associated to the shifted module $R(d)$; $\mathcal{O}_X(d) = \widetilde{R(d)}$. If $\mathcal{F}$ is a $\mathcal{O}_X$-module, then we write $\mathcal{F}(d)$ for the sheaf $\mathcal{F} \otimes_{\mathcal{O}_X} \mathcal{O}_X(d)$ (cf. 6.11).*

The sections of $\mathcal{O}_X(d)$ over the open set $D^+(f)$ are therefore the degree d elements of $R(f)$, i.e., elements of the form a/f^r, where $\deg a - r \deg f = d$.

Remarks 9.8.

0) Warning. The sheaves $\mathcal{O}_X(d)$ depend fundamentally on the grading on R, and hence on the chosen embedding of X in $\mathbf{P}^n$.

1) If M is a graded R-module, then $\widetilde{M(d)} = \widetilde{M} \otimes_{\mathcal{O}_X} \mathcal{O}_X(d) = \widetilde{M}(d)$. (This follows from the fact that tensor products commute.)

2) We now seek an inverse functor. Since sheaves on projective varieties generally have too few functions, sections of the sheaves $\mathcal{F}(d)$ (for all d in $\mathbf{Z}$) will play the role in projective theory played by global sections of $\mathcal{F}$ in affine theory. However, the correspondence between sections and sheaves is less perfect in projective geometry. More precisely, if $\mathcal{F}$ is a $\mathcal{O}_X$-module, then we define a graded R-module $\Gamma_*(\mathcal{F})$ by the formula

$$\Gamma_*(\mathcal{F}) = \bigoplus_{d \in \mathbf{Z}} \Gamma(X, \mathcal{F}(d)).$$

(The module structure comes from the natural map

$$\Gamma(X, \mathcal{O}_X(p)) \otimes_k \Gamma(X, \mathcal{F}(q)) \longrightarrow \Gamma(X, \mathcal{F}(p+q)).)$$

We can check that if $\mathcal{F}$ is a sheaf, then $\mathcal{F} \simeq \widetilde{\Gamma_*(\mathcal{F})}$: in other words, Γ_* is a right inverse of $\sim$ (*cf.* [H] Chapter II, 5.15). In particular, a module is quasi-coherent (resp. coherent) if and only if it is of the form $\widetilde{M}$ (resp. $\widetilde{M}$ for some M of finite type).

3) NB: this functor is not, however, a left inverse. If the given sheaf $\mathcal{F}$ is of the form $\widetilde{M}$, then the modules M and $\Gamma_*(\mathcal{F})$ are not necessarily isomorphic. More precisely, there is a natural homomorphism $r_d : M_d \to \Gamma(X, \widetilde{M}(d))$ for each d which associates to $x \in M_d$ the element $x/1$ in $M_{(f)}(d)$ (note that these sections glue together to form a global section). This gives a homomorphism of degree 0, $r : M \to \Gamma_*(\mathcal{F})$, but we have to be careful because r is not generally either injective or surjective (*cf.* Chapter VII or consider the case where X is not connected, $M = R$ and $d = 0$). However, we will show in Chapter VII that if M is of finite type, then r_d is an isomorphism for large enough d. (The moral is that the sheaf $\widetilde{M}$ actually only depends on the terms of large degree in M.)

For $\mathbf{P}^n$, however, the global sections are what we expect.

Proposition 9.9. *Let R_d be the vector space of homogeneous polynomials of degree d in $X_0, \ldots, X_n$. Then*

$$\Gamma(\mathbf{P}^n, \mathcal{O}_{\mathbf{P}^n}(d)) = \begin{cases} 0 & \text{if } d < 0; \\ R_d & \text{if } d \geqslant 0. \end{cases}$$

In particular, $\Gamma(\mathbf{P}^n, \mathcal{O}_{\mathbf{P}^n}) = k$ (cf. 8.8).

Proof. Consider $f \in \Gamma(\mathbf{P}^n, \mathcal{O}_{\mathbf{P}^n}(d))$, $f \neq 0$. By definition its restriction to the open set $D^+(X_i)$ is a rational function of the form $P_i(X_0, \ldots, X_n)/X_i^r$, where P_i is homogeneous of degree $d + r$. After simplifying if necessary we can assume that X_i does not divide P. Likewise, $f|_{D^+(X_j)}$ is of the form $P_j(X_0, \ldots, X_n)/X_j^s$, where P_j is homogeneous of degree $d + s$ and X_j does not divide P_j. Since these elements are restrictions of f, they coincide on the intersection $D^+(X_i X_j)$ and hence are equal in the localised ring $k[X_0, \ldots, X_n]_{(X_i X_j)}$ or in the field of rational fractions $k(X_0, \ldots, X_n)$. It follows that $X_j^s P_i = X_i^r P_j$, but since X_i does not divide P_i, this is only possible if $r = 0$ and likewise $s = 0$, so $P_i = P_j$. The section f is therefore given by a homogeneous polynomial P_i of degree d which is independent of i.

Corollary 9.10. *We have*

$$\dim_k \Gamma(\mathbf{P}^n, \mathcal{O}_{\mathbf{P}^n}(d)) = \binom{n+d}{n}.$$

Remark. Unlike affine varieties, whose spaces of sections are rarely finite dimensional (this only happens for finite varieties *cf.* Chapter I, 4.8), the space of sections over projective space of a sheaf $\mathcal{O}_{\mathbf{P}^n}(d)$ is finite dimensional. This is a general phenomenon: we will show in Chapter VII that if $\mathcal{F}$ is a coherent sheaf on a projective algebraic variety X, then the space $\Gamma(X, \mathcal{F})$ is a finite-dimensional k-vector space.

10 Two important exact sequences

a. The exact sequence of a hypersurface

We work in projective space $\mathbf{P}^n$ and we consider a homogeneous polynomial of degree $d > 0$, $F \in R = k[X_0, \ldots, X_n]$, which is non-zero and has no multiple factors. Let $X = V_p(F)$ be the projective hypersurface defined by F. We know that the ideal $I(X)$ is equal to (F) by the Nullstellensatz. Multiplication by F induces an isomorphism of graded R-modules $R(-d) \to I(X)$. (Be careful when dealing with shifts: here we have to make sure that the constant 1, whose image is F, has the same degree, namely d, on both sides of the equation.) Passing to sheaves, we have the following exact sequence

$$0 \longrightarrow \mathcal{O}_{\mathbf{P}^n}(-d) \xrightarrow{\cdot F} \mathcal{O}_{\mathbf{P}^n} \longrightarrow \mathcal{O}_X \longrightarrow 0.$$

b. The exact sequence of a complete intersection

Proposition 10.1. *Let $F, G \in R = k[X_0, \ldots, X_n]$ be two homogeneous polynomials of degrees s and t without common factors and set $I = (F, G)$. We have an exact sequence of graded R-modules*

$$0 \longrightarrow R(-s-t) \xrightarrow{\varphi} R(-s) \oplus R(-t) \xrightarrow{\psi} I \longrightarrow 0,$$

where $\varphi(C) = (-CG, CF)$ and $\psi(A, B) = AF + BG$.

Proof. We note that the maps are indeed homogeneous of degree 0. It is clear that ψ is a surjective map onto I. Let us find its kernel: if $AF + BG = 0$, then $AF = -BG$, but since F and G are coprime, F divides B and hence $B = CF$ and it follows that $A = -CG$. And, finally, since R is an integral domain, it is clear that φ is injective.

Corollary 10.2. *With the above notation we consider* $V = V_p(F, G)$. *We assume that* $I_p(V) = I = (F, G)$ *(in other words, we assume this ideal is radical). We then have the following exact sequence:*

$$0 \longrightarrow \mathcal{O}_{\mathbf{P}^n}(-s-t) \longrightarrow \mathcal{O}_{\mathbf{P}^n}(-s) \oplus \mathcal{O}_{\mathbf{P}^n}(-t) \longrightarrow \mathcal{J}_V \longrightarrow 0.$$

When F and G have no common factors, we say that the variety $V = V_p(F, G)$ is a *complete intersection* of the hypersurfaces $V(F)$ and $V(G)$ and we express the fact that $I(V) = (F, G)$ by saying that V is a *scheme-theoretic complete intersection* of the hypersurfaces in question. NB: a set-theoretic complete intersection is not always a scheme-theoretic complete intersection (*cf.* the space cubic in Exercise II,4).

For more examples of calculations of "resolutions", see Exercise II.

11 Examples of morphisms

a. Some remarks on morphisms

Let $(X, \mathcal{O}_X)$ and $(Y, \mathcal{O}_Y)$ be algebraic varieties and let $\varphi : X \to Y$ be a map. We wish to know under what conditions φ is a morphism.

Proposition 11.1. *The fact of being a morphism is a local condition: given an open cover V_i of Y and for each i an open cover U_{ij} of $\varphi^{-1}(V_i)$, then φ is a morphism if and only if for all i, j, $\varphi|_{U_{ij}} : U_{ij} \to V_i$ is a morphism.*

Proof. Only the "if" part needs to be proved. We note that φ is continuous. If V is an open set of Y and $f \in \Gamma(V, \mathcal{O}_Y)$, then the restriction f_i of f to V_i is a section, and hence $f_i\varphi$ restricted to $U_{i,j}$ is a section, so $f\varphi$, obtained by gluing, is also a section.

Proposition 11.2. *If Y is affine, it is enough to consider global sections: φ is a morphism if and only if for any $f \in \Gamma(Y, \mathcal{O}_Y)$, $f\varphi \in \Gamma(X, \mathcal{O}_X)$.*

Proof.
   1) Assume that X is affine. Then φ is regular and we are done by 3.5.

   2) To prove the result in general, take an open affine covering of X: the proposition follows from 11.1 plus the case where X is affine.

Remarks 11.3.

1) It can be proved that the natural projection from $k^{n+1} - \{0\}$ to $\mathbf{P}^n$ is a morphism.

2) If $\varphi : X \to Y$ is a morphism and V and W are subvarieties of X and Y respectively such that $\varphi(V) \subset W$, then the restriction $\varphi|_V : V \to W$ is a morphism.

b. Applications to morphisms from $\mathbf{P}^n$ to $\mathbf{P}^m$

Consider $m + 1$ polynomials $F_0, \ldots, F_m \in k[X_0, \ldots, X_n]$, homogeneous of the same degree d. We then define a map

$$\varphi : \Omega = \mathbf{P}^n - V_p(F_0, \ldots, F_m) \longrightarrow \mathbf{P}^m$$

by setting $\varphi(x) = \varphi(x_0, \ldots, x_n) = (F_0(x), \ldots, F_m(x))$.

Don't forget that 1) the coordinates of a point in projective space are not all 0, which is why φ is not defined on the whole of $\mathbf{P}^n$, and 2) these coordinates are homogeneous, which is why all the polynomials F_i must be of the same degree.

Proposition 11.4. *With the above notations φ is a morphism of varieties.*

Proof. We apply 11.1 and 11.2 using the affine cover $D^+(X_i)$ of $\mathbf{P}^m$. The inverse image of $D^+(X_i)$ is the open set $D^+(F_i)$, and these open sets cover Ω. If g is in $\Gamma(D^+(X_i), \mathcal{O}_{\mathbf{P}^m})$, then $g = G(X_0, \ldots, X_m)/X_i^r$ and

$$g\varphi = G(F_0(X_0, \ldots, X_n), \ldots, F_m(X_0, \ldots, X_n))/F_i(X_0, \ldots, X_n)^r$$

is indeed an element of $\Gamma(D^+(F_i), \mathcal{O}_{\mathbf{P}^n})$.

Remark 11.5. If V is the closure of the image of φ, then Remark 11.3 shows that φ also defines a morphism from $\mathbf{P}^n - V_p(F_0, \ldots, F_m)$ into V.

Examples 11.6.

a) Parameterising conics. We consider the morphism $\varphi : \mathbf{P}^1 \to \mathbf{P}^2$ given by

$$\varphi(u, v) = (u^2, uv, v^2).$$

This morphism is defined on the whole of $\mathbf{P}^1$ since the polynomials u^2, uv, v^2 have no common zeros on $\mathbf{P}^1$. It is clear that its image is contained in the conic $C = V_p(XT - Y^2)$ and it is easy to see that $\varphi : \mathbf{P}^1 \to C$ is bijective: we will show that it is in fact an isomorphism.

To do this we note that C is covered by affine open sets $D^+(X)$ and $D^+(T)$ in $\mathbf{P}^2$. We know that these open sets are isomorphic to k^2. For $D^+(X)$, for example, the isomorphism is given by $(1, y, t) \mapsto (y, t)$. This induces an isomorphism j from $C \cap D^+(X)$ onto C_b, an affine conic defined by $y^2 - t = 0$.

We also have an isomorphism i from k onto $D^+(u) \subset \mathbf{P}^1$ given by $v \mapsto (1, v)$. Composing these maps, we get maps

$$k \xrightarrow{\ i\ } D^+(u) \xrightarrow{\ \varphi\ } D^+(X) \cap C \xrightarrow{\ j\ } C,$$

such that $v \mapsto (1, v) \mapsto (1, v, v^2) \mapsto (v, v^2)$, and since the composition of these three morphisms is clearly an isomorphism, φ is an isomorphism from $D^+(u)$ to $D^+(X) \cap C$. A similar argument with respect to v and T completes the proof of this result.

Warning: in this example, even though $\mathbf{P}^1$ and C are isomorphic, their associated graded rings are not the same. The morphism φ induces a homomorphism

$$\varphi^* : \Gamma_h(C) = k[X, Y, T]/(XT - Y^2) \longrightarrow \Gamma_h(\mathbf{P}^1) = k[U, V],$$

given by the three polynomials U^2, UV, V^2. But this homomorphism is not an isomorphism (its image is the subring $k[U^2, UV, V^2]$) and does not preserve the grading. Indeed, the two rings are not isomorphic (the localisation of $\Gamma_h(C)$ at the ideal (X, Y, T) is not regular because it corresponds to the vertex of the affine cone $V(XT - Y^2)$, cf. Chapter V).

b) *The space cubic.* We now consider the morphism $\varphi : \mathbf{P}^1 \to \mathbf{P}^3$ given by $\varphi(u, v) = (u^3, u^2 v, uv^2, v^3)$ (cf. Exercises II). A similar argument shows that φ is an isomorphism from $\mathbf{P}^1$ to the curve

$$C = V(XT - YZ, \ Y^2 - XZ, \ Z^2 - YT).$$

c) *The Veronese map.* This generalises the above example. Let d be a positive integer and denote by $M_0, M_1, \ldots, M_N$ the monomials of degree d in variables $X_0, \ldots, X_n$, so that $N = \binom{n+d}{n} - 1$. The Veronese map (associated to this ordering on the set of monomials) is the morphism $\varphi_d : \mathbf{P}^n \to \mathbf{P}^N$ given by

$$\varphi_d(x_0, \ldots, x_n) = (M_0(x), \ldots, M_N(x)).$$

We consider the homomorphism

$$\theta : k[Y_0, \ldots, Y_N] \longrightarrow k[X_0, \ldots, X_n]$$

sending Y_i onto M_i. We let I be its kernel. We then have the following result.

Proposition 11.7. *The Veronese map φ_d is an isomorphism from $\mathbf{P}^n$ to the projective algebraic variety $V = V_p(I)$ (called the Veronese variety).*

The proof of this proposition is not difficult (modulo a suitable choice of notation, cf. Exercise III, B.3).

The Veronese map is useful because it transforms a hypersurface in $\mathbf{P}^n$ into a hyperplane in $\mathbf{P}^N$, which sometimes allows us to reduce to this case. For example:

Proposition 11.8. *Consider $F \in k[X_0, \ldots, X_n]$, a homogeneous polynomial of degree d. The open set $D^+(F)$ is then an open affine set in $\mathbf{P}^n$. (It follows that in a projective algebraic variety V all the sets $D^+(f)$ are open affine sets.)*

Proof. We write $F = \sum_i a_i M_i$ (where the terms M_i are the degree d monomials). Under the Veronese morphism, the monomials M_i correspond to coordinates Y_i and hence $\varphi_d(D^+(F)) = D^+(H) \cap V$, where $H = \sum_i a_i Y_i$ is now of degree 1. But $D^+(H)$ is isomorphic to $D^+(Y_i)$ via a homography, and this latter is an open affine set. Since $D^+(F)$ is isomorphic to a closed subset of $D^+(H)$, it is also an affine variety.

Exercises A

1 Sheaves and sheaves of functions

Let X be a topological space and let $\mathcal{F}$ be a sheaf on X. Consider $P \in X$ and let E_P be the set of pairs (U, s), where U is an open set containing P and $s \in \mathcal{F}(U)$. We check that we can define an equivalence relation on E_P as follows: $(U, s) \sim (V, t)$ if and only if there is an open set W containing P, $W \subset U \cap V$ such that $s|_W = t|_W$.

The equivalence class of (U, s) is called the *germ* of s at P. We denote it by s_P. We denote the set of germs by $\mathcal{F}_P$: this is the *fibre* of $\mathcal{F}$ over P. We say that $\mathcal{F}_P$ is the inductive limit of the spaces $\mathcal{F}(U)$ for $P \in U$.

We set $K = \coprod_{P \in X} \mathcal{F}_P$ (the disjoint union of the spaces $\mathcal{F}_P$). Show that it is possible to define an injection i_U from $\mathcal{F}(U)$ to the set of functions from U to K by setting $i_U(s)(P) = s_P$. Show that the maps i_U are compatible with restrictions and hence $\mathcal{F}$ is a subsheaf of the sheaf of functions from X to K.

2 Sections over an open set

Let V be an affine algebraic variety. We assume that $\Gamma(V)$ is a factorial ring (this is the case if $V = k^n$, for example).

a) Let $f_1, \ldots, f_n \in \Gamma(V)$ be non-zero elements and let h be their gcd. Prove that $D(f_1) \cup \cdots \cup D(f_n) \subset D(h)$ and the natural restriction homomorphism

$$r : \Gamma(D(h), \mathcal{O}_V) \longrightarrow \Gamma(D(f_1) \cup \cdots \cup D(f_n), \mathcal{O}_V)$$

is an isomorphism. (Argue by induction on n.)

b) Deduce that if U is an open set in V and U is not contained in an open set $D(f)$ different from V, then $\Gamma(U, \mathcal{O}_V) = \Gamma(V, \mathcal{O}_V)$. An example of this is the case $V = k^2$, $U = k^2 - \{(0,0)\}$. More generally, $\Gamma(U) = \Gamma(V)$ whenever $V - U$ is of codimension $\geqslant 2$ in V.

3 Sections and quotients

Consider $Q \subset k^4$ given by $Q = V(XY - ZT)$ with its algebraic variety structure. Consider U_Y and U_Z, the open sets of Q defined by $y \neq 0$ and $z \neq 0$ and set $U = U_Y \cup U_Z$.

a) Prove that the function f from U to k defined by $f(x, y, z, t) = x/z$ (resp. $= t/y$) if $P = (x, y, z, t) \in U_Z$ (resp. if $P \in U_Y$) is an element of $\Gamma(U, \mathcal{O}_Q)$.

b) ¶ Prove that f is not the restriction to U of a quotient G/H, where G, H are elements of $k[X, Y, Z, T]$ such that $H(P) \neq 0$ for all $P \in U$.

(Note that this implies that $V(Y, X)$, which is of codimension 1 in Q, cannot be defined by a unique equation (*cf.* Chapter IV).)

4 To be or not to be affine

Prove that $k^2 - \{(0,0)\}$ with the open subvariety structure inherited from k^2 is not an affine variety. (Consider its inclusion into k^2 and use Exercise 2 and the full faithfulness of the functor $V \mapsto \Gamma(V)$ on affine varieties.)

5 To be or not to be isomorphic

Prove that k is not isomorphic to $k - \{x_1, \ldots, x_n\}$ for $n \geqslant 1$. (Look at the invertible elements in their rings.)

6 A local criterion

Let $\varphi : X \to Y$ be a morphism of varieties. Prove that φ is an isomorphism if and only if the following two conditions are satisfied.

i) φ is a homeomorphism,
ii) For any $x \in X$, $\varphi^* : \mathcal{O}_{Y,\varphi(x)} \to \mathcal{O}_{X,x}$ is an isomorphism.

7 Semicontinuity of rank

Let X be an affine variety with associated ring $A = \Gamma(X, \mathcal{O}_X)$, let M be an A-module of finite type and let $\mathcal{F} = \widetilde{M}$ be the corresponding sheaf on X. Consider a point $x \in X$ corresponding to a maximal ideal m_x in A and let $k(x)$ be its residue field (which is isomorphic to the base field k). We set $\mathcal{F}(x) = M \otimes_A k(x)$. (Be careful not to confuse this object with $\mathcal{F}_x$, the sheaf fibre, which is the tensor product with the local ring $M \otimes_A A_{m_x}$. The link between the two is given by the formula $\mathcal{F}(x) = \mathcal{F}_x / m_x \mathcal{F}_x$.)

a) Prove that $\mathcal{F}(x)$ is a finite-dimensional k-vector space.
b) Set $U_n = \{x \in X \mid r(x) = \text{rank}(\mathcal{F}(x)) \leqslant n\}$. Prove that U_n is open in X or, in other words, that the function r is upper semi-continuous. (Use Nakayama, cf. Summary 2.)
c) Generalise the above result to a not necessarily affine variety X and a coherent sheaf $\mathcal{F}$.
d) Assume that X is connected and $\mathcal{F}$ is locally free (i.e., for any $x \in X$ there is an affine open set U in X containing x over which $\mathcal{F}$ is isomorphic to $\mathcal{O}_U^n$). Prove that the function r is constant on X.
e) ¶ Conversely, suppose that r is constant. Prove that $\mathcal{F}$ is locally free. (Start by lifting a basis $\overline{x_1}, \ldots, \overline{x_r}$ of $\mathcal{F}(x)$ over k to $x_1, \ldots, x_r$ in $\mathcal{F}_x$ or even to $\mathcal{F}(U)$ for some open affine set U containing x and prove that these elements form a basis for $\mathcal{F}(U)$ over $\Gamma(U, \mathcal{O}_X)$.)

8 Direct and inverse images of sheaves

The ideas introduced in this exercise are of fundamental importance in algebraic geometry. However, they will not be used in the rest of this book. More details can be found in [H].

Let $\varphi : X \to Y$ be a morphism of varieties and let $\mathcal{F}$ be a sheaf on X. We defined in 6.9 the direct image sheaf $\varphi_* \mathcal{F}$ via the formula $\varphi_* \mathcal{F}(V) = \mathcal{F}(\varphi^{-1} V)$, where V is an open set of Y. The aim of this exercise is to define an inverse map φ^{-1} and study its properties.

1) Let $\mathcal{G}$ be a sheaf on Y and let U be an open set on X. Prove that we can define an equivalence relation on pairs (V, s), where V is an open set of Y containing $\varphi(U)$ and s is an element of $\mathcal{G}(V)$, by setting $(V, s) \sim (V', s')$ if and only if s and s' coincide on an open set V'' containing $\varphi(U)$ and contained in $V \cap V'$. (This definition is analogous to the definition of germs of functions in 5.1.) The set of equivalence classes for this relationship is denoted by $\varphi_0^{-1} \mathcal{G}(U)$ (in sophisticated language we call it the inductive limit of the spaces $\mathcal{G}(V)$ for $V \supset \varphi(U)$). This defines a presheaf on X and the associated sheaf on X is called the *inverse image* of $\mathcal{G}$ and denoted $\varphi^{-1} \mathcal{G}$, *i.e.*, $\varphi^{-1} \mathcal{G} = \varphi_0^{-1} \mathcal{G}^+$. Prove that the operation φ^{-1} is functorial.
2) Prove that the fibres $\varphi^{-1} \mathcal{G}_P$ and $\mathcal{G}_{\varphi(P)}$ correspond bijectively (*cf.* Exercise 1).
3) Prove that if φ is open, then, for every open set U in X, $\varphi^{-1} \mathcal{G}(U) = \mathcal{G}(\varphi U)$. Prove that if X is an open set of Y and φ is the inclusion, then $\varphi^{-1} \mathcal{G} = \mathcal{G}|_X$ (and in particular, $\varphi^{-1} \mathcal{O}_Y = \mathcal{O}_X$).
4) Prove that the above formula is not true in general even if X is a closed subset of Y. (Take X to be a point and use 2).)
5) Let $\mathcal{F}$ (resp. $\mathcal{G}$) be a sheaf over X (resp. over Y). Prove that there exist natural morphisms $\lambda : \varphi^{-1} \varphi_* \mathcal{F} \to \mathcal{F}$ and $\mu : \mathcal{G} \to \varphi_* \varphi^{-1} \mathcal{G}$. Deduce the following identity (called the adjunction formula):

$$\mathrm{Hom}_X(\varphi^{-1} \mathcal{G}, \mathcal{F}) \simeq \mathrm{Hom}_Y(\mathcal{G}, \varphi_* \mathcal{F}).$$

9 Direct and inverse images of $\mathcal{O}_Y$-modules

Let $\varphi : X \to Y$ be a morphism of varieties.

1) Prove there is a sheaf morphism $j : \varphi^{-1} \mathcal{O}_Y \to \mathcal{O}_X$. (Use 8.5 above.)
2) Let $\mathcal{G}$ be a $\mathcal{O}_Y$-module. Prove that we can define a $\mathcal{O}_X$-module, called the inverse image of $\mathcal{G}$ and written $\varphi^* \mathcal{G}$, via the formula

$$\varphi^* \mathcal{G} = \varphi^{-1} \mathcal{G} \otimes_{\varphi^{-1} \mathcal{O}_Y} \mathcal{O}_X.$$

Prove that this operation is functorial. Prove that $\varphi^* \mathcal{O}_Y = \mathcal{O}_X$.
3) Let $\mathcal{F}$ (resp. $\mathcal{G}$) be a $\mathcal{O}_X$-module (resp. a $\mathcal{O}_Y$-module). Prove that there are natural morphisms $\lambda : \varphi^* \varphi_* \mathcal{F} \to \mathcal{F}$ and $\mu : \mathcal{G} \to \varphi_* \varphi^* \mathcal{G}$. Deduce the following formula (called the adjunction formula):

$$\mathrm{Hom}_{\mathcal{O}_X}(\varphi^* \mathcal{G}, \mathcal{F}) \simeq \mathrm{Hom}_{\mathcal{O}_Y}(\mathcal{G}, \varphi_* \mathcal{F}).$$

4) Assume that X is a (closed or open) subvariety of Y and that φ is the canonical injection.

a) Let $\mathcal{F}$ be a $\mathcal{O}_X$-module. Prove that $\varphi^*\varphi_*\mathcal{F} \simeq \mathcal{F}$.

b) Let $\mathcal{F}$ and $\mathcal{F}'$ be two $\mathcal{O}_X$-modules. Prove that

$$\mathrm{Hom}_{\mathcal{O}_Y}(\varphi_*\mathcal{F}, \varphi_*\mathcal{F}') \simeq \mathrm{Hom}_{\mathcal{O}_X}(\mathcal{F}, \mathcal{F}').$$

c) Prove that φ_* and φ^* provide an equivalence of categories between $\mathcal{O}_X$-modules and $\mathcal{O}_Y$-modules of the form $\varphi_*\mathcal{F}$. This equivalence of categories is the theoretical justification for our identification of $\mathcal{F}$ and $\varphi_*\mathcal{F}$ in Chapter III, 6.10.

5) Assume that X and Y are affine with associated rings A and B respectively. There is therefore a ring homomorphism $f : A \to B$ associated to φ. Let M (resp. N) be an A-module (resp. a B-module). Prove the following formulae: $\varphi^*\widetilde{M} = \widetilde{M \otimes_A B}$ and $\varphi_*\widetilde{N} = \widetilde{N_{[A]}}$, where $N_{[A]}$ denotes the A-module obtained from N by reduction of scalars to A.

Exercises B

1 Homogenisation and dehomogenisation

In projective space $\mathbf{P}^n(k)$ with homogeneous coordinates $x_0, \ldots, x_n$ we identify affine space $\mathbf{A}^n(k)$ with the open set U_0 defined by $x_0 \neq 0$. The hyperplane "at infinity" $x_0 = 0$ is denoted by H_0. We will study the relationships between the algebraic subsets of $\mathbf{A}^n(k)$ and $\mathbf{P}^n(k)$.

We recall the following definitions.

For any $P \in k[X_0, X_1, \ldots, X_n]$, $P_\flat(X_1, \ldots, X_n) = P(1, X_1, \ldots, X_n)$.

For any $p \in k[X_1, \ldots, X_n]$, $p^\sharp(X_0, X_1, \ldots, X_n) = X_0^{\deg p} p(X_1/X_0, \ldots, X_n/X_0)$.

The notations below will be used throughout the following.

If I is an ideal of $k[X_1, \ldots, X_n]$, then $I^\sharp$ is the ideal in $k[X_0, X_1, \ldots, X_n]$ generated by the elements $\{p^\sharp \mid p \in I\}$.

If J is an ideal in $k[X_0, \ldots, X_n]$, then $J_\flat$ is the ideal in $k[X_1, \ldots, X_n]$ generated by the elements $\{P_\flat \mid P \in J\}$.

If V (resp. W) is an algebraic set in $\mathbf{A}^n(k)$ (resp. $\mathbf{P}^n(k)$) of the form $V = V(I)$ (resp. $W = V(J)$), then $V^\sharp$ (resp. $W_\flat$) is the algebraic set in $\mathbf{P}^n(k)$ (resp. $\mathbf{A}^n(k)$) defined by $I^\sharp$ (resp. $J_\flat$).

1) Prove that the operations $\flat$ and $\sharp$ are increasing on algebraic sets.

2) a) Prove that $(V^\sharp)_\flat = V$.

  b) Prove that $V^\sharp$ is the closure of V in $\mathbf{P}^n$.

  c) Prove that if $V = V_1 \cup \cdots \cup V_r$ is the decomposition of V into irreducible components, then the sets $V_i^\sharp$ are the irreducible components of $V^\sharp$.

3) a) Prove that $W_\flat = W \cap U_0$.

  b) Assume that no irreducible component W_i of W is contained in H_0. Prove that $(W_\flat)^\sharp = W$ and the components of $W_\flat$ are the sets $W_{i\flat}$.

4) Prove that I is radical if and only if $I^\sharp$ is radical. Prove that if J is radical, then $J_\flat$ is radical. What can you say about the converse? Prove that $I(V^\sharp) = I(V)^\sharp$ and $I(W_\flat) = I(W)_\flat$.

5) Let I be the ideal of $k[X, Y, Z]$ generated by F and G, where $F = Y - Z^2$ and $G = X - Z^3$. Is the ideal $I^\sharp$ generated by $F^\sharp$ and $G^\sharp$? (cf. Exercise II, 4.)

2 Resolution of a graded module

Let R be a graded Noetherian ring, $R = \oplus_{n \geqslant 0} R_n$ and M a graded R-module of finite type, $M = \oplus_{n \in \mathbf{Z}} M_n$. We denote by $M(n)$ the shifted module defined by $M(n)_p = M_{n+p}$.

1) Prove that M is generated by a finite number of homogeneous elements.
2) Prove there is a surjective homomorphism $p : L_0 \to M$ such that L_0 is a graded R-module of the form $\oplus_{i=1}^r R(n_i)$ and p is homogeneous of degree 0, i.e., p sends an element of degree n to an element of degree n.
3) Prove that M has a graded resolution, i.e., there exists an exact sequence

$$\cdots \longrightarrow L_d \xrightarrow{\;u_d\;} L_{d-1} \longrightarrow \cdots \longrightarrow L_0 \xrightarrow{\;p\;} M \longrightarrow 0$$

such that the spaces L_i are of the above form and the maps u_i are degree zero homogeneous morphisms. (Start by considering the kernels of p and the maps u_i.)

3 The Veronese map

Our aim is to prove Theorem III, 11.7. Let n and d be numbers > 0 and set

$$A = \{\alpha = (\alpha_0, \ldots, \alpha_n) \in \mathbf{N}^{n+1} \mid \alpha_i \geqslant 0 \text{ and } \sum_{i=0}^n \alpha_i = d\}.$$

We note that $|A| = \binom{n+d}{d}$. We set $|A| = N + 1$.

The support of $\alpha \in A$ is the set of indices i such that $\alpha_i \neq 0$. Its cardinal is called the breadth of α.

Consider distinct integers $i, j \in [0, n]$. We denote by (i, j) (resp. (i)) the element α in A defined by $\alpha_k = 0$ for $k \neq i, j$; $\alpha_i = d - 1$; $\alpha_j = 1$ (resp. $\alpha_k = 0$ for $k \neq i$ and $\alpha_i = d$).

If $X_0, \ldots, X_n$ are variables, then for any $\alpha \in A$ we set $X^\alpha = X_0^{\alpha_0} \cdots X_n^{\alpha_n}$.

We consider the map $\varphi : \mathbf{P}^n \to \mathbf{P}^N$ defined by the formula

$$\varphi(x_0, \ldots, x_n) = ((x^\alpha)_{\alpha \in A}) = ((x_0^{\alpha_0} \cdots x_n^{\alpha_n})).$$

1) Prove that φ is an injective map.
2) Consider the ring homomorphism

$$\theta : k[(Y_\alpha)]_{(\alpha \in A)} \longrightarrow k[X_0, \ldots, X_n]$$

defined by $\theta(Y_\alpha) = X^\alpha$. Set $I = \operatorname{Ker} \theta$ and consider $V = V_p(I)$ (the Veronese variety).
   Prove that I is a homogeneous ideal and $\varphi(\mathbf{P}^n) \subset V$.
3) Consider $\alpha, \beta, \gamma, \delta \in A$. We assume that $\alpha + \beta = \gamma + \delta$. Prove that $Y_\alpha Y_\beta - Y_\gamma Y_\delta$ is in the ideal I.
4) Prove that the open sets $D^+(Y_{(i)})$ cover V. (Consider $y = (y_\alpha) \in V$. Our aim is to prove that one of the elements $y_{(i)}$ is $\neq 0$. Argue by contradiction, by considering a non-zero y_α of minimal breadth and using 3.)
5) We define $\psi : D^+(Y_{(i)}) \cap V \to D^+(X_i)$ by the formula

$$\psi((y_\alpha)) = (y_{(i,0)}, y_{(i,1)}, \ldots, y_{(i)}, \ldots, y_{(i,n)}).$$

Prove that φ and ψ are mutually inverse morphisms on the open sets in question.
6) Prove that φ gives an isomorphism from $\mathbf{P}^n$ to the Veronese variety V.

4 Local rings in projective space

Our aim is to prove Proposition III, 8.7.

Consider $x = (x_0, \ldots, x_n) \in \mathbf{P}^n$. We consider the ideals m_x and I_x in $k[X_0, \ldots, X_n]$ defined as follows: $m_x = (X_0 - x_0, \ldots, X_n - x_n)$ and I_x is the ideal generated by the polynomials $x_i X_j - x_j X_i$ for $0 \leqslant i < j \leqslant n$.

1) Assume $x_0 \neq 0$. Prove that I_x is generated by polynomials $X_i - \frac{x_i}{x_0} X_0$ for $i = 1, \ldots, n$. Prove that $I_x \subset m_x$.
2) Prove that I_x is the ideal $I_p(\{x\})$ of polynomials vanishing at x. Deduce that I is homogeneous and prime.
3) Prove there is a canonical isomorphism

$$k[X_0, \ldots, X_n]_{(m_x)} \simeq k[X_0, \ldots, X_n]_{(I_x)},$$

where the brackets indicate that in the localised ring we restrict ourselves to elements of degree 0.
4) Assume $x_0 = 1$ and set

$$\xi = (x_1, \ldots, x_n) \quad \text{and} \quad n_\xi = (X_1 - x_1, \ldots, X_n - x_n) \subset k[X_1, \ldots, X_n].$$

Prove that the homomorphism $\flat : k[X_0, \ldots, X_n] \to k[X_1, \ldots, X_n]$ sending X_0 to 1 induces an isomorphism

$$k[X_0, \ldots, X_n]_{(m_x)} \simeq k[X_1, \ldots, X_n]_{n_\xi}.$$

5) Complete the proof of the theorem.

5 An example of a non quasi-coherent module

Let X be an affine irreducible variety, let a be a point of X and let $\mathcal{F}$ be the presheaf defined over X by the formula

$$\Gamma(U, \mathcal{F}) = \begin{cases} \Gamma(U, \mathcal{O}_X) & \text{if } a \notin U, \\ 0 & \text{if } a \in U. \end{cases}$$

Prove that $\mathcal{F}$ is a sheaf, and that it is a non-zero $\mathcal{O}_X$-module such that $\Gamma(X, \mathcal{F}) = 0$. Deduce that $\mathcal{F}$ is not quasi-coherent.

IV

Dimension

Throughout this chapter we work over an algebraically closed base field k.

0 Introduction

Dimension is the first and most natural invariant of an algebraic variety. We will finally be able to talk about varieties of dimension 0 (points), 1 (curves) and 2 (surfaces)... We will give a very natural topological definition of dimension, which is not always easy to work with, followed by other definitions which are easier to work with but which depend on results from algebra.

1 The topological definition and the link with algebra

a. Definition

The basic idea we are going to formalise is that any closed irreducible subset of an irreducible algebraic variety is of smaller dimension than the initial variety.

Definition 1.1. *Let X be a set. A* chain *of subsets of X is a sequence $X_0 \subset X_1 \subset \cdots \subset X_n$ such that the sets X_i are distinct. Such a chain is said to be of length n.*

Definition 1.2. *Let X be a topological space. The* dimension *of X is the maximum of the lengths of chains of irreducible closed subsets of X. It is either a positive integer or $+\infty$. We denote it by $\dim X$.[1]*

---

[1] Of course, this definition is only useful for topologies such as the Zariski topology. If X is separated, for example, then $\dim X$ is always 0.

b. Some topological remarks

Proposition-Definition 1.3. *If Y is a topological subspace of X, then* $\dim Y \leqslant \dim X$. *If X is finite dimensional, then we define the codimension of Y in X to be the number* $\dim X - \dim Y$. *If moreover X is irreducible and of finite dimension and Y is a closed subset different from X, then* $\dim Y < \dim X$.

Proof. Let $F_1 \subset \cdots \subset F_n$ be a chain of closed irreducible subsets of Y. There is then a sequence $\overline{F}_1 \subset \cdots \subset \overline{F}_n$ of closed irreducible subsets of X (*cf.* Chapter I, 3.7). These closed sets are distinct since, for every i, $F_i = \overline{F}_i \cap Y$, since the sets F_i are closed in Y. The theorem follows. The second statement is obvious (simply add X to a maximal chain in Y).

Proposition 1.4. *Let X be a topological space. Assume $X = \bigcup_{i=1}^{n} X_i$, where the sets X_i are closed. Then $\dim X = \sup \dim X_i$.*

Proof. Given 1.3 it is clear that $\dim X \geqslant \sup \dim X_i$. Conversely, let p be the sup in question. If p is infinite, the theorem is trivial. Assume not and take a chain in X of length $p+1$, $F_0 \subset \cdots \subset F_{p+1}$. Then $F_{p+1} = \bigcup_{i=1}^{n}(X_i \cap F_{p+1})$, but since F_{p+1} is irreducible, it is included in one of the sets X_i, which contradicts $\dim X_i \leqslant p$.

In particular, the above proposition applies when the sets X_i are the irreducible components of an algebraic variety X (*cf.* Chapter III, 4.4). The problem of dimension is thus more or less reduced to the problem of dimension of irreducible varieties.

c. Relation with Krull dimension

We recall the definition of the Krull dimension of a ring A.

Definition 1.5. *The Krull dimension of A is the maximum of the lengths of chains of prime ideals of A. We denote it by $\dim_K A$.*

Example 1.6. A principal ring which is not a field is of dimension 1 (since every non-zero prime ideal is maximal); $k[X_1, \ldots, X_n]$ is a ring of dimension $\geqslant n$, as the chain $(0) \subset (X_1) \subset (X_1, X_2) \subset \cdots \subset (X_1, \ldots, X_n)$ shows. In fact, this ring is of dimension n (*cf.* 1.9 below).

Proposition 1.7. *Let V be an affine algebraic variety and let $\Gamma(V) = \Gamma(V, \mathcal{O}_V)$ be the algebra of regular functions on V. Then*

$$\dim V = \dim_K \Gamma(V).$$

Proof. This follows from the decreasing bijection (Corollary I, 4.9 of the Nullstellensatz) between closed irreducible subsets of V and prime ideals of $\Gamma(V)$.

d. A fundamental theorem in algebra

The key result of this chapter is the following one, whose proof may be found in Problem III.

Theorem 1.8. *Let A be an integral domain which is a k-algebra of finite type and let $K = \mathrm{Fr}(A)$ be its fraction field. The Krull dimension of A is equal to the transcendence degree of K over k: $\dim_K A = \partial_k K$.*

Corollary 1.9. *We have $\dim_K k[X_1, \ldots, X_n] = n$. It follows that the affine space k^n is of dimension n.*

Proof. Indeed, the fraction field of $k[X_1, \ldots, X_n]$ is $k(X_1, \ldots, X_n)$, whose transcendence degree over k is n (*cf.* Summary 3.2.b).

Corollary 1.10. *Let V be an irreducible affine algebraic variety, let $\Gamma(V)$ be the algebra of regular functions on V and let $K(V)$ be the field of rational functions on V (cf. Chapter I, 6.15). Then $\dim V = \dim_K \Gamma(V) = \partial_k K(V)$. In particular, the dimension of an affine algebraic variety V is finite.*

Proof. The first claim follows from 1.7 and 1.8. For the second, 1.4 allows us to assume that V is irreducible and hence $\Gamma(V)$ is an integral domain. As $\Gamma(V)$ is a k-algebra of finite type, the field $K(V)$ is a k-extension of finite type and hence its transcendence degree is finite (*cf.* Summary 3.c).

e. Passing from a variety to an open subset

Proposition 1.11. *Let X be a irreducible algebraic variety and let U be a non-empty open subset of X. Then $\dim X = \dim U$, and this dimension is finite.*

Proof.

1) We treat first the case where X is affine with associated ring $\Gamma(X)$. The open set U then contains an open standard set $D(f)$, where $f \in \Gamma(X)$ is non-zero and $\dim D(f) \leqslant \dim U \leqslant \dim X$. As the ring of $D(f)$ is a localised ring of the ring of X, these two rings have the same fraction field and hence X and $D(f)$ have the same dimension by 1.10, hence so does U.

2) The above shows that the non-empty affine subsets of X (which are irreducible) all have the same finite dimension r (consider the intersection of two of them).

3) Assume $\dim X > r$. There is therefore a chain $F_0 \subset \cdots \subset F_n$ such that $n > r$. Consider $x \in F_0$ and let U be an open affine set containing x. We consider the closed sets $U \cap F_i$ in U. Being non-empty open subsets of a irreducible set, they are irreducible: they are distinct because $\overline{F_i \cap U} = F_i$ (because $F_i \cap U$ is a non-empty open subset of the irreducible set F_i). There is therefore a chain of length n in U, which is absurd.

4) Finally, if U is an arbitrary open set, then U contains an open affine subset and we are done.

Comment 1.12. The above gives us a method for calculating the dimension of an arbitrary algebraic variety X.

1) After decomposing X as a finite union of irreducible subsets if necessary, we reduce to the case where X is irreducible.

2) If X is irreducible, we can, after possibly passing to an affine open set, assume that X is affine and irreducible.

3) And finally, if X is irreducible and affine, we calculate the transcendence degree of $K(X)$.

Examples 1.13.

1) We have $\dim \mathbf{P}^n = n$. (Reduce to affine space.)

2) An affine variety V of dimension 0 is finite. (Reduce to the irreducible affine case: the result is then clear.)

Definition 1.13. *An algebraic variety of dimension 1 (resp. 2) is called a curve (resp. a surface).*

Note that for the moment we allow our varieties to have components of smaller dimension.

2 Dimension and counting equations

a. The Hauptidealsatz

Let V be an affine algebraic variety of dimension d and consider $f \in \Gamma(V)$. The aim of this paragraph is to put the intuitive idea that the closed subvariety $V(f)$ in V should have dimension $d-1$ on a firm footing. This idea is suggested in particular by the linear model—*i.e.*, hyperplanes in vector spaces.

We start by noting two extreme cases to be avoided.

Proposition 2.1.

a) $V(f)$ empty $\Leftrightarrow$ f is invertible in $\Gamma(V)$,

b) $V(f)$ contains an irreducible component $\Leftrightarrow$ f is a zero divisor.

Proof. Claim a) follows from the weak Nullstellensatz (I, 4.1). To prove b), note that if f is a zero divisor, then $fg = 0$ for some $g \neq 0$ and hence $V = V(f) \cup V(g)$, where $V(g) \neq V$. If V_i is a component of V, then $V_i = (V(f) \cap V_i) \cup (V(g) \cap V_i)$ and hence V_i is contained in $V(f)$ or in $V(g)$. As not all the components can be contained in $V(g)$ (since $V(g) \neq V$), $V(f)$ contains at least one component. Conversely, if $V(f)$ contains a component V_i, then consider some $g \in \Gamma(V)$ which is non-zero and vanishes on all the other components of V (*cf.* Chapter I, 2.2.2). We see that $fg = 0$.

We have the following definition.

Definition 2.2. *An algebraic variety X is said to be equidimensional if all its irreducible components are of the same dimension.*

Of course, an irreducible algebraic variety is equidimensional.

The following theorem is a geometric version of an algebraic result (Krull's Hauptidealsatz, or principal ideal theorem).

Theorem 2.3. *Let V be an equidimensional affine algebraic variety of dimension n and let $f \in \Gamma(V)$ be an element which is neither invertible nor a zero-divisor. Then $V(f)$ is an equidimensional affine algebraic variety of dimension $n - 1$.*

Proof. See [M] Chapter I, § 7 Theorem 2 for an excellent proof. We will restrict ourselves to the trivial case where $V = k^n$: f is then a non-constant polynomial (which we assume to be of degree > 0 in X_n) and we can assume that f is irreducible (*cf.* Chapter I, 4.12). Then

$$\Gamma(V(f)) = k[X_1, \ldots, X_n]/(f),$$

and this ring is an integral domain. We will show that the images $x_1, \ldots, x_{n-1}$ of the variables X_i in this ring form a transcendence basis in the field of fractions, which will be enough to show that $V(f)$ is of dimension $n-1$. Indeed, the last variable x_n is algebraic over $k(x_1, \ldots, x_{n-1})$ because it satisfies the equation $F(x_1, \ldots, x_{n-1}, x_n) = 0$. On the other hand, the variables $x_1, \ldots, x_{n-1}$ are algebraically independent, since otherwise we would have a polynomial equation $g(X_1, \ldots, X_{n-1}) \in (f)$, but on considering the degree of g with respect to the variable X_n we see that this is impossible.

Corollary 2.4. *Let V be an equidimensional affine algebraic variety of dimension n and consider $f_1, \ldots, f_r \in \Gamma(V)$. If W is an irreducible component of $V(f_1, \ldots, f_r)$, then $\dim W \geqslant n - r$.*

Proof. We proceed by induction on r. We note that for a given r it will be enough to prove the result for irreducible V (since W is contained in a component of V). Assume $r = 1$ and that V is irreducible. If f_1 is neither invertible nor zero we are done by 2.3. Let us note that f_1 cannot be invertible (since otherwise $V(f_1) = \varnothing$, but this variety contains W which is irreducible and hence non-empty). If f_1 vanishes, then $V(f_1) = V$ and the result follows. And finally, to pass from $r - 1$ to r we apply the induction hypothesis to a component of $V(f_1)$ containing W.

The reader should be aware of the fact that if f is a zero divisor, then $V(f)$ is not necessarily equidimensional or of dimension n or of dimension $n - 1$. For example, if $V = V(XY)$ and $f = x(x + y + 1)$, then we can prove that $V(f)$ is the union of a line and a point.

Of course, for $r > 1$ equations we cannot improve on this result. (Consider the case where the functions f_i are all the same.)

b. The intersection theorem

Once again, the inspiration for this result comes from linear algebra—namely, the lemma on the dimension of the intersection of two subspaces. However, in the context of algebraic varieties we only get an inequality.

Proposition 2.5. *Let X and Y be two irreducible affine algebraic sets contained in k^n of dimensions r and s respectively. Then every irreducible component of $X \cap Y$ is of dimension $\geqslant r + s - n$.*

Proof.

1) Assume first that $X = V(F)$ is a hypersurface. In this case, if f is the image of F in $\Gamma(Y)$, then $X \cap Y = V_Y(f)$, and we are done by 2.4.

2) The general case can be reduced to the above by identifying $X \cap Y$ with the intersection of the product with the diagonal in the product space $k^n \times k^n$ (*cf.* Exercise IV, 1 or [H] Chapter I, 7.1).

c. Converses and systems of parameters

We saw in 2.3 and 2.4 that if we take r equations in an affine variety of dimension n, then we obtain a subvariety of dimension $\geqslant n - r$ and probably equal to $n - r$ in good cases. Conversely, we might wonder whether or not any subvariety of dimension $n - r$ (or codimension r) can be defined by r equations. This is clearly too much to ask for but we will give some partial converses. We start with the case $r = 1$. In this case, we will be able to establish a good converse, provided we make (strong) assumptions on the original variety.

Proposition 2.6. *Let V be an irreducible affine algebraic variety. Assume that the ring $\Gamma(V)$ is factorial. Let W be a closed irreducible subset of V of codimension 1. Then there is an $f \in \Gamma(V)$ such that $W = V(f)$.*

Proof. We consider $I(W)$, which is a prime ideal of $\Gamma(V)$. This ideal is of height 1, which is to say it is minimal in the set of non-zero prime ideals of $\Gamma(V)$. (This follows from the bijection between irreducible subsets and prime ideals.) It is enough to check that I is principal since if $I(W) = (f)$, then $W = V(f)$. Consider $g \in I(W)$, $g \neq 0$ and decompose g as a product of irreducible elements $g = f_1 \cdots f_r$. Since $I(W)$ is prime one of the functions f_i is in $I(W)$ and hence $(f_i) \subset I(W)$. But now since $\Gamma(V)$ is factorial, the ideal f_i is prime (by Euclid's lemma, *cf.* Summary 1.5) and non-zero, and hence is equal to $I(W)$.

In general we have to make do with the following result.

Proposition 2.7. *Let V be an irreducible affine algebraic variety and let W be an irreducible affine algebraic subvariety of codimension $r > 1$. For every s such that $1 \leqslant s \leqslant r$, there exist $f_1, \ldots, f_s \in \Gamma(V)$ such that:*
  1) $W \subset V(f_1, \ldots, f_s)$ and
  2) All the components of $V(f_1, \ldots, f_s)$ are of codimension s.

In particular, there are elements $f_1, \ldots, f_r \in \Gamma(V)$ such that W is an irreducible component of $V(f_1, \ldots, f_r)$. We say that the functions f_i are a system of parameters for W.

Proof. We proceed by induction on s. For $s = 1$ we take $f \in I(W)$, $f \neq 0$ ($I(W) \neq (0)$ since $V \neq W$). We have $W \subset V(f)$ and we are done by 2.3.

We now show how to pass from $s - 1$ to s. The induction hypothesis for $s - 1$ says that $W \subset V(f_1, \ldots, f_{s-1})$, and if $Y_1, \ldots, Y_n$ are the components of $V(f_1, \cdots, f_{s-1})$, then the sets Y_i are of codimension $s - 1$. As $s - 1 < r$, none of the components Y_i are contained in W and hence $I(W)$ is contained in none of the ideals $I(Y_i)$. By the avoidance of prime ideals lemma (*cf.* Summary 4.1) $I(W) \not\subset \bigcup_{i=1}^{n} I(Y_i)$, and there is therefore an $f_s \in I(W)$ such that $f_s \notin I(Y_i)$ for all $i = 1, \ldots, n$. It follows that $W \subset V(f_1, \ldots, f_s)$. Moreover, if Z is a component of $V(f_1, \ldots, f_s)$, then Z is of codimension $\leqslant s$. Furthermore, Z is contained in $V(f_1, \ldots, f_{s-1})$, and is hence contained in one of the components Y_i, and is not equal to this Y_i (since $f_s \notin I(Y_i)$). It follows that Z is of codimension $\geqslant s$ (*cf.* 1.3) and hence is of codimension s. QED.

Corollary 2.8. *Let A be an integral k-algebra of finite type and of Krull dimension n and let I be a prime ideal in A. We assume $\dim_K(A/I) = r$. Then there is a sequence of prime ideals of A of length n "passing" through I:*

$$(0) = I_0 \subset I_1 \subset \cdots \subset I_{n-r} = I \subset \cdots \subset I_n.$$

(We say that the ring A is catenary.).

Proof. The existence of suitable ideals I_k for $k \geqslant n - r$ follows from the definition of the Krull dimension for A/I. Moreover, A is the ring of an irreducible affine variety V and I corresponds to an irreducible subvariety of dimension r. The theorem follows on applying 2.7 in the form stated above.

Corollary 2.9. *Let V be an irreducible algebraic variety and consider $x \in V$. We have $\dim V = \dim_K \mathcal{O}_{V,x}$. If V is not irreducible, then the Krull dimension of $\mathcal{O}_{V,x}$ is the maximum of the dimensions of the irreducible components of V containing x. We denote this number by $\dim_x(V)$.*

Proof. We reduce immediately to the case where V is irreducible and affine with associated ring A and the point x corresponds to the maximal ideal m_x in A. By the description of the prime ideals of $\mathcal{O}_{V,x} = A_{m_x}$ given in Summary 1.6 we know that $\dim_K \mathcal{O}_{V,x} \leqslant \dim_K A = \dim V$. Once again, the converse follows from 2.7.

d. Projective versions

We now state rapidly the projective versions of some of the above results. Most of the proofs are immediate using affine open sets and cone varieties (*cf.* Chapter II, 4.5.C). The advantage of projective space is that in certain cases we can be sure that the varieties obtained are non-empty.

Proposition 2.10. *Let $V \subset \mathbf{P}^n$ be an irreducible projective algebraic variety and consider $f \in \Gamma_h(V)$, a non-constant homogeneous element.*

1) Every irreducible component of $V(f)$ is of codimension 1 in V.

2) If $\dim V > 0$, then $V(f)$ is non-empty.

Proof. To prove the first statement we consider the affine open sets $U_i = D^+(X_i)$. The components of $V(f)$ are non-empty and hence not contained in all the sets $V(X_i)$. Let us look for example at those components of $V(f)$ which are not contained in $V(X_0)$. Then $V \cap U_0$ is an irreducible affine algebraic variety and $V(f) \cap U_0 = V(f_\flat) \neq \varnothing$, where $f_\flat \in \Gamma(V \cap U_0)$ is defined as in Chapter III, 8.b. The element $f_\flat$ is neither invertible (since $V(f_\flat)$ is non-empty) nor 0 (since otherwise $V(f)$ would contain $V \cap U_0$ and hence V and f would be 0). It follows that $\dim V(f_\flat) = \dim(V \cap U_0) - 1 = \dim V - 1$, and moreover this holds for all the irreducible components not contained in $V(X_0)$. The result follows.

To prove 2) we work with cones in k^{n+1}. We consider $C(V)$, the cone over V (*cf.* Chapter II, 4.5). The (affine) ideal of $C(V)$ is none other than $I_p(V)$. It is therefore prime, so $C(V)$ is irreducible. Moreover, $\dim C(V) = \dim V + 1$. (The simplest way to see this is to use the dimension of fibres theorem (4.7) applied to the natural projection from $C(V) - \{0\}$ to V whose fibres are lines with the origin removed. We can also apply the Hauptidealsatz in order to pass from the ring $k[X_0, \ldots, X_n]/I(V)$ to the affine ring $k[X_1, \ldots, X_n]/I(V)_\flat$ by quotienting by the element $X_0 - 1$.) We then consider the affine subvariety $V_a(f) \subset C(V)$. This is non-empty (since f is homogeneous, it contains 0) and is of codimension 1 in $C(V)$, and is hence of dimension $\dim V > 0$. It therefore does not consist only of the point 0 and hence its image in $\mathbf{P}^n$, which is simply $V(f)$, is non-empty.

We leave the proofs of the following two propositions, which are similar to the above, as an exercise for the reader.

Corollary 2.11. *Let V be an irreducible projective algebraic variety and consider non-constant homogeneous elements $f_1, \ldots, f_r \in \Gamma_h(V)$.*

1) Every irreducible component of $V(f_1, \ldots, f_r)$ is of codimension $\leqslant r$ in V.

2) If $r \leqslant \dim V$, then $V(f_1, \ldots, f_r)$ is non-empty.

We can also define homogeneous systems of parameters.

Proposition 2.12. *Let V be an irreducible projective algebraic variety and let W be an irreducible projective algebraic subvariety of codimension r. There are homogeneous elements $f_1, \ldots, f_r \in \Gamma_h(V)$ such that W is an irreducible component of $V(f_1, \ldots, f_r)$.*

3 Morphisms and dimension

a. Examples and discussion

Let $\varphi : X \to Y$ be a morphism of algebraic varieties. We assume that X and Y are irreducible. (To reduce to this case we decompose Y into irreducible components $Y = Y_1 \cup \cdots \cup Y_r$ and then consider the inverse images of the components Y_i, which are themselves closed subsets covering X, which we then further decompose, cf. 3.8 below.)

Let y be a point in Y and denote its fibre by $\varphi^{-1}(\{y\})$, or rather $\varphi^{-1}(y)$. As $\{y\}$ is closed in Y (exercise: prove this) this fibre is a closed subset of X which we equip with its subvariety structure.[2]

Our aim is to compare the dimensions of X, Y and the fibres.

Example 3.1. Consider the projection $\varphi : k^{n+d} \to k^n$ given by the first n coordinates. The fibre of an arbitrary point of k^n is isomorphic to k^d. Here, we have the relationship

$$(*) \qquad \dim X = \dim Y + \dim \varphi^{-1}(y).$$

This is the relationship we expect, but we have to be a bit careful.

Example 3.2. If $\operatorname{Im}\varphi$ is too small, then no formula of type $(*)$ can hold. For example, if φ is constant, $\varphi(x) = b$ for all $x \in X$, say, then

$$\varphi^{-1}(y) = \begin{cases} \varnothing, & \text{if } y \neq b, \\ X, & \text{if } y = b, \end{cases}$$

and formula $(*)$ is not true (if only because Y does not appear).

We are therefore going to have to make some assumptions about φ. The natural hypothesis, that φ should be surjective, is too strong (cf. 3.3 and 3.4); the right condition is that φ should be dominant, i.e., (cf. Chapter I, 6.10) $\overline{\varphi(X)} = Y$.

Example 3.3. Take $V = V(XY - 1) \subset k^2$, $W = k$ and take φ to be the projection $\varphi(x,y) = x$. We have $\varphi(V) = k - \{0\}$, and the projection is dominant but not surjective. In this case all the fibres except for the fibre over 0, which is empty, have the correct dimension predicted by $(*)$: zero.

Example 3.4. Consider $V = V(XZ - Y) \subset k^3$ and let $\varphi : V \to k^2$ be given by $\varphi(x,y,z) = (x,y)$. If $x \neq 0$, then $\varphi^{-1}(x,y) = \{(x,y,y/x)\}$; if $y \neq 0$, then $\varphi^{-1}(0,y) = \varnothing$ and $\varphi^{-1}(0,0) = \{(0,0,z) \mid z \in k\}$. The image of φ is hence $(k^2 - V(X)) \cup \{(0,0)\}$ and the non-empty fibres all satisfy $(*)$, which predicts that their dimension should be 0 except the fibre over the point $(0,0)$ which is of dimension 1.

---

[2] When dealing with dimension, which is a fairly coarse invariant in the final analysis, the variety structure is sufficient. On the other hand, for more delicate invariants (such as degree) we will have to define a scheme structure on the fibre. See the appendix on schemes 4.b.

About images. We note that the image of a morphism is not necessarily open, or closed, or even locally closed (*i.e.*, the intersection of an open set and a closed set). However, it is always constructible, that is to say, a finite union of locally closed sets (*cf.* [H] Exercise II, 3.18 and 3.19 or [M] Chapter I, 8 Corollary 2).

The moral of this story is that the "general" fibre has to have the expected dimension $\dim X - \dim Y$ but that "special" fibres may be larger.

b. Reduction to the affine case

Let $\varphi : X \to Y$ be a dominant morphism of irreducible algebraic varieties and let $y \in Y$ be in the image of $\varphi(X)$. Let Z be an irreducible component of $\varphi^{-1}(y)$.

Lemma 3.5. *There are non-empty affine open sets $U \subset X$ and $V \subset Y$ such that*
 1) $\varphi(U) \subset V$,
 2) $\varphi|_U : U \to V$ *is dominant,*
 3) $y \in V$,
 4) $Z \cap U \neq \emptyset$.

Proof. Take an open affine set V in Y containing y. We consider $\varphi^{-1}(V)$. This is an open subset of X containing Z. We take $z \in Z$ and we consider U, an affine open set of $\varphi^{-1}(V)$ containing z. It will be enough to show that $\varphi|_U : U \to V$ is dominant. Let Ω be a non-empty open subset of V. As φ is dominant, Ω meets $\varphi(X)$, so $\varphi^{-1}(\Omega)$ is a non-empty open set in X. But as X is irreducible this open set meets U in a point x and hence $\varphi(x) \in \Omega \cap \varphi|_U(U)$. QED.

3.6. Consequences.
   a) Under the hypotheses of the lemma, $\dim X = \dim U$, $\dim Y = \dim V$, $\dim Z = \dim Z \cap U$ and $Z \cap U$ is an irreducible component of $\varphi|_U^{-1}(y)$. We are therefore in exactly the same situation, but all our algebraic varieties are affine. This enables us to reduce most of our claims to the affine case.
   b) It follows from the lemma that if $\varphi : X \to Y$ is dominant (and X, Y are irreducible), then $\dim Y \leqslant \dim X$. Indeed, we can assume that X, Y are affine and we then know that the map $\varphi^* : \Gamma(Y) \to \Gamma(X)$ associated to φ is injective. This induces an injective map on fraction fields, $\varphi^* : K(Y) \to K(X)$, and hence the transcendence degree of $K(Y)$ over k is smaller than that of $K(X)$. The result follows from 1.8.

c. The dimension theorem

Theorem 3.7. *Let $\varphi : X \to Y$ be a dominant morphism of irreducible algebraic varieties.*

1) Let y be a point of Y. Every irreducible component of $\varphi^{-1}(y)$ has dimension at least $\dim X - \dim Y$.

2) There is a non-empty open set $U \subset Y$ such that
a) $U \subset \varphi(X)$,
b) $\forall y \in U$, $\dim \varphi^{-1}(y) = \dim X - \dim Y$. More precisely, every irreducible component of $\varphi^{-1}(y)$ is of dimension $\dim X - \dim Y$.

Proof. By 3.5 and 3.6 we may assume that X and Y are affine. We set $A = \Gamma(Y)$, $B = \Gamma(X)$: φ induces an injective map $\varphi^* : A \to B$.

1) We set $p = \dim Y$. Consider $y \in Y$. By 2.7, there are elements $f_1, \ldots, f_p \in A$ such that $V(f_1, \ldots, f_p)$ is finite and y is a point of this set. Replacing Y (as in 3.5) with an open affine set containing y and no other point of $V(f_1, \ldots, f_p)$ we may assume $V(f_1, \ldots, f_p) = \{y\}$. Set $g_i = \varphi^*(f_i)$. We then have $\varphi^{-1}(y) = V(g_1, \ldots, g_p)$. Indeed, the statement that $x \in \varphi^{-1}(y)$, and hence $\varphi(x) = y$, means exactly that, for every i, $f_i\varphi(x) = 0$, but $f_i\varphi$ is simply g_i.

But it then follows by 2.4 that every component of $\varphi^{-1}(y) = V(g_1, \ldots, g_p)$ is of codimension $\leqslant p$ and is hence of dimension $\geqslant \dim X - \dim Y$. (NB: this fibre may however be empty, in which case it has no irreducible component.)

2) Set $p = \dim Y$, $q = \dim X$ and $r = q - p$. There is an injection φ^* from A to B. As B is a k-algebra of finite type, it is also an A-algebra of finite type: $B = A[b_1, \ldots, b_n]$. (NB: this notation does not mean that B is a polynomial algebra cf. Problem III, 0). Let K and L be the fraction fields of A and B; we have $L = K(b_1, \ldots, b_n)$. Moreover, since $\partial_k K = p$ and $\partial_k L = q$ it follows that $\partial_K L = r$ (cf. Summary 3.c). We can therefore pick r elements $b_1, \ldots, b_r$ amongst the elements b_i which form a transcendence basis for L over K and, conversely, $r+1$ of the elements b_i are always algebraically dependent over K.

To prove the theorem it will be enough to deal with claims a) and b) separately and then take the intersection of the non-empty open sets thus obtained (which will be non-empty because Y is irreducible).

Let us first prove a). Let C be a A-subalgebra of B generated by the subsets $b_1, \ldots, b_r$, $C = A[b_1, \ldots, b_r]$. As the elements b_i are algebraically independent over K, this is a polynomial algebra and we have $C \simeq A \otimes_k k[T_1, \ldots, T_r]$ (cf. Summary 2.4), so C is the algebra of an affine variety Z, isomorphic to $Y \times k^r$ (cf. Problem I, 1.b), such that the projection $\pi : Z \to Y$ corresponds to the canonical injection $j : A \to C$. The other injection $i : C \to B$ corresponds to a dominant map $\varphi' : X \to Z$, where $f = \pi\varphi'$.

To prove a) it will be enough to show that $\varphi'(X)$ contains a non-empty open set Ω in Z. Indeed, $\varphi(X)$ will then contain $\pi(\Omega)$, which is a non-empty open set in Y (cf. Problem I, 1.c). We consider $M = K(b_1, \ldots, b_r)$. We have $L = M(b_{r+1}, \ldots, b_n)$ and as $b_1, \ldots, b_r$ is a transcendence basis for L over K, the elements b_i, $i > r$, are algebraic over M. There are therefore equations $c_{n_i,i}b_i^{n_i} + \cdots + c_{0,i} = 0$ for every $i = r+1, \ldots, n$, and (after possibly clearing denominators) we can assume that $c_{i,j} \in C$ and $c_{n_i,i} \neq 0$.

We then set $f = \prod_{i=r+1}^{n} c_{n_i,i}$. We have $f \in C$ and $f \neq 0$. Let Ω be the non-empty open set $D_Z(f)$. Its inverse image under φ' is simply $D_X(\varphi'^*(f))$, and these open sets correspond to the local algebras C_f and $B_{\varphi'^*(f)}$ respectively. The second algebra is generated as a C_f-algebra by the elements b_i for $i > r$, but (and this is what we have gained) since f is invertible in C_f, so is $c_{n_i,i}$, and hence the elements b_i are *integral* over C_f. In other words, the ring $B_{\varphi'^*(f)}$ is integral over C_f, or, alternatively, the restricted morphism $\varphi' : D_X(\varphi'^*(f)) \rightarrow D_Z(f)$ is finite and hence surjective (*cf.* Annex 4, Theorem 4.2 below) and the open set Ω is indeed contained in $\varphi'(X)$.

We now turn to point b). If we consider $r + 1$ elements $b_{i_1}, \ldots, b_{i_{r+1}}$ amongst the elements $b_1, \ldots, b_n$, we know they are algebraically dependent over K and hence they satisfy a non-trivial equation $F_{\underline{i}}(b_{i_1}, \ldots, b_{i_{r+1}}) = 0$, $(\underline{i} = \{i_1, \ldots, i_{r+1}\})$ whose coefficients can be assumed to lie in A. We choose a non-zero coefficient $a_{\underline{i}}$ of this equation and set $a = \prod a_{\underline{i}}$, the product being taken over all subsets of $\{1, \ldots, n\}$ containing $r+1$ elements. We then consider the non-empty open set $D_Y(a) \subset Y$. We will show that over this open set all irreducible components of the fibres are of the right dimension.

Consider $y \in D_Y(a)$. It corresponds to a maximal ideal m in A, $m = (f_1, \ldots, f_s)$, and, on setting $g_i = \varphi^*(f_i)$, it is easy to check that $\varphi^{-1}(y) = V(g_1, \ldots, g_s) = V(mB)$, as in the proof of point 1). By the Nullstellensatz we know that $I(\varphi^{-1}(y)) = \mathrm{rac}(mB)$, and if W is an irreducible component of $\varphi^{-1}(y)$, then W corresponds to a prime ideal $\mathfrak{q}$ in B which contains mB and is a minimal ideal satisfying this condition. We then have $\Gamma(W) = B/\mathfrak{q}$ and, by 1), it will be enough to show that $\dim B/\mathfrak{q} \leqslant r$, or, alternatively, $\partial_k \mathrm{Fr}(B/\mathfrak{q}) \leqslant r$.

But $\mathfrak{q} \cap A = m$: indeed, $m \subset mB \subset \mathfrak{q}$ hence $m \subset \mathfrak{q} \cap A$, and since $\mathfrak{q} \cap A$ is prime (and therefore $\neq A$) and m is maximal, they are equal. It follows that $B/\mathfrak{q}$ is generated as a $k = A/m$-algebra by the images $\bar{b}_i$ of the elements b_i. Likewise, $\mathrm{Fr}(B/\mathfrak{q}) = k(\bar{b}_1, \ldots, \bar{b}_n)$. Assume $\partial_k \mathrm{Fr}(B/\mathfrak{q}) > r$. We would then have $r + 1$ elements $\bar{b}_{i_1}, \ldots, \bar{b}_{i_{r+1}}$ amongst the elements $\bar{b}_i$ which are algebraically independent over k. But we know that the corresponding elements b_i satisfy the equation

$$F_{\underline{i}}(b_{i_1}, \ldots, b_{i_{r+1}}) = \sum a_{i,\alpha} b_{i_1}^{\alpha_1} \cdots b_{i_{r+1}}^{\alpha_{r+1}} = 0,$$

where $a_{i,\alpha} \in A$. The same equation holds upon passing to the quotient by $\mathfrak{q}$ and gives us an equation satisfied by the elements $\bar{b}_i$, with coefficients $\overline{a_{i,\alpha}} \in k$. As $\mathfrak{q} \cap A = m$ is the maximal ideal corresponding to y, $\overline{a_{i,\alpha}} = a_{i,\alpha}(y)$. But as $y \in D_Y(a)$, one of the elements $\overline{a_{i,\alpha}}$ does not vanish by construction, and hence the equation satisfied by the $\bar{b}_{i_1}, \ldots, \bar{b}_{i_{r+1}}$ is non-trivial, and hence these elements are algebraically dependent, which gives us a contradiction.

d. Some corollaries

Our first corollary takes the reducible case into account.

Corollary 3.8. *Let $\varphi : X \to Y$ be a morphism of algebraic varieties.*

1) Assume that all the fibres of φ are of dimension $\leqslant r$. Then $\dim X \leqslant r + \dim Y$.

2) Assume that φ is dominant and all the non-empty fibres of φ are of dimension r. Then $\dim X = r + \dim Y$.

Proof.

1) We decompose Y into irreducible components, $Y = Y_1 \cup \cdots \cup Y_n$, and similarly decompose $\varphi^{-1}(Y_i) = \bigcup_j X_{i,j}$. We consider the restriction $\varphi : X_{i,j} \to Y_i$ and set $Z_i = \overline{\varphi(X_{i,j})}$. We apply 3.7 to $\varphi : X_{i,j} \to Z_i$, and we get $\dim X_{i,j} \leqslant \dim Z_i + \dim \varphi^{-1}(z) \leqslant \dim Y + r$. The result follows since X is the union of the sets $X_{i,j}$.

2) Let Y_i be a component of Y of dimension $\dim Y$. The restriction $\varphi : \varphi^{-1}(Y_i) \to Y_i$ is dominant (consider the non-empty open set $Y_i - \bigcup_{j \neq i} Y_j$). We then decompose $\varphi^{-1}(Y_i) = \bigcup_j X_{i,j}$ into components. After possibly removing certain components we can assume that all the sets $X_{i,j}$ dominate Y_i. If the dimensions of all the $X_{i,j}$ are $< r + \dim Y$, then 2.b of 3.7 applied to the restriction $\varphi : X_{i,j} \to Y_i$ shows that the general fibre of $\varphi^{-1}Y_i \to Y_i$ is of dimension $> r$, which is impossible.

Corollary 3.9. *Let $\varphi : X \to Y$ be a closed morphism (i.e., a morphism which sends closed sets into closed sets: this is the case if, for example, X and Y are both projective varieties, cf. Problem II). For any $i \in \mathbf{N}$ we set*

$$Y_i = \{y \in Y \mid \dim \varphi^{-1}(y) \geqslant i\}.$$

Then the sets Y_i are closed in Y.

Proof. We proceed by induction on $\dim Y$. If $\dim Y = 0$, this is clear. Otherwise, set $p = \dim Y$, $q = \dim X$. We may assume that X and Y are irreducible and φ is surjective and we note that the sets Y_i are decreasing in i. Theorem 3.7 shows that $Y_{q-p} = Y$ (and hence this is also the case for Y_i, $i \leqslant q - p$) and that there is a closed set $Y' \neq Y$ (namely the complement of the open set U constructed in 3.7) such that $Y_i \subset Y'$ for $i \geqslant q - p + 1$. We then apply the induction hypothesis to Y and its inverse image under φ.

Remark 3.10. The closed sets Y_i form what is known as a stratification of Y. For each i we additionally construct a locally closed set $Y_i^0 = Y_i - Y_{i+1}$ over which the dimension of the fibre is constant and equal to i (these sets are the strata: they are disjoint and their union is Y). As these are locally closed subsets of Y, they are equipped with variety structures. The use of this type of stratification is a standard technique in algebraic geometry, as the following corollary, which is often useful in dimension problems, shows.

Corollary 3.11. *Let $\varphi : X \to Y$ be a closed surjective morphism. With the notations of 3.9 the following formula holds; $\dim X = \sup(i + \dim Y_i^0)$, where the sup is taken over all the $i \in \mathbf{N}$ such that Y_i^0 is non-empty.*

Proof. We can assume that X and Y are irreducible. Set $p = \dim X$, $q = \dim Y$. As Y^0_{q-p} contains a non-empty open set, we have $\dim Y^0_{q-p} = \dim Y = p$, so $\dim X \leqslant \sup(i + \dim Y^0_i)$. Conversely, consider a non-empty Y^0_i and set $X^0_i = \varphi^{-1}(Y^0_i)$. Corollary 3.8 then says that $\dim Y^0_i + i \leqslant \dim X^0_i \leqslant \dim X$, and the other inequality follows.

Counter-example 3.12. Corollary 3.9 is only valid if we assume φ is closed. Indeed, the image of φ is simply Y_0 and this is not generally closed, *cf.* 3.4. We note that the assumption that φ is surjective is not enough, as is shown by the example of the morphism $\varphi : k^3 \to k^3$ given by $\varphi(x, y, z) = (x, (xy-1)y, (xy-1)z)$, which is surjective, but for which Y_1 is not closed.[3]

4 Annex: finite morphisms

We recall (*cf.* Summary 1.7) that if B is an A-algebra of finite type, then B is integral over A if and only if B is an A-module of finite type: we then say that B is a finite A-algebra.

Definition 4.1. *Let* $\varphi : X \to Y$ *be a dominant morphism of irreducible affine algebraic varieties and let* $\varphi^* : \Gamma(Y) \to \Gamma(X)$ *be the associated morphism of algebras turning* $\Gamma(X)$ *into a* $\Gamma(Y)$-*algebra. We say that* φ *is a finite morphism if* $\Gamma(X)$ *is a finite* $\Gamma(Y)$-*algebra.*

Theorem 4.2. *Let* $\varphi : X \to Y$ *be a finite morphism. Then* φ *is surjective.*

Proof. We work with rings. We denote by $\operatorname{Max} A$ (resp. $\operatorname{Spec} A$) the set of maximal (resp. prime) ideals in A. Theorem 4.2 then follows from the following result.

Theorem 4.3. *Let* $A \subset B$ *be integral domains such that* B *is integral over* A. *We consider* $m \in \operatorname{Max} A$. *There is an* $n \in \operatorname{Max} B$ *such that* $m = n \cap A$.

Proof (of 4.3). We start by proving a lemma.

Lemma 4.4. *With the notations of 4.3, consider* $\mathfrak{q} \in \operatorname{Spec} B$ *such that* $\mathfrak{p} = \mathfrak{q} \cap A \in \operatorname{Spec} A$. *Then* $\mathfrak{q}$ *maximal* $\Leftrightarrow$ $\mathfrak{p}$ *maximal.*

Proof (of 4.4). As B is integral over A, the same is true of $B/\mathfrak{q}$ over $A/\mathfrak{p}$ (take the integral dependence equation of $b \in B$ and project it to the quotient). We then only have to prove the following.

Lemma 4.5. *Let* $A \subset B$ *be integral domains such that* B *is integral over* A. *Then* A *is a field* $\Leftrightarrow$ B *is a field.*

---

[3] This example was pointed out to me by Nicusor Dan.

Proof (of 4.5). Suppose that A is a field. Consider $b \in B$, $b \neq 0$. This element satisfies an equation $b^n + a_{n-1}b^{n-1} + \cdots + a_1 b + a_0 = 0$ and we may assume $a_0 \neq 0$ (after possibly dividing by b^i, which is possible since B is a domain). But a_0 is then invertible in A and we have

$$b(a_0^{-1}b^{n-1} + a_0^{-1}a_{n-1}b^{n-2} + \cdots + a_0^{-1}a_1) = -1,$$

and hence b is invertible.

Assume that B is a field and consider $a \in A$, $a \neq 0$. There is an inverse a^{-1} in B which is therefore integral over A, $a^{-n} + a_{n-1}a^{-n+1} + \cdots + a_0 = 0$, or, alternatively, multiplying by a^n, $1 + a(a_{n-1} + \cdots + a_0 a^{n-1}) = 0$. This is an equality in A and shows that a is invertible in A. This completes the proof of 4.5 and 4.4.

The following result is a consequence of 4.4.

Corollary 4.6. *With the notations of 4.3, if A is a local ideal with maximal ring m, there is an $n \in \operatorname{Max} B$ such that $m = n \cap A$.*

Proof (of 4.6). We have $A \subset B$ and hence $B \neq 0$, so B contains a maximal ideal n which must satisfy the condition given by 4.4.

We can now complete the proof of 4.3. We consider $m \in \operatorname{Max} A$ and consider the localised rings A_m and B_m obtained on inverting the elements of the multiplicative set $S = A - m$. B_m is then a domain containing A_m and is still integral over A_m. (It is enough to write down an equation for the numerator.) By 4.6 it follows that there is an ideal n' in B_m lying over mA_m (*i.e.*, such that $n' \cap A_m = mA_m$). Set $n = n' \cap B$: this is a prime ideal of B and we have $n \cap A = n' \cap A_m \cap A = m$. But n is then maximal by 4.4 and we are done.

For other results on finite morphisms, *cf.* Problem III, Midterm 1991 and Chapter IX.

For applications of the results contained in this chapter, see the exam problems from June 1993 and February 1994.

Exercises

Throughout the following exercises, we work over an algebraically closed base field k.

1 Affine intersections

Let X and Y be two irreducible algebraic subsets of k^n of respective dimensions r and s. Our aim is to prove that any irreducible component of $X \cap Y$ is of dimension $\geqslant r + s - n$.

a) Prove that this result is true if X is a hypersurface in k^n.
b) Let Δ be the diagonal in $k^n \times k^n$ (*cf.* Problem I, 4). Prove that the variety $X \cap Y$ is isomorphic to $(X \times Y) \cap \Delta$. Show that $X \times Y$ is of dimension $r + s$ (use the theorem on dimensions of fibres).
c) Using b) and explicit equations for Δ, finish the problem by reducing to the case of a hypersurface.

2 Projective intersections

Let X and Y be two irreducible algebraic subsets of $\mathbf{P}^n$ of respective dimensions r and s.

a) Show that any irreducible component of $X \cap Y$ is of dimension $\geqslant r + s - n$ (use the affine cones of X and Y to reduce to 1).

b) Show that if in addition $r + s - n \geqslant 0$, then $X \cap Y$ is non-empty.

3 Matrices of rank at most r

Let p, q be integers > 0 and r an integer such that $0 \leqslant r \leqslant \inf(p, q)$. We denote by $\mathbf{M}_{p,q}$ the set of matrices $p \times q$ with coefficients in k. We endow this set with its natural affine space structure of dimension pq. We set

$$C_r = \{A \in \mathbf{M}_{p,q} \mid \operatorname{rank}(A) \leqslant r\} \quad \text{and} \quad C_r' = \{A \in \mathbf{M}_{p,q} \mid \operatorname{rank}(A) = r\}.$$

a) Prove that C_r is a closed set in $\mathbf{M}_{p,q}$ and C_r' is open in C_r.

b) We set

$$J = \begin{pmatrix} I_r & 0 \\ 0 & 0 \end{pmatrix} \in \mathbf{M}_{p,q},$$

where I_r is the $r \times r$ identity matrix. Show that the map $\varphi : \mathbf{M}_{p,p} \times \mathbf{M}_{q,q} \to C_r$ given by $\varphi(P, Q) = PJQ$ is a surjective morphism of varieties. Deduce that C_r is irreducible.

c) We consider φ', the restriction of φ to $\mathrm{GL}(p, k) \times \mathrm{GL}(q, k)$. Prove that the image of φ' is equal to C_r'. Prove that all the fibres of φ' are isomorphic. Calculate the dimension of $\varphi'^{-1}(J)$.

d) Deduce from b) and c) the dimension of C_r and the codimension of C_r in $\mathbf{M}_{p,q}$.

4 Dimension of the orthogonal group

We denote by $O(n, k)$ the set of $n \times n$ matrices with coefficients in k such that ${}^t\!A A = I_n$.

a) Prove that $O(n, k)$ is both a subgroup of $\mathrm{GL}(n, k)$ and an affine variety. Prove that $\dim O(n, k) \geqslant n(n-1)/2$.

b) Set $S = \{x = (x_1, \ldots, x_n) \in k^n \mid \sum_1^n x_i^2 = 1\}$ and $e = (1, 0, \ldots, 0)$, $e \in S$. Prove that the map $\varphi : O(n, k) \to S$ given by $\varphi(A) = Ae$ is a surjective morphism. Determine the fibres of φ and deduce the dimension of $O(n, k)$ by induction[4] on n.

---

[4] We can prove, using Problem V, 6, that all the irreducible components of $O(n, k)$ have the same dimension. We can also prove that these components are $O^+(n, k) = \{A \in O(n, k) \mid \det(A) = 1\}$ and $O^-(n, k)$ (corresponding to $\det(A) = -1$).

5 Dimension of the commutation variety

Our aim is to calculate the dimension of the affine variety

$$C = \{(A, B) \in M_n(k) \times M_n(k) \mid AB = BA\}.$$

We recall the following theorems on matrices. (We assume the reader is familiar with Jordan normal form.)

1) If $A \in M_n(k)$, then the commutant $C(A) = \{B \in M_n(k) \mid AB = BA\}$ is a k-vector space of dimension $\geqslant n$.
2) We have $\dim C(A) = n$ if and only if the Jordan reduction is complete, $i.e.$, is for every eigenvalue λ the corresponding block is of the form

$$\begin{pmatrix} \lambda & 1 & 0 & \cdots & 0 & 0 \\ 0 & \lambda & 1 & \cdots & 0 & 0 \\ \vdots & \vdots & \ddots & \ddots & \vdots & \vdots \\ \vdots & \vdots & \ddots & \ddots & \ddots & \vdots \\ 0 & 0 & 0 & \cdots & \lambda & 1 \\ 0 & 0 & 0 & \cdots & 0 & \lambda \end{pmatrix}.$$

We then say that A is a generic matrix.
  a) Prove that the generic matrices form an open set U in $M_n(k)$.
  b) Let A be an arbitrary matrix. Show that $C(A)$ contains a generic matrix. (Reduce to the case where A is in Jordan normal form.)
  c) Let p_1 and p_2 be the projections from C to $M_n(k)$. Prove that $\Omega = p_1^{-1}(U) \cup p_2^{-1}(U)$ is an open set of C of dimension $n^2 + n$.
  d) Prove that Ω is dense in C (if $(A, B) \in C$, then "approximate" (A, B), using b), by pairs $(A, B + \lambda B')$ such that $\lambda \in k$ and B' is generic).
  e) Calculate the dimension of C.

6 Dimensions of Grassmannians

Let E be a k-vector space of dimension n and let $G_{n,p}$ be the set of subspaces of E of dimension p. We will admit the result that $G_{n,p}$ is a projective algebraic variety (called a Grassmannian).

Prove that the map φ, associating to a p-tuplet of independent vectors of E the subspace they generate, induces a surjection from an open set of E^p to $G_{n,p}$. Determine the fibres of φ and calculate $\dim G_{n,p}$. (You may use the fact that φ is a morphism.)

(See also the June 1993 exam).

7 An irreducibility theorem

Our aim is to prove the following theorem.

Theorem. *Let* $\varphi : X \to Y$ *be a dominant morphism of* projective *varieties. Assume that 1)* Y *is irreducible and 2) all the fibres* $\varphi^{-1}(y)$ *for* $y \in Y$ *are irreducible and of constant dimension* n. *Then* X *is irreducible.*

1) Prove that φ is surjective and closed (*cf.* Problem II). Prove that $\dim X = n + \dim Y$.

2) Let $X = X_1 \cup \cdots \cup X_r$ be the decomposition of X into irreducible components. Prove that there exists a component X_i such that $\varphi(X_i) = Y$.

In what follows we assume the components X_i such that $\varphi(X_i) = Y$ are those of index $i = 1, \ldots, s$, $1 \leqslant s \leqslant r$. We denote the restriction of φ to X_i by φ_i.

3) Prove there is an $i \leqslant s$ such that $\dim X_i = \dim X$. Prove that for such i all the fibres of the the maps φ_i are of dimension $\geqslant n$, and finally prove that $X = X_i$ (compare the fibres of φ and φ_i). Complete the proof of the theorem.

We note that the theorem does not hold if the varieties are not assumed to be projective: there exist surjective morphisms between affine varieties $p : X \to Y$ such that Y is irreducible and all the fibres of p are irreducible and of the same dimension without X being irreducible or even equidimensional. It is enough to take X to be the union of the origin in k^2 and the hyperbole $xy = 1$ and p to be the projection onto the x-axis.

V

Tangent spaces and singular points

Throughout this chapter we work over an algebraically closed base field k.

0 Introduction

We start with a little differential geometry. Let $f(x_1, \ldots, x_n) = 0$ be a hypersurface $S \subset \mathbf{R}^n$. We assume that f is C^∞. Consider $a = (a_1, \ldots, a_n) \in S$. What is the tangent space to S at a?

To get this tangent space we expand f at a point $x = a + h$ close to a using Taylor's formula

$$f(x_1, \ldots, x_n) = \sum_{i=1}^n h_i \frac{\partial f}{\partial x_i}(a_1, \ldots, a_n) + \frac{1}{2} \sum_{i,j} h_i h_j \frac{\partial^2 f}{\partial x_i \partial x_j}(a_1, \ldots, a_n) + \cdots$$

(since $f(a_1, \ldots, a_n) = 0$). The basic idea is that, in a neighbourhood of a, $h_i = x_i - a_i$ is small and the products $h_i h_j$ are even smaller, so S is approximated by the "tangent" hyperplane

$$\sum_{i=1}^n (x_i - a_i) \frac{\partial f}{\partial x_i}(a_1, \ldots, a_n) = 0,$$

which we obtain on writing the equations $(a_1 + h_1, \ldots, a_n + h_n) \in S$ and neglecting the terms of order $\geqslant 2$.

In algebraic geometry, we do not, in general, have any notion of smallness. We replace it with the idea of a first-order deformation (the analogue of $(a_1 + h_1, \ldots, a_n + h_n)$) by considering deformations of the form $(a_1 + b_1\varepsilon, \ldots, a_n + b_n\varepsilon)$, where $a_i, b_i \in k$, but ε is "infinitesimally small to first order," which means that $\varepsilon \neq 0$, but $\varepsilon^2 = 0$.

In other words, to get the tangent space to an algebraic variety, we will solve the equations defining it not in the field k, but in the ring of dual numbers

$$k[\varepsilon] = k[X]/(X^2) = \{a + b\varepsilon \mid a, b \in k; \ \varepsilon^2 = 0\}.$$

For hypersurfaces we recover the usual tangent hyperplane: this method is most useful, however, when dealing with varieties for which we have some global description. Here are two example of this type.

Examples 0.1.

1) We calculate the tangent space at A to the algebraic variety of matrices $O(n, \mathbf{R})$ defined by the equation ${}^t\!AA = I$. A deformation of A is a matrix of the form $A + \varepsilon B$; it is in $O(n, \mathbf{R})$ if and only if ${}^t(A + \varepsilon B)(A + \varepsilon B) = I$, which, expanding, gives us ${}^t\!AA + \varepsilon({}^tBA + {}^t\!AB) + \varepsilon^2({}^tBB) = I$: since $\varepsilon^2 = 0$ and ${}^t\!AA = I$, it follows that ${}^tBA + {}^t\!AB = 0$, which gives us the tangent space we are looking for. We note that for $A = I$ we obtain the equation ${}^tB + B = 0$: the tangent space to $O(n, \mathbf{R})$ at the origin is the vector space of anti-symmetric matrices.

2) It is easy to see using the same method that the tangent space to the group $SL(n, k)$ at the origin is the space of trace zero matrices.

1 Tangent spaces

a. Definitions and examples

As tangent spaces are local objects, we can essentially restrict ourselves to affine algebraic varieties (*cf.* 1.11).

Let V be an affine algebraic variety and consider $x \in V$. We saw in Chapter I that a point of V is equivalent to a maximal ideal in $\Gamma(V)$: $m_x = \{f \in \Gamma(V) \mid f(x) = 0\}$, or, alternatively, a homomorphism of algebras (a character) $\chi_x : \Gamma(V) \to k$, given by $f \mapsto f(x)$, the link between the two being clear: $m_x = \operatorname{Ker} \chi_x$.

Another way of considering a point x in V is to introduce the "point scheme": this is the affine variety P, consisting of one point P, such that $\Gamma(P, \mathcal{O}_P) = k$. This will also be denoted by $\operatorname{Spec} k$. Giving a character $\chi_x : \Gamma(V) \to k = \Gamma(P)$ is therefore equivalent to giving a morphism, which we simply denote $x : P \to V$, sending the point scheme into V.

A deformation of V is the same thing, but we replace the point scheme P by the fat point[1] P_ε (which is also denoted by $\operatorname{Spec} k[\varepsilon]$). This thickened point is not a variety (it is our first scheme!) but a ringed space consisting of one unique point such that $\Gamma(P_\varepsilon) = k[\varepsilon]$. There are non-constant functions on this point, such as the nilpotent element ε, and the ring of functions is not reduced, as it always is in the case of varieties. There is an obvious morphism $i : P \to P_\varepsilon$ which corresponds on rings to the projection $p : k[\varepsilon] \to k$ sending ε to 0.

---

[1] This fat point is fat in one direction only. It is also possible to create multi-dimensional fattenings by replacing, for example, $k[\varepsilon]$ by $k[X, Y]/(X^2, Y^2)$.

Definition 1.1. *Let V be an affine algebraic variety and consider $x \in V$. A deformation[2] of V at x is a morphism of ringed spaces $t : P_\varepsilon \to V$ (in the sense of Chapter III, 1.9.b) such that $ti = x$. (In other words, P_ε is sent set-theoretically to x.) We denote by $\mathrm{Def}(V, x)$ the set of deformations of V at x.*

This is equivalent to giving a k-algebra homomorphism $t^* : \Gamma(V) \to k[\varepsilon]$ such that $pt^* = \chi_x$. Such a homomorphism will be called a deformation of $\Gamma(V)$ at x. The set of such morphisms will be denoted by $\mathrm{Def}(\Gamma(V), x)$.

Consider a deformation t^* of $\Gamma(V)$. Since $pt^* = \chi_x$ it can be written in the form $t^*(f) = f(x) + \varepsilon v_t(f)$, where $v_t(f) \in k$. The deformation t^* therefore defines a map $v_t : \Gamma(V) \to k$ given by $t^* = \chi_x + \varepsilon v_t$. The fact that t^* is a homomorphism of algebras implies that v_t is linear and the fact that $\varepsilon^2 = 0$ implies that multiplication is respected if and only if for all f, g

$$v_t(fg) = f(x)v_t(g) + g(x)v_t(f).$$

We therefore introduce the following definition.

Definition 1.2. *Let A be a k-algebra and let M be an A-module. A map $D : A \to M$ is called a derivation if it satisfies the following conditions:*
1) D is k-linear;
2) For all $a, b \in A$, $D(ab) = aD(b) + bD(a)$.

The space of derivations from A to M is denoted by $\mathrm{Der}_k(A, M)$. It is a k-vector space. We note that if $\lambda \in k$, then $D(\lambda) = 0$ (calculate $D(1^2)$). An obvious example of a derivation is the map from $k[X_1, \dots, X_n]$ to itself given by $F \mapsto \partial F / \partial X_j$.

Example and Definition 1.3. *Let V be an affine algebraic variety and consider $x \in V$. Set $A = \Gamma(V)$ and $M = k$. Then M inherits a $\Gamma(V)$-module structure via χ_x: if $f \in \Gamma(V)$ and $\lambda \in k$, we set $f \cdot \lambda = f(x)\lambda$. A derivation from $\Gamma(V)$ to k is therefore a k-linear map $v : \Gamma(V) \to k$ such that, for all $f, g \in \Gamma(V)$, $v(fg) = f(x)v(g) + g(x)v(f)$. We say that v is a tangent vector to V at x; the space of tangent vectors to V at x (which is simply $\mathrm{Der}_k(\Gamma(V), k)$) is also denoted by $T_x(V)$ and is called the tangent space to V at x. It is a k-vector space.*

We see that the map v_t associated to a deformation t is a tangent vector. The following proposition summarises the above.

Proposition 1.4. *Let V be an affine algebraic variety and consider $x \in V$. There are canonical bijections*

$$\mathrm{Def}(V, x) \simeq \mathrm{Def}(\Gamma(V), x) \simeq T_x(V) = \mathrm{Der}_k(\Gamma(V), k)$$

given by $t \mapsto t^* \mapsto v_t$, where we set $t^* = \chi_x + \varepsilon v_t$.

---
[2] It would in fact be more correct to say an infinitesimal deformation.

Remark 1.5. The zero tangent vector $v_t = 0$ corresponds to the trivial deformation $t = xi$ which sends the fat point P_ε onto the ordinary point P and then sends the latter into V at x.

Definition 1.6. *Let $\varphi : V \to W$ be a morphism of affine algebraic varieties and consider $x \in V$. We set $\varphi(x) = y$. The map φ induces a map $\mathrm{Def}(\varphi, x)$: $\mathrm{Def}(V, x) \to \mathrm{Def}(W, y)$ given by $\mathrm{Def}(\varphi, x)(t) = \varphi t$. This map induces a linear map $T_x(\varphi) : T_x(V) \to T_y(W)$ given by the formula $T_x(\varphi)(v_t) = v_{\varphi t} = v_t \varphi^*$. We call it the linear tangent map associated to φ at x, or the differential of φ at x.*

Remark 1.7. The above correspondence is functorial (*i.e.*, $T_x(\psi\varphi) = T_y(\psi)T_x(\varphi)$ and the identity is sent to the identity). The reader may wish to calculate for example's sake the differential of some classical maps, such as the product of two matrices or the inverse maps.

Examples 1.8.

1) Let V be k^n and set $a = (a_1, \ldots, a_n) \in V$. The tangent space of V at a consists of deformations $t^* : k[X_1, \ldots, X_n] \to k[\varepsilon]$ such that $\chi_a = pt^*$. Such a deformation is given by the images of the variables $t^*(X_i) = a_i + \varepsilon b_i$, that is to say, by a vector $(b_1, \ldots, b_n)$. The tangent space is therefore the vector space k^n.

2) Let V be an affine algebraic variety, embedded in k^n, and assume $I(V) = (F_1, \ldots, F_r)$. We set $a = (a_1, \ldots, a_n) \in V$ and seek the tangent space of V at a.

To do this we consider a deformation

$$t^* : k[X_1, \ldots, X_n]/(F_1, \ldots, F_r) \longrightarrow k[\varepsilon]$$

such that $\chi_a = pt^*$, determined by the images of the variables X_i, $t^*(X_i) = a_i + \varepsilon b_i$, which must satisfy the relations $F_j(a + \varepsilon b) = 0$ for all $j = 1, \ldots, r$. In other words, we have the Taylor formula

$$\sum_{i=1}^n b_i \frac{\partial F_j}{\partial X_i}(a_1, \ldots, a_n) = 0.$$

The corresponding tangent vector v_t is the vector $b = (b_1, \ldots, b_n)$. If $d_a(F_1, \ldots, F_r)$ is the Jacobian matrix of the functions F_j at the point a, which is a linear map from k^n to k^r, then

$$T_a(V) = \mathrm{Ker}\, d_a(F_1, \ldots, F_r).$$

In particular, $T_a(V)$ is finite dimensional.

b. Relation with local rings

Proposition 1.9. *Let V be an affine algebraic variety, let x be a point of V and let m_x be the maximal ideal of $\Gamma(V)$ corresponding to x. There is an isomorphism of vector spaces $T_x(V) \simeq (m_x/m_x^2)^*$, where the star denotes the dual vector space.*

Proof. Let $v : \Gamma(V) \to k$ be a tangent vector to V at x, *i.e.*, a derivation from $\Gamma(V)$ to k. We also denote its restriction to m_x by v. This linear map vanishes on m_x^2 since, on the one hand, for all $f, g \in m_x$, $v(fg) = f(x)v(g) + g(x)v(f)$ and, on the other hand, as f, g are contained in m_x, $f(x)$ and $g(x)$ are zero. This map therefore factorises through $\overline{v} : m_x/m_x^2 \to k$, which is an element of the dual of m_x/m_x^2.

Conversely, let θ be an element of $(m_x/m_x^2)^*$. We reconstruct v via the formula $v(f) = \theta(\overline{f - f(x)})$ (where we denote the image in m_x/m_x^2 of $a \in m_x$ by $\overline{a}$).

Corollary 1.10. *Let V be an affine algebraic variety, x a point of V, $\mathcal{O}_{V,x}$ the local ring of V at x and $m_{V,x}$ its maximal ideal. There is an isomorphism $T_x(V) \simeq (m_{V,x}/m_{V,x}^2)^*$. In particular, the tangent space only depends on the local ring of V at x.*

Proof. Set $A = \Gamma(V)$ and $m = m_x$, so that $\mathcal{O}_{V,x} = A_m$ and $m_{V,x} = mA_m$. The natural homomorphism $m \to mA_m$ given by $x \mapsto x/1$ factorises through $\theta : m/m^2 \to mA_m/(mA_m)^2$ and our aim is to show that θ is an isomorphism of k-vector spaces.

a) θ is injective. Consider $x \in m$ such that $x \in (mA_m)^2$. This means there is an $s \notin m$ such that $sx \in m^2$. But if $s \notin m$, since A/m is a field, there is a $t \in A$ such that $st = 1 - a$, $a \in m$. It follows that $x = stx + ax \in m^2$.

b) θ is surjective. Consider $x/s \in mA_m$, where $x \in m$ and $s \notin m$. Using the same notation as in a), $x/s = \theta(tx)$, since $x/s - tx = txa/(1 - a) \in (mA_m)^2$.

Remark 1.11. We can now define the tangent space to an arbitrary algebraic variety at a point x as being the tangent space at this point of an open affine subset. By 1.10, this space does not depend on the choice of open set. We could also have used the formula given in 1.10 directly or used the definition via deformations of X given in 1.1.

2 Singular points

Definition 2.1. *Let V be an irreducible algebraic variety and let x be a point of V. We say that x is a regular (or smooth) point of V (or that V is non-singular at x) if $\dim V = \dim_k T_x(V)$. We say that V is non-singular (or smooth or regular) if it is non-singular at every point.*

Remarks 2.2.
1) If V is not irreducible, we ask for $\dim_x V = \dim T_x(V)$, where $\dim_x V$ is the sup of the dimensions of all the irreducible components of V passing through x. (In fact, it is possible to show, *cf.* 3.6, that if x lies on more than one component, then it is a singular point.)
2) It is always the case that $\dim T_x(V) \geqslant \dim_x(V)$ (*cf.* Problem V).

Theorem 2.3 (Jacobian criterion). *Let $V \subset k^n$ be an irreducible affine algebraic variety of dimension d. Assume $I(V) = (F_1, \ldots, F_r)$. Then*

$$V \text{ non-singular at } x \iff \operatorname{rank} d_x(F_1, \ldots, F_r) = n - d \quad (cf. \ 1.8).$$

Proof. This follows immediately from 1.8.

Remark 2.4. If we assume only that $V = V(F_1, \ldots, F_r)$, and

$$\operatorname{rank} d_x(F_1, \ldots, F_r) = n - d,$$

then V is non-singular at x. Indeed, $I(V)$ is generated by the functions F_i plus (possibly) some other polynomials, but the presence of these extra polynomials cannot increase the dimension of the kernel of the Jacobian matrix, and we are done by 2.2.2. On the other hand, if the rank is $> n - d$, then we cannot conclude anything, *cf.* the example of $V(X^2)$ in k^2. (The variety $V(X^2) = V(X)$ is non-singular, but the scheme $V(X^2)$, *cf.* the appendix on schemes, is singular.)

Examples 2.5.

a) Plane curves. If $F(X, Y)$ is a polynomial without multiple factors and $F(a, b) = 0$, then the point (a, b) is singular in $V(F)$ if and only if $\partial F / \partial X(a, b) = \partial F / \partial Y(a, b) = 0$. (Indeed, our assumption implies $I(V(F)) = (F)$, *cf.* Chapter I, 4.11.) Hence, for example, $V(Y^2 - X^3)$ and $V(X^3 + X^2 - Y^2)$ are singular at $(0, 0)$. (We note that $V(F)$ is singular at the point $(0, 0)$ if and only if F has no terms of degree < 2.) On the other hand, $V(Y^2 - X(X - 1)(X - \lambda))$ is smooth whenever $\lambda \neq 0, 1$ (in characteristic other than 2). We note that if F is not irreducible, $F = F_1 \cdots F_r$, then the points at the intersections of F_i and F_j are singular.

b) Hypersurfaces. An identical result holds for affine hypersurfaces: $a = (a_1, \ldots, a_n)$ is a singular point of $V(F)$ if all the partial derivatives of F (and F itself!) vanish at a.

Proposition 2.6 (the projective case). *Let $V \subset \mathbf{P}^n$ be an irreducible projective algebraic variety and consider $x = (x_0, \ldots, x_n) \in V$. We assume $I(V) = (F_1, \ldots, F_r)$, where the polynomials F_i are homogeneous. Let A be the matrix whose general term is $\frac{\partial F_i}{\partial X_j}(x)$, where $i = 1, \ldots, r$ and $j = 0, \ldots, n$. Then V is non-singular in x if and only if $\operatorname{rank}(A) = n - \dim V$.*

Proof. We can assume $x_0 \neq 0$ or even $x_0 = 1$: x is a smooth point of V if and only if $\xi = (x_1, \ldots, x_n)$ is a smooth point of the affine open set $V_\flat = V \cap D^+(X_0)$ in V. We know that $I(V_\flat) = (F_{1\flat}, \ldots, F_{r\flat})$ (*cf.* Exercise III, B.1), ξ is smooth in $V_\flat$ if and only if the matrix B of partials $(\partial F_{i\flat} / \partial X_j)(\xi)$, where $j = 1, \ldots, n$ and $i = 1, \ldots, r$, is of rank $n - \dim V$. However, if F is a homogeneous polynomial, then, for all $j \geqslant 1$, $(\partial F / \partial X_j)(x) = (\partial F_\flat / \partial X_i)(\xi)$, so the matrix B is obtained from A by suppressing the first column of partials $(\partial F_i / \partial X_0)(x)$. It follows that $\operatorname{rank}(B) \leqslant \operatorname{rank}(A)$.

If $F \in k[X_0, \ldots, X_n]$ is a homogeneous polynomial of degree d, then Euler's formula

$$dF = \sum_{j=0}^{n} X_j \frac{\partial F}{\partial X_j}$$

holds, and for F_i of degree d_i it follows that at x

$$\frac{\partial F_i}{\partial X_0}(x) = d_i F_i(x) - \sum_{j=1}^{n} x_j \frac{\partial F_i}{\partial X_j}(x),$$

and since $F_i(x) = 0$, this implies that the first column of A is a linear combination of the others, so $\text{rank}(A) = \text{rank}(B)$.

Examples 2.7.

1) The projective elliptic plane curve $V(Y^2T - X(X - T)(X - \lambda T))$ is smooth for $\lambda \neq 0, 1$.

2) The space cubic C in $\mathbf{P}^3$ given by

$$I(C) = V(XT - YZ, Y^2 - XZ, Z^2 - YT)$$

(*cf.* Exercise II, 4) is smooth. Indeed, the Jacobian matrix

$$d_x = \begin{pmatrix} T & -Z & -Y & X \\ -Z & 2Y & -X & 0 \\ 0 & -T & 2Z & -Y \end{pmatrix}$$

is of rank 2 at every point (X^2 and T^2 appear in its minors and at every point of C, X or T is non-zero). It is left to the reader to check that the 3-minors are all in $I(C)$.

3) If the characteristic of k does not divide d, then the Fermat curve $V(X^d + Y^d - T^d)$ is a smooth curve in $\mathbf{P}^2$. Exercise: find a smooth plane curve of degree d in characteristic p when p divides d. (Answer: $X^d + Y^{d-1}T + XT^{d-1}$.)

Remark 2.8. As in the affine case, if we assume only that $V = V(F_1, \ldots, F_r) \subset \mathbf{P}^n$, where the polynomials F_i are homogeneous and the matrix of partial derivatives at the point x is of rank $n - \dim V$, then the point x is smooth in V.

3 Regular local rings

We start by recalling an algebraic result.

Proposition-Definition 3.1. *Let A be a local Noetherian ring, let m be its maximal ideal and set $k = A/m$. The quotient $m/m^2 \simeq m \otimes_A k$ is a k-vector space and $\dim_k m/m^2 \geqslant \dim_K A$ (the Krull dimension of A, cf. Chapter IV, 1.5).*

We say that A is regular if $\dim_k m/m^2 = \dim_K A$.

Proof. The fact that m/m^2 is a k-vector space is immediate. Consider $\lambda \in A$ and $x \in m$ and denote by $\overline{\lambda}$ and $\overline{x}$ their images modulo m and m^2 respectively. We set $\overline{\lambda}\overline{x} = \overline{\lambda x}$ and we check that this definition does not depend on the choice of representatives (or, alternatively, use the tensor product description of m/m^2).

When A is a local ring at a point of an algebraic variety, the inequality we seek follows from 2.2.2. For a proof of this inequality in the general case, see [Ma].

Proposition 3.2. *Let V be an algebraic variety and let x be a point in V. We have the equivalence*

$$V \text{ smooth at } x \Longleftrightarrow \mathcal{O}_{V,x} \text{ regular.}$$

Proof. We know that $T_x(V) \simeq (m_{V,x}/m_{V,x}^2)^*$ (*cf.* 1.10). On the other hand, $\dim_x V = \dim_K \mathcal{O}_{V,x}$ for any x (*cf.* Chapter IV, 2.9) and the result follows (*cf.* 2.2.1).

The following proposition gives a handy interpretation of the dimension of m/m^2.

Proposition 3.3. *Let A be a local Noetherian ring, let m be its maximal ideal and set $k = A/m$. The dimension of m/m^2 as a k-vector space is equal to the minimal number of generators of m.*

Proof. (This is a special case of Nakayama's lemma, *cf.* Summary 2.)

1) Assume $m = (x_1, \ldots, x_d)$. Then the classes $\overline{x}_i$ generate m/m^2 over k. Indeed, if $x \in m$, then $x = \sum_{i=1}^d a_i x_i$, where $a_i \in A$, and hence, on reducing modulo m^2, $\overline{x} = \sum_{i=1}^d \overline{a_i x_i} = \sum_{i=1}^d \overline{a}_i \overline{x}_i$ by definition of the vector space structure. It follows that $\dim_k m/m^2 \leqslant d$.

2) Conversely, let $\overline{x}_1, \ldots, \overline{x}_d$ be a basis for m/m^2 over k. Then $m = (x_1, \ldots, x_d)$. Indeed, we can complete the elements x_i to a system of generators $x_1, \ldots, x_d, \ldots, x_n$ of m and we complete the proof by induction, using the following lemma.

Lemma 3.4. *If $m = (x_1, \ldots, x_n)$ and $\overline{x}_1, \ldots, \overline{x}_{n-1}$ generate m/m^2 over k, then $m = (x_1, \ldots, x_{n-1})$.*

Proof (of 3.4). We write $\overline{x}_n = \sum_{i=1}^{n-1} \overline{a}_i \overline{x}_i$, where $a_i \in A$. We deduce that $x_n - \sum_{i=1}^{n-1} a_i x_i \in m^2$. As m^2 is generated by the products $x_i x_j$, we have $x_n = \sum_{i=1}^{n-1} a_i x_i + \sum_{i \leqslant j} b_{ij} x_i x_j$. We gather the terms containing x_n:

$$x_n \left(1 - \sum_{i \leqslant n} b_{in} x_i\right) = \sum_{i=1}^{n-1} a_i x_i + \sum_{i \leqslant j < n} b_{ij} x_i x_j.$$

As the elements x_i are in m, the coefficient of x_n is not in m and hence is invertible in A, and it follows that $x_n \in (x_1, \ldots, x_{n-1})$.

Corollary 3.5. *Let A be a local Noetherian ring and let m be its maximal ideal. Then A is regular if and only if m is generated by $\dim A$ elements.*

Remark 3.6. A regular local ring is factorial (*cf.* [Ma]) and in particular is an integral domain (which proves, *cf.* 2.2.1, that a point at the intersection of two irreducible components is necessarily singular).

4 Curves

In this chapter we will use the word curves only for *equidimensional* varieties of dimension 1. Our aim is to study the detailed structure of local rings of curves. The proofs are contained in Problem IV.

Proposition 4.1. *Let C be a curve and x a point of C. We have an equivalence*

$$x \text{ non-singular} \Longleftrightarrow \mathcal{O}_{C,x} \text{ is a discrete valuation ring}$$
$$\text{(in other words a local principal ring).}$$

Proof. Here, the non-singularity condition means that the maximal ideal is principal, which is equivalent to the local ring being a discrete valuation ring (*cf.* Problem IV).

Besides this local characterisation, there is a very simple global characterisation of smooth affine curves (loc. cit.).

Theorem 4.2. *Let C be an irreducible affine curve. Then C is smooth if and only if $\Gamma(C)$ is integrally closed.*

Example 4.3. The curve $V(Y^2 - X^3)$ is singular at the origin. We check that its ring $A = k[X,Y]/(Y^2 - X^3)$ is not integrally closed. The element Y/X in the fraction field of A is not in A but is integral over A since it satisfies the equation $(Y/X)^2 - X = 0$.

When x is a singular point of C, it is possible to define its multiplicity $\mu_x(C)$ (which is, of course, 1 for a regular point). We restrict ourselves to plane curves (*cf.* 4.7.2 for a generalisation).

Consider $F \in k[X,Y]$ (a priori, we do not suppose that F has no multiple factors) and consider $P = (a,b) \in k^2$. After translation, we may assume $P = (0,0)$. We write $F = F_0 + F_1 + \cdots + F_d$, where F_i is homogeneous of degree i.

Definition 4.4. *The multiplicity of F at P, which we denote by $\mu_P(F)$, is the smallest integer i such that $F_i \neq 0$. If F has no multiple factor and $C = V(F)$, this is also the multiplicity of C at P.*

Remark 4.5. The statement $\mu_P(F) = 0$ is exactly equivalent to $P \notin V(F)$; the statement $\mu_P(F) = 1$ is exactly equivalent to the statement that P is a smooth point of $V(F)$ since if $F_1 = \alpha X + \beta Y$, then $\alpha = \frac{\partial F}{\partial X}(P)$ and $\beta = \frac{\partial F}{\partial Y}(P)$. If F has no multiple factors, then $\mu_P(F) \geq 2$ if and only if P is a singular point of $V(F)$.

We note that if $\mathcal{O}_P$ is the local ring of k^2 at P and m_P is its maximal ideal, then $\mu_P(F)$ is the largest integer r such that $F \in m_P^r$. It is also the smallest integer r such that there is a partial derivative of rth order of F which is non-zero at P.

This definition of multiplicity has the apparent defect of depending on the choice of embedding of the curve, or the choice of coordinates. The following proposition shows that in fact it is an intrinsic property.

Proposition 4.6. *Consider $F \in k[X,Y]$: let $F = F_r + \cdots + F_p$ be its decomposition into non-zero homogeneous polynomials such that $r \leq p$. Consider $P = (0,0) \in k^2$, so $r = \mu_P(F)$. Let $\mathcal{O}_P$ be the local ring of k^2 at P, let $m = m_P$ be its maximal ideal, let A be the quotient ring $A = \mathcal{O}_P/(F)$ and let $\overline{m}$ be the maximal ideal of A. Then, for large enough n, $\dim_k A/\overline{m}^n = rn + c$, where c is an integral constant.*

Proof. We have $A/\overline{m}^n \simeq \mathcal{O}_P/(F, m^n)$. For $n \geq r$ we have the following exact sequence:

$$0 \longrightarrow \mathcal{O}_P/m^{n-r} \overset{F}{\longrightarrow} \mathcal{O}_P/m^n \longrightarrow \mathcal{O}_P/(F, m^n) \longrightarrow 0.$$

Indeed, considering the map $\mathcal{O}_P \overset{F}{\longrightarrow} \mathcal{O}_P$ given by multiplication by F we see that, for all $x \in m^{n-r}$, $Fx \in m^n$ since $F \in m^r$. This produces the required factorisation. The only thing that needs to be proved is that the map F is injective. Consider $a \in \mathcal{O}_P$ and assume $aF \in m^n$: we aim to prove that $a \in m^{n-r}$. We write $a = a'/s$, where $a', s \in k[X,Y]$ and $s(P) \neq 0$. We set $a' = a_l + \cdots + a_q$, where a_i is homogeneous of degree i, $a_l \neq 0$ and $l \leq q$. As s does not vanish at P, the fact that Fa is in m^n means that the valuation at P of the polynomial $a'F$ is $\geq n$, but this valuation is $r+l$ (since $a_l F_r \neq 0$). It follows that $l \geq n - r$, and hence $a \in m^{n-r}$.

It follows from this exact sequence that

$$\dim A/\overline{m}^n = \dim \mathcal{O}_P/m^n - \dim \mathcal{O}_P/m^{n-r}.$$

Denoting by x, y the images of the variables, there is an obvious basis for $\mathcal{O}_P/m^n$:

$$1, x, y, x^2, xy, y^2, \ldots, x^{n-1}, x^{n-2}y, \ldots, y^{n-1}$$

of size $n(n+1)/2$, and hence $\dim A/\overline{m}^n = n(n+1)/2 - (n-r)(n-r+1)/2 = rn - r(r-1)/2$.

Remarks 4.7.

1) Curves are a special case of the above proposition (namely the case where F has no multiple factors). The ring A is then simply the local ring of $C = V(F)$ at P and the multiplicity of C only depends on this local ring. We have given above a more general form of the proposition so as to leave open the possibility of curves with multiple components (*cf.* Chapter VI).

2) The notion of multiplicity can be generalised as follows. If A is a local Noetherian ring with residue field $k = A/m$, then we can show that for n large enough the dimension of the k-vector space A/m^n is a polynomial function of n (called the Hilbert-Samuel function of A) of degree $d = \dim_K A$ and whose dominant coefficient is of the form $\mu_A/d!$, where μ_A is an integer called the multiplicity of A (*cf.* [H] Chapter V, Exercise 3.4).

4.8. Tangent lines of a plane curve in a point. For a plane curve we can also define the *tangent lines* of the curve at a possibly singular point. As above, we use a translation to reduce to the case where $P = (0,0)$ and $F(P) = 0$.

If $\mu_P(F) = 1$, then the tangent line is given by $F_1 = 0$ and is simply the tangent space (*cf.* 1.8).

If $\mu_P(F) = r \geqslant 2$, then the tangent space is the whole of k^2, but since k is algebraically closed, we can decompose the homogeneous polynomial F_r as a product of factors of degree 1: $F_r = \prod_{i=1}^{r}(\alpha_i X + \beta_i Y)^{r_i}$. The lines T_i given by the equations $\alpha_i X + \beta_i Y = 0$ are called the *tangent lines* of F at P, and the number r_i is called the *multiplicity* of T_i. (We will also call the set of all such lines the *tangent cone*.) When all the integers r_i are equal to 1, we say that the multiple point is *ordinary*.

Examples 4.9.

If $F = Y^2 - X^3$, then the origin is a double point with a double tangent line. It is not an ordinary double point (but a cusp).

If $F = X^3 + X^2 - Y^2$, then the origin is an ordinary double point with tangent lines $Y = \pm X$.

If $F = (X^2 + Y^2)^2 + 3X^2Y - Y^3$ (trefoil), then the origin is an ordinary triple point with tangent lines $Y = 0$ and $Y = \pm\sqrt{3}X$.

If $F = (X^2 + Y^2)^3 - 4X^2Y^2$ (quadrifoil), the origin is a (non-ordinary) quadruple point with two double tangent lines $X = 0$ and $Y = 0$.

Exercises

In what follows we work over an algebraically closed base field k. We refer to the exercises from Chapter II for the definition of a homography. A homography of the projective plane $\mathbf{P}^2$ is simply a linear change of coordinates, ${}^t(X', Y', T') = A^t(X, Y, T)$, where A is an invertible 3×3 matrix.

1 Three remarks

a) Let $F \in k[X_0, \ldots, X_n]$ be a homogeneous polynomial of degree d. Assume that F is reducible. Prove that $V(F)$ is not smooth. (If $F = GH$ consider a common point of $V(G)$ and $V(H)$, which exists by Exercise IV, 2b.)

b) Let $F \in k[X, Y, Z, T]$ be homogeneous of degree 3. Assume that $V(F)$ contains two distinct singular points a and b. Prove that the line (ab) is contained in $V(F)$.

c) Let $F \in k[X, Y, T]$ be homogeneous of degree 3. Assume that $V(F)$ has three distinct singular points. Show that $V(F)$ is the union of three lines.

2 Some examples

Find the singular points of the following varieties and say whether or not they are irreducible.

a) In $\mathbf{P}^2$:

$$V(XY^4 + YT^4 + XT^4), \quad V(X^2Y^3 + X^2T^3 + Y^2T^3),$$
$$V(X^n + Y^n + T^n), \ n > 0,$$
$$V((X^2 - YT)^2 + Y^3(Y - T)), \quad V(2X^4 + Y^4 - TY(3X^2 + 2Y^2) + Y^2T^2),$$
$$V(Y^2T^2 + T^2X^2 + X^2Y^2 - 2XYT(X + Y + T)).$$

If possible, try to draw the affine (relative to the line at infinity $T = 0$) real part of these curves.

b) In $\mathbf{P}^3$:

$$V(X^2 + Y^2 - Z^2), \quad V(XYT + X^3 + Y^3),$$
$$V(XY^2 - Z^2T), \quad V(XYZ + XYT + XZT + YZT),$$
$$V(XT - YZ, \ YT^2 - Z^3, \ ZX^2 - Y^3).$$

3 Conics

Let $F \in k[X, Y, T]$ be an irreducible homogeneous polynomial of degree 2 and let $C = V(F) \subset \mathbf{P}^2$ be the "conic" defined by F.

a) Prove that up to homography we may assume $F = YT - X^2$. (Consider a non-singular point in C, P, and reduce by homography to the case $P = (0, 0, 1)$ and the tangent line to C at P is the line $Y = 0$. Now show that F is of the form $aX^2 + bY^2 + dYT + fXY$. Perform another homography of the form $X' = X, T' = T, Y' = uX + vY + wT$ to reduce to the required form. Another possibility would be to use the reduction of quadratic forms.)

b) Prove that any irreducible conic is smooth.

4 Cuspidal cubics

Let $F \in k[X, Y, T]$ be an irreducible homogeneous polynomial of degree 3 and let $C = V(F) \subset \mathbf{P}^2$ be the "cubic" defined by F. Assume that C has a cusp at P.

a) Prove that up to homography we may assume $P = (0, 0, 1)$ and that the double tangent line is $Y = 0$.

b) Prove that up to homography the curve is given by the equation $F = Y^2T - X^3$. Determine the singular points of C.

5 Nodal cubics

The same exercise as above for a cubic C with a double ordinary point (*i.e.*, a node).
(Reduce to curves of the form $X^3 + Y^3 - XYT$.)

6 A space elliptic quartic

Set $C = V(X^2 - XZ - YT, \ YZ - XT - ZT) \subset \mathbf{P}^3$ and let P be the point $(0,0,0,1)$.

a) Consider the projection φ with centre P onto the plane $T = 0$ (*i.e.*, the map
   sending $Q = (x,y,z,t) \in \mathbf{P}^3$, $Q \neq P$ to $(x,y,z) \in \mathbf{P}^2$). Show that φ induces an
   isomorphism of $C - \{P\}$ onto the plane cubic $\Gamma = V(Y^2 Z - X^3 + XZ^2)$ minus
   the point $(1,0,-1)$. Prove that Γ is smooth.
b) Prove that C is smooth and irreducible. Prove that the number of intersection
   points of C with any plane is $\leqslant 4$ and that this number is "generally" equal to 4
   (we say that C is of degree 4 or that C is a quartic).

7 Products

Study how the smoothness of the point $(x,y) \in X \times Y$ depends on the nature of x
and y.

8 Linear subvarieties

Let $V(F)$ be a hypersurface of $\mathbf{P}^n$ of degree $\geqslant 2$. Assume that $V(F)$ contains a
linear subvariety L (that is to say, a projective subspace) of dimension $r \geqslant n/2$.
Prove that $V(F)$ is singular. (Look for singular points on L, which we may assume
is given by the equations $X_{r+1} = 0, \ldots, X_n = 0$.)

9 Quadrics

Assume that k is not of characteristic 2. By a quadric of $\mathbf{P}^n$ we mean a variety of
the form $Q = V(F)$, where F is homogeneous of degree 2 (*i.e.*, a quadratic form).

a) Prove that up to homography we may assume $F = X_0^2 + \cdots + X_r^2$, where
   $0 \leqslant r \leqslant n$.
b) Prove that Q is irreducible if and only if $r \geqslant 2$.
c) Assume $r \geqslant 2$. Prove that the singular locus of Q is a linear subvariety of
   dimension $n - r - 1$. In particular, Q is smooth if and only if $r = n$.

VI

Bézout's theorem

0 Introduction

Our aim is to show that two plane curves of degrees s and t have exactly st intersection points. We saw in the introduction that we need to take care when stating this result.

1) We have to assume the curves have no common components.
2) We have to assume the base field k is algebraically closed.
3) We have to work in projective space.
4) We have to count intersections with multiplicities.

We still need to deal with this last point. As multiplicity is a local concept we start with affine spaces.

1 Intersection multiplicities

a. Finite schemes

We start with a very simple example.

We intersect the parabola $C = V(Y - X^2)$ and the line $D_\lambda = V(Y - \lambda)$. The intersection variety is $C \cap D_\lambda = V(Y - X^2, Y - \lambda)$. Consider the ideal $I_\lambda = (Y - X^2, Y - \lambda) = (Y - \lambda, X^2 - \lambda)$ and let us calculate the quotient ring $A_\lambda = k[X, Y]/I_\lambda$. Sending Y to λ we see that this ring is isomorphic to $k[X]/(X^2 - \lambda)$. There are two cases to consider.

a) If $\lambda \neq 0$, set $\lambda = \alpha^2$. Then A_λ is isomorphic to the product ring $k \times k$ via the homomorphism sending X to $(\alpha, -\alpha)$. This ring is reduced, so I_λ is equal to the ideal $I(C \cap D_\lambda)$ and the variety $C \cap D_\lambda$ has ring A_λ. It is formed of two distinct points and at each of these points the local ring is k.

b) $\lambda = 0$: D_0 is tangent to C. We have $A_0 = k[X]/(X^2) = k[\varepsilon]$ (the ring of dual numbers). The quotient is not reduced and hence $I(C \cap D_0)$ is the radical of I_0, or, alternatively, the ideal (X, Y). The variety $C \cap D_0$ is a point with local ring k. Of course, to get the intersection multiplicity of D_0 and C—namely 2—we simply don't pass to the radical. It is the ideal $I_0 = (Y, X^2)$, the limit of the ideals I_λ when λ tends to 0, not its radical, which contains the information we seek. Our solution is to define $C \cap D_0$ not as a variety, but as a scheme. In other words, we equip the unique point $P = (0,0)$ of $C \cap D_0$ with the ring $\Gamma(P) = k[X]/(X^2) = k[\varepsilon]$, not k. The point P then possesses an unusual "function" ε, which is non-zero but nilpotent. The above suggests the following definition.

Definition 1.1 (Finite schemes). *A finite scheme* $(Z, \mathcal{O}_Z)$ *is a ringed space such that Z is a finite discrete set and the ring $\mathcal{O}_Z(\{P\})$ at each (open) point $P \in X$ is a local k-algebra which is finite dimensional as a k-vector space (i.e., a finite k-algebra). This dimension is called the* multiplicity *of Z at P. We denote it by $\mu_P(Z)$.*

Remarks 1.2.

1) The ring $\mathcal{O}_Z(\{P\})$ is also the local ring of Z at P, $\mathcal{O}_{Z,P}$ (in the sense given in Chapter III, 5.1).

2) The unique prime ideal of $\mathcal{O}_{Z,P}$ is its maximal ideal m_P (indeed, if I is a prime ideal, then its quotient is an integral domain which is finite dimensional over k and is hence a field, so I is maximal and equal to m_P). It follows that m_P is the nilradical of $\mathcal{O}_{Z,P}$ (*cf.* Summary 1.2.d) and hence its elements are nilpotent and, being of finite type, m_P itself is nilpotent, *i.e.*, there is an integer n such that $m_P^n = 0$.

3) A finite variety is a finite scheme such that all the local rings are equal to k and hence all the points are of multiplicity 1.

Proposition 1.3. *Let $(Z, \mathcal{O}_Z)$ be a finite scheme. For any subset V in Z, $\Gamma(V, \mathcal{O}_Z) = \prod_{P \in V} \mathcal{O}_{Z,P}$. Conversely, if we assign to every point of a finite set Z a local finite k-algebra, then the above formula defines a finite scheme structure on Z.*

Proof. This is clear: we associate to any section over V its restriction to each of the (open) points $P \in V$, and the gluing condition is empty in this case.

Definition 1.4. *If Z is a finite scheme and $A = \Gamma(Z, \mathcal{O}_Z)$, then we write $Z = \operatorname{Spec} A$.*

We note that A is a finite k-algebra (*i.e.*, a k-vector space of finite dimension) since it is a product of local rings $\mathcal{O}_{Z,P}$ which are finite dimensional (*cf.* 1.11.4). The scheme Z is a variety if and only if A is reduced.

Example 1.5. Even a single point has many possible scheme structures, such as $k[X]/(X^n)$ or $k[X,Y]/(X^2, XY, Y^2)$, for example.

b. The finite scheme structure on the intersection of two plane affine curves

Let $F, G \in k[X, Y]$ be two non-zero polynomials *without common factors*. We recall (*cf.* Chapter I,5) that the set $Z = V(F, G)$ is finite and the algebra $k[X, Y]/(F, G)$ is finite. We seek to define a finite scheme structure on Z. To do this, we recall that given a closed subset Z in an affine variety X such that $R = \Gamma(X)$ and $I = I(Z)$, we have $\mathcal{O}_Z = \widetilde{R/I}$ (*cf.* Chapter III, 7.4). We will therefore define $\mathcal{O}_Z$ using a similar formula, replacing $I(Z)$ with (F, G), as the above example of $C \cap D_0$ suggests. We will need a detailed description of the sheaf $\widetilde{R/I}$, which we give in full generality.

Proposition-Definition 1.6. *Let X be an irreducible affine algebraic variety and set $R = \Gamma(X)$. Let I be an ideal of R and let Z be the closed set $V(I)$ (we do not assume $I = I(Z)$). Let i be the inclusion of Z in X and set $\mathcal{F} = \widetilde{R/I}$. Let $D(f)$ be a standard open set in X. The following results hold.*

1) If $D(f) \cap Z = \varnothing$, then $\Gamma(D(f), \mathcal{F}) = 0$.

2) If $D(f) \cap Z = \{x\}$, then $\Gamma(D(f), \mathcal{F}) = \mathcal{O}_{X,x}/I\mathcal{O}_{X,x}$.

3) If $D(f) \cap Z = \{x_1, \ldots, x_n\}$, then

$$\Gamma(D(f), \mathcal{F}) = \prod_{i=1}^{n} \mathcal{O}_{X,x_i}/I\mathcal{O}_{X,x_i}.$$

4) The set $Z = V(I)$ is finite if and only if R/I is a finite k-algebra; Z is then discrete and we can define a sheaf of rings $\mathcal{O}_Z$ on Z by setting, for any $U \subset Z$:

$$\Gamma(U, \mathcal{O}_Z) = \prod_{x \in U} \mathcal{O}_{X,x}/I\mathcal{O}_{X,x}.$$

We then have $\mathcal{F} = i_*\mathcal{O}_Z$. In particular,

$$\Gamma(Z, \mathcal{O}_Z) = \Gamma(X, \mathcal{F}) = R/I = \prod_{x \in Z} \mathcal{O}_{X,x}/I\mathcal{O}_{X,x}.$$

The ringed space $(Z, \mathcal{O}_Z)$ is a finite scheme, denoted by $\mathrm{Spec}(R/I)$ as in 1.4.

Proof. We certainly have

$$\Gamma(D(f), \mathcal{F}) = (R/I)_f = R_f/IR_f.$$

1) We have $Z \subset V(f)$, and hence f, which vanishes on Z, is in the radical of I, so $f^r \in I$. But f^r is then both invertible and zero in the ring R_f/IR_f and hence this ring is the zero ring.

2) The point x corresponds to the maximal ideal m_x in R. Saying that $x \in D(f)$ means exactly that $f \notin m_x$ and hence m_x defines a maximal ideal of R_f. Saying that $x \in Z$ means exactly that $I \subset m_x$ and hence m_x gives

us a maximal ideal of R_f/IR_f. Saying that x is the only point in $Z \cap D(f)$ means that the ideal corresponding to m_x is the only maximal ideal in the ring R_f/IR_f, which is therefore local and hence $R_f/IR_f = (R_f/IR_f)_{m_x} = R_{m_x}/IR_{m_x}$, which is the result we seek.

3) There are elements $f_1, \ldots, f_n \in R$ such that, for all i, $f_i(x_i) \neq 0$ and $f_i(x_j) = 0$ for $j \neq i$. We therefore have $x_i \in D(ff_i)$, but $x_i \notin D(ff_j)$ for $j \neq i$. We calculate $\Gamma(D(f), \mathcal{F})$ by covering $D(f)$ with the sets $D(ff_i)$ and $D(g_i)$, which are disjoint from Z. We deduce an injective homomorphism

$$\Gamma(D(f), \mathcal{F}) \longrightarrow \prod_{i=1}^{n} \Gamma(D(ff_i), \mathcal{F}) \times \prod_j \Gamma(D(g_j), \mathcal{F}).$$

The last rings vanish (cf. 1) and so only the sets $D(ff_i)$ remain. As $D(ff_if_j) \cap Z = \varnothing$, there is no gluing condition and φ is an isomorphism, which, together with 2), proves the result.

4) We start by proving the first statement. Assume first that R/I is finite dimensional. Then, since $I \subset I(Z)$, the ring $\Gamma(Z) = R/I(Z)$ is finite dimensional, and hence Z is finite (cf. Chapter I, 4.8).

Conversely, if Z is finite the ring $\Gamma(Z)$, which is the reduced ring associated to R/I, is finite dimensional by Chapter I, 4.8. Moreover, the formula from 3) applied to $f = 1$ shows that R/I is a product of local rings, and hence the same is true of $\Gamma(Z)$, but in the latter all the local rings are equal to k (cf. 1.2). The local rings of R/I therefore all have nilpotent maximal ideals, and we will be done if we can prove the following lemma.

Lemma 1.7. *Let A be a local k-algebra with maximal ideal m and residue field k. We assume $m^n = 0$. Then A is finite dimensional over k.*

Proof. It is enough to consider the sequence of subspaces

$$(0) = m^n \subset m^{n-1} \subset \cdots \subset m \subset A$$

and note that the quotients m^i/m^{i+1} are all finite dimensional over $k = A/m$ (since the ideals m^i are of finite type).

Let us return to 4). The fact that the sheaves $i_*\mathcal{O}_Z$ are equal to $\mathcal{F}$ follows from 3). And finally, $(Z, \mathcal{O}_Z)$ is a finite scheme if and only if the local rings $\mathcal{O}_{X,x}/I\mathcal{O}_{X,x}$ are finite dimensional over k and this follows from the finiteness of R/I.

Remark 1.8. The sheaf of rings $\mathcal{O}_Z$ defined above is called the inverse image of the sheaf $\mathcal{F}$ under the map i and is denoted by $i^*(\mathcal{F})$.

Returning to the beginning of paragraph b), we define a finite scheme structure on $V(F, G)$ as a special case of 1.6 in the following way.

Proposition-Definition 1.9. *Consider two polynomials* $F, G \in k[X, Y]$ *without common factors and let* Z *be the finite set* $V(F, G)$. *We set* $R = k[X, Y]$ *and* $I = (F, G)$. *We equip* Z *with a ringed space structure by defining* $\mathcal{O}_Z$ *as in 1.6* $(\mathcal{O}_Z = i^*(\widetilde{R/I})$ *in the notation of 1.8). The ringed space* $(Z, \mathcal{O}_Z)$ *is then a finite scheme. The local ring* $\mathcal{O}_{Z,P}$ *of* Z *at* P *is equal to* $\mathcal{O}_{k^2,P}/(F, G)$ *and*

$$k[X, Y]/(F, G) \simeq \prod_{P \in Z} \mathcal{O}_{k^2,P}/(F, G) = \Gamma(Z, \mathcal{O}_Z),$$

so $Z = \operatorname{Spec}(R/I) = \operatorname{Spec} k[X, Y]/(F, G)$.

Definition 1.10. *With the notation of 1.9, we define the intersection multiplicity* $\mu_P(F, G)$ *of* F *and* G *at* P *to be the multiplicity of the finite scheme* $Z = V(F, G)$ *at* P: *alternatively*

$$\mu_P(F, G) = \dim_k \mathcal{O}_{k^2,P}/(F, G).$$

Corollary 1.11. *With the notation of 1.9,*

$$\sum_{P \in V(F,G)} \mu_P(F, G) = \dim_k k[X, Y]/(F, G) = \dim_k \Gamma(Z, \mathcal{O}_Z).$$

We note that this formula already contains a large part of the information we seek, namely the sum of the multiplicities of the intersection points. However, we need to be careful: this formula may miss out some points at infinity—think of the case $F = X$, $G = X - 1$.

Remarks 1.12.

1) In definition 1.10 the polynomials F and G can have multiple factors. In other words, we allow ourselves to consider intersection multiplicities of curves which may themselves have multiple factors (this is important for certain calculations, *cf.* Problem VII).

2) Definition 1.10 is meaningful even if $P \notin V(F, G)$. The multiplicity is then zero because either F or G is non-zero at P and is therefore invertible in the local ring, so the quotient is zero.

3) For an axiomatic definition of multiplicity and an algorithm for calculating it (which is indispensable in practice), *cf.* Problem VII or [F] Chapter 3.

4) We saw in Proposition 1.6 that any finite k-algebra is isomorphic to a product of finite local k-algebras. From this it is easy to deduce that the functor from the category of finite schemes to the category of finite k-algebras given by $Z \mapsto \Gamma(Z, \mathcal{O}_Z)$ is an equivalence of categories.

Example 1.13. Let us calculate the intersection multiplicity of $X^3 - X^2 - Y$ and Y at the point $P = (0, 0)$. Let $\mathcal{O}$ be the local ring $\mathcal{O}_{k^2,P}$. The ideal $I\mathcal{O}$ is equal to $(X^3 - X^2 - Y, Y) = (X^3 - X^2, Y) = (X^2(X - 1), Y) = (X^2, Y)$, since $X - 1$ is invertible in $\mathcal{O}$. The quotient $\mathcal{O}/I\mathcal{O}$ therefore has a basis given by the images of 1 and X and the multiplicity is 2.

2 Bézout's theorem

Throughout the following, we consider two homogeneous non-zero polynomials F, G in $k[X, Y, T]$ without common factors and of degrees s and t respectively.

a. Intersection multiplicity in projective space

Consider $P = (x, y, t) \in \mathbf{P}^2$. One of its coordinates is non-zero: we can assume that t is non-zero or even that $t = 1$. We then consider the dehomogenised (with respect to T) polynomials $F_\flat$ and $G_\flat$ (that is, $F_\flat(X, Y) = F(X, Y, 1)$, *cf.* Chapter III, 8.b), and we define the intersection multiplicity of F and G in P by $\mu_P(F, G) = \mu_{(x,y)}(F_\flat, G_\flat)$.

We now show that this definition does not depend on the choice of line at infinity ($T = 0$ in this case). Let $I(P)$ be the (prime, homogeneous) ideal of the point P in $\mathbf{P}^2$. The local ring $\mathcal{O}_{\mathbf{P}^2, P}$ can be obtained as the subring of the localised ring $k[X, Y, T]_{I(P)}$ consisting of elements A/B, where A, B are homogeneous of equal degree and $B(P) \neq 0$ (*cf.* Chapter III, 8.7). If J is an ideal of $k[X, Y, T]$, we denote by J_P the restriction to $\mathcal{O}_{\mathbf{P}^2, P}$ of the ideal generated by J in $k[X, Y, T]_{I(P)}$.

The fact that multiplicity is well defined follows from the following lemma.

Lemma 2.1. *With the above notations, there is a ring isomorphism*

$$\mathcal{O}_{\mathbf{P}^2, P}/(F, G)_P \simeq \mathcal{O}_{k^2, P}/(F_\flat, G_\flat).$$

Proof. Note that in the case at hand (*i.e.,* $P = (x, y, 1)$) the ideal $(F, G)_P$ is the ideal generated by F/T^s and G/T^t. We consider the homomorphism $\flat : k[X, Y, T] \to k[X, Y]$ sending T to 1. This homomorphism induces a map $\Phi : k[X, Y, T]_{I(P)} \to k[X, Y]_{m_P} = \mathcal{O}_{k^2, P}$ (here, m_P is the ideal of the point P in k^2). The restriction of Φ to $\mathcal{O}_{\mathbf{P}^2, P}$ is an isomorphism sending $\mathcal{O}_{\mathbf{P}^2, P}$ to $\mathcal{O}_{k^2, P}$ (*cf.* Chapter III, 8.7). Since the image under Φ of (F, G) is the ideal $(F_\flat, G_\flat)$, the result follows on passing to the quotient.

b. Statement of the theorem

Theorem 2.2 (Bézout). *Let $F, G \in k[X, Y, T]$ be two homogeneous polynomials without common factors of degrees s and t respectively. We have*

$$\sum_{P \in V(F) \cap V(G)} \mu_P(F, G) = st.$$

(We could also take this sum over all points of $\mathbf{P}^2$ since if $P \notin V(F) \cap V(G)$, then its multiplicity is zero, *cf.* 1.11.2.) The proof of Bézout's theorem takes up the rest of this chapter.

c. The link with the affine case

Lemma 2.3. *The set $V(F) \cap V(G) \subset \mathbf{P}^2$ is finite.*

Proof. Take $T = 0$ to be the line at infinity and denote it by D_∞. The restriction of $V(F)$ (resp. $V(G)$) to D_∞ is a projective algebraic subset of D_∞ and is hence either finite or equal to D_∞. If it is equal to D_∞, then T divides F (resp. T divides G). Since F and G have no common factors, the set $V(F) \cap V(G) \cap D_\infty$ is finite.

We identify $\mathbf{P}^2 - D_\infty$ with k^2 in the usual way. We then have $V(F) \cap k^2 = V(F_\flat)$ (*cf.* Chapter III, 8.b) and likewise $V(G) \cap k^2 = V(G_\flat)$. But $F_\flat$ and $G_\flat$ still have no common factors and hence $V(F) \cap V(G) \cap k^2 = V(F_\flat) \cap V(G_\flat)$, which is finite. The intersection is therefore finite.

Lemma 2.4. *There is a projective line D which does not meet the intersection $V(F) \cap V(G)$.*

Proof. Let $Z \subset \mathbf{P}^2$ be a finite set. We want to show that there is a line which does not meet Z. Take $a \notin Z$ and consider the lines passing through a. There are infinitely many such lines, and only a finite number of them meet Z. QED.

d. Description of the structure sheaf of the intersection

Let D be a projective line which does not meet $Z = V(F, G)$. After homography (which does not change the intersection multiplicities of the polynomials F and G or their degrees) we can assume that D is the line $T = 0$ and take it to be the line at infinity. On identifying $\mathbf{P}^2 - D_\infty$ and k^2 we have $V(F) \cap V(G) = V(F_\flat) \cap V(G_\flat) \subset k^2$. We equip $Z = V(F_\flat, G_\flat)$ with its finite scheme structure as in 1.9. We know that

$$\sum_{P \in V(F,G)} \mu_P(F, G) = \sum_{P \in V(F_\flat, G_\flat)} \mu_P(F_\flat, G_\flat) = \dim_k \Gamma(Z, \mathcal{O}_Z)$$

and it remains to show that

$$\dim_k \Gamma(Z, \mathcal{O}_Z) = \dim_k k[X, Y]/(F_\flat, G_\flat) = st.$$

We set $S = k[X, Y, T]$, $J = (F, G)$ (a homogeneous ideal) and $B = S/J$ (a graded S-module). In affine space, we set $R = k[X, Y]$ and $I = (F_\flat, G_\flat)$. We denote by i the inclusion of Z in k^2 and by j the inclusion of Z in $\mathbf{P}^2$. We then have the following description of $\mathcal{O}_Z$.

Proposition 2.5. *There are isomorphisms $i_*(\mathcal{O}_Z) \simeq \widetilde{R/I}$ and $j_*(\mathcal{O}_Z) \simeq \widetilde{S/J} = \widetilde{B}$. In particular, $\Gamma(Z, \mathcal{O}_Z) = \Gamma(\mathbf{P}^2, \widetilde{S/J})$.*

Proof. The first isomorphism was proved in 1.6. The second can be proved by the same method, replacing the open set $D(f)$ by $D^+(f)$ everywhere. (We will reconstruct this isomorphism using another method in 2.11.)

Given 2.5, it is enough to prove the formula $\Gamma(\mathbf{P}^2, \widetilde{S/J}) = st$. In fact, we do not really need to calculate this space of sections (*cf.* 2.13), but this notation enables us to better understand what is going on. The great advantage of working in projective space is that all the spaces of sections of sheaves are finite dimensional (*cf.* Chapter VII, 4.4). As we will see in Chapter VII, the following arguments contain a certain amount of hidden cohomology.

e. Description of $B = S/J$

Lemma 2.6. *There is an exact sequence of graded S-modules,*

$$0 \longrightarrow S(-s-t) \xrightarrow{\ \psi\ } S(-s) \oplus S(-t) \xrightarrow{\ \varphi\ } S \longrightarrow S/J \longrightarrow 0,$$

where $\psi(W) = (WG, -WF)$ and $\varphi(U, V) = UF + VG$.

Proof. This was proved in Chapter III, 10.1.

Corollary 2.7. *We have the following exact sequences of sheaves.*

$$0 \longrightarrow \mathcal{O}_{\mathbf{P}^2}(-s-t) \longrightarrow \mathcal{O}_{\mathbf{P}^2}(-s) \oplus \mathcal{O}_{\mathbf{P}^2}(-t) \longrightarrow \tilde{J} \longrightarrow 0,$$
$$0 \longrightarrow \tilde{J} \longrightarrow \mathcal{O}_{\mathbf{P}^2} \longrightarrow \widetilde{S/J} \longrightarrow 0.$$

The second exact sequence is to be understood as follows. Using j_*, we can identify the sheaves $\widetilde{S/J}$ and $\mathcal{O}_Z$. The sheaf $\tilde{J}$ is therefore a sheaf of ideals of $\mathcal{O}_{\mathbf{P}^2}$ which we can denote by $\mathcal{J}_Z$, and the second sequence is the analogue for the subscheme Z in $\mathbf{P}^2$ of the fundamental sheaf associated to a closed subset of a variety (*cf.* Chapter III, 9.5).

The problem with calculating $\Gamma(\mathbf{P}^2, \tilde{B})$ is that, although we have an exact sequence

$$0 \longrightarrow \Gamma(\mathbf{P}^2, \tilde{J}) \longrightarrow \Gamma(\mathbf{P}^2, \mathcal{O}_{\mathbf{P}^2}) \longrightarrow \Gamma(\mathbf{P}^2, \tilde{B}),$$

the last map is certainly not surjective: the dimension of $\Gamma(\mathbf{P}^2, \mathcal{O}_{\mathbf{P}^2})$ is 1 and we hope that the dimension of $\Gamma(\mathbf{P}^2, \tilde{B})$ will turn out to be st. We would have had exactly the same problem with the other exact sequence: remember that surjectivity of sheaves does not imply surjectivity of global sections (*cf.* Chapter III, 6.7).

The idea for getting round this problem—due to the lack of sections of $\mathcal{O}_{\mathbf{P}^2}$—is to work with the shifted sheaves $\mathcal{O}_{\mathbf{P}^2}(d)$ (*cf.* Chapter III, 9.7), which have many sections for large d.

f. Comparison of $\widetilde{B}$ and $\widetilde{B(d)}$ (or $\mathcal{O}_Z$ and $\mathcal{O}_Z(d)$)

The fact that Z does not meet the line at infinity will imply that shifting does not alter $j_*(\mathcal{O}_Z) = \widetilde{B}$.

Proposition 2.8. *Multiplication by T induces an injection α of graded S-modules from $B(-1)$ to B. Moreover, for $n \geqslant s + t - 1$ the map $\alpha_n : B_{n-1} \rightarrow B_n$ is surjective.*

Proof.

1) Injectivity. Consider $\overline{H} \in B(-1)$ such that $\alpha(\overline{H}) = 0$. This means that $TH = UF + VG$ in S. We set $T = 0$ and obtain $U(X,Y,0)F(X,Y,0) + V(X,Y,0)G(X,Y,0) = 0$. But $F(X,Y,0)$ and $G(X,Y,0)$ are coprime since F and G have no common point at infinity and hence $G(X,Y,0)$ divides $U(X,Y,0)$, $U(X,Y,0) = G(X,Y,0)C(X,Y)$, and hence $V(X,Y,0) = -F(X,Y,0)C(X,Y)$. Returning to S we see that $U = GC + TU'$, $V = -FC + TV'$, so $TH = T(U'F + V'G)$ and $H = U'F + V'G$ therefore vanishes in the quotient S/J.

2) Surjectivity of α_n follows from its injectivity and the following lemma.

Lemma 2.9. *For $d \geqslant s + t - 2$, $\dim_k B_d = st$.*

Proof. In degree d, the exact sequence 2.6 shows that

$$0 \longrightarrow S_{d-s-t} \longrightarrow S_{d-s} \oplus S_{d-t} \longrightarrow S_d \longrightarrow B_d \longrightarrow 0,$$

and therefore $\dim B_d = \dim S_d - \dim S_{d-s} - \dim S_{d-t} + \dim S_{d-s-t}$. A straightforward calculation using the formula

$$\dim S_n = \binom{n+2}{2} = (n+2)(n+1)/2,$$

which is valid for $n \geqslant -2$ (which explains the condition $d \geqslant s + t - 2$) then gives us the result.

(If we wanted to avoid the calculation, we could also have noted that the previous formula depends only on s and t and calculate the dimension of B_d in the special case where $F = X^s$, $G = Y^t$. In this case, there is a basis of B_d consisting of the images of the monomials $X^i Y^j T^{d-i-j}$, where $0 \leqslant i \leqslant s - 1$, $0 \leqslant j \leqslant t - 1$, and $d - i - j \geqslant 0$. For $d \geqslant s + t - 2$ the last condition is unnecessary, so the required dimension is indeed st.)

Corollary 2.10. *Multiplication by T induces a sheaf isomorphism $\widetilde{B(-1)} \simeq \widetilde{B}$. (It follows that for every integer d there is an isomorphism $\widetilde{B(d)} \simeq \widetilde{B}$.)*

Proof. This follows from 2.8 and Chapter III, 9.4 (since surjectivity for n large is enough).

Remark 2.11. Using 2.8 we can give another proof of 2.5. Let $H \in S$ be a homogeneous polynomial of degree > 0. Then $D^+(H) \cap D^+(T) = D(H_\flat)$ and $Z \cap D^+(H) = Z \cap D(H_\flat)$. We want to prove there is an isomorphism

$$\Gamma(D^+(H), \widetilde{S/J}) \simeq \Gamma(D(H_\flat), \widetilde{R/I}).$$

(By 1.6 this last group is isomorphic to $\Gamma(Z \cap D^+(H), \mathcal{O}_Z)$.) The careful reader will easily establish the existence of such an isomorphism using the flat (*i.e.*, $\flat$) operation and the ideas of the proof of Proposition 2.13 below.

Identifying $j_*(\mathcal{O}_Z)$ and $\widetilde{B}$, Corollary 2.10 can be stated as follows.

Corollary 2.12. *Multiplying by T induces a sheaf isomorphism $\mathcal{O}_Z(-1) \simeq \mathcal{O}_Z$, and hence for all integers d we have an isomorphism $\mathcal{O}_Z \simeq \mathcal{O}_Z(d)$.*

g. Conclusion

It only remains to establish the equalities

$$\dim_k \Gamma(Z, \mathcal{O}_Z(d)) = \dim_k \Gamma(\mathbf{P}^2, \widetilde{B(d)}) = \dim k[X, Y]/(F_\flat, G_\flat) = st.$$

In fact, as $\dim_k B_d = st$ for d large enough, it will be enough to find an isomorphism of B_d with $k[X, Y]/(F_\flat, G_\flat)$. (We can also, *cf.* Chapter VII, 4.3, prove that for d large enough there is an isomorphism $\Gamma(\mathbf{P}^2, \widetilde{B(d)}) \simeq B_d$, but this is not useful here.)

We consider the ring homomorphism

$$\flat : k[X, Y, T] = S \longrightarrow k[X, Y] = R$$

sending T to 1. As the image of the ideal J under this homomorphism is I, it factors through a homomorphism which we also denote by $\flat$, $\flat : B = S/J \to A = R/I$. After restriction we obtain a linear map $v_d : B_d \to R/I$. It remains to prove the following result.

Proposition 2.13. *The map $v = v_d : B_d \to R/I = k[X, Y]/(F_\flat, G_\flat)$ is an isomorphism for $d \geqslant s + t - 2$.*

Proof.

1) We show first that v is injective. Let $\overline{P} \in B_d$ be the image of $P \in S_d$ and assume $v(\overline{P}) = 0$. Then $P_\flat \in I$, *i.e.*, $P_\flat = aF_\flat + bG_\flat$, $a, b \in k[X, Y]$. Applying the sharp map we get $T^\alpha P = T^\beta a^\sharp F + T^\gamma b^\sharp G$ (*cf.* Chapter III, 8.b) and hence $T^\alpha P \in J$, but as multiplication by T is injective in S/J (*cf.* 2.8), it follows that $P \in J$, and hence $\overline{P} = 0$.

2) We now prove that v is surjective. Consider $\overline{f} \in k[X, Y]/I$, the image of the polynomial f. We consider $f^\sharp$ which is homogeneous of a certain degree n, and we consider its image $\overline{f^\sharp}$ in B_n. If $n \leqslant d$, then the element $T^{d-n}\overline{f^\sharp}$ in B_d is sent to f by v. If $n > d \geqslant s + t - 2$, then we know that multiplication by T^{n-d} gives an isomorphism between B_d and B_n and hence $\overline{f^\sharp} = T^{n-d}\overline{P}$, where v sends $\overline{P} \in B_d$ to f.

This completes the proof of Bézout's theorem.

Exercises

The following exercises deal with projective plane curves, particularly applications of Bézout's theorem.

1 Linear systems of curves

Let d be a strictly positive integer. We denote by V_d the vector space of homogeneous polynomials of degree d in X, Y, T.

a) Prove that $\dim_k(V_d) = (d+1)(d+2)/2$.

b) For all $P \in \mathbf{P}^2$ we set $V_d(P) = \{F \in V_d \mid F(P) = 0\}$. Prove that $V_d(P)$ is a hyperplane in V_d.

c) For all $r \in \mathbf{N}$ and $P \in \mathbf{P}^2$ we set $V_d(rP) = \{F \in V_d \mid \mu_P(F) \geqslant r\}$. Calculate $\dim_k(V_d(rP))$.

d) For all $r_1, \ldots, r_n \in \mathbf{N}$ and $P_1, \ldots, P_n \in \mathbf{P}^2$ we set

$$V_d(r_1 P_1, \ldots, r_n P_n) = \{F \in V_d \mid \forall i = 1, \ldots, n \quad \mu_{P_i}(F) \geqslant r_i\}.$$

Prove that

$$\dim_k V_d(r_1 P_1, \ldots, r_n P_n) \geqslant (d+1)(d+2)/2 - \sum_{i=1}^{n} r_i(r_i + 1)/2.$$

e) Using the above notation, prove that if

$$\sum_{i=1}^{n} r_i(r_i + 1)/2 \leqslant d(d+3)/2,$$

then there exists a curve (in the most general sense of the word: a polynomial, which may have multiple factors) with multiplicity $\geqslant r_i$ at each point P_i.

2 Degree and singular points

Let F be a homogeneous polynomial of degree $d \geqslant 1$ in X, Y, T. We denote by $\mu_P(F)$ (or simply μ_P) the multiplicity of F at P.

a) Assume that F has no multiple factors. Prove that

$$\sum_{P \in \mathbf{P}^2} \mu_P(\mu_P - 1) \leqslant d(d-1).$$

(Apply Bézout's theorem to F and $\partial F/\partial X$.)

b) Assume that F is irreducible. Prove the following strengthening of the above inequality:

$$\sum_{P \in \mathbf{P}^2} \mu_P(\mu_P - 1) \leqslant (d-1)(d-2).$$

(Prove as in 1.e that there is a polynomial G of degree $d-1$ which has multiplicity $\mu_P - 1$ at each of the the singular points of F and which vanishes at a certain number of suitably chosen points elsewhere on F. Apply Bézout's theorem.)

The (positive or zero) integer

$$g = (d-1)(d-2)/2 - \sum_{P \in \mathbf{P}^2} \mu_P(\mu_P - 1)/2$$

is called the *genus* of the curve $V(F)$ (whenever $V(F)$ has no multiple points with multiple tangents).

3 Applications

a) Re-prove the fact that an irreducible conic is non-singular and an irreducible cubic has at most one double point (*cf.* Exercise V, 1).

b) Prove that a quartic with more than three double points or a more-than-triple point is reducible.

c) Prove that a quartic with a unique double point with distinct tangent lines is irreducible.

d) Set $F(X,Y,T) = \sum_{i+j+k=4} X^i Y^j T^k$. Prove that $V(F)$ is smooth (and hence irreducible). (Note that $V(F)$ is stable under the action of the symmetric group S_3 which permutes the coordinates. Calculate the cardinal of the orbits of points of $\mathbf{P}^2$ under the action of this group, paying careful attention to the special cases, and examine the orbits of the singular points.)

4 The tricuspidal quartic

We set $F(X,Y,T) = Y^2T^2 + T^2X^2 + X^2Y^2 - 2XYT(X+Y+T)$.

a) Prove that $C = V(F)$ has three cusp points at $P = (0,0,1), Q = (0,1,0), R = (1,0,0)$.

b) Consider a conic Γ passing through P, Q, R and tangent to C at P. Prove that C and Γ have at most one other point in common.

c) Prove that all conics Γ of the above form have equations of the form $\mu(YT - XT) + \lambda XY$ with $(\lambda, \mu) \in \mathbf{P}^1(k)$. Calculate explicitly the intersection points of C and Γ. (Carry out your calculations in affine space.)

d) Prove that this gives us a birational parameterisation of C. We first calculate this parameterisation in affine space. We then homogenise and get a morphism $\varphi : \mathbf{P}^1 \to C$ given by the formula $\varphi(\lambda, \mu) = (4\mu^2(\lambda+\mu)^2, 4\mu^2(\lambda-\mu)^2, (\lambda^2-\mu^2)^2)$. Prove that this morphism is an isomorphism except at the points $(1,0), (-1,1)$ and $(1,1)$ whose images are the points P, Q and R and that the inverse of φ outside of P, Q, R is given by the map $\varphi^{-1}(x,y,t) = ((x-y)t, xy)$.

e) Construct the real part of C in the affine plane $T \neq 0$.

VII

Sheaf cohomology

0 Introduction

We return for a moment to the proof of Bézout's theorem. Given $Z = V(F, G)$ we had to calculate the dimension of $\Gamma(Z, \mathcal{O}_Z)$. The method used was to consider the exact sequences

$$0 \longrightarrow \mathcal{J}_Z \longrightarrow \mathcal{O}_{\mathbf{P}^2} \longrightarrow \mathcal{O}_Z \longrightarrow 0,$$
$$0 \longrightarrow \mathcal{O}_{\mathbf{P}^2}(-s-t) \longrightarrow \mathcal{O}_{\mathbf{P}^2}(-s) \oplus \mathcal{O}_{\mathbf{P}^2}(-t) \longrightarrow \mathcal{J}_Z \longrightarrow 0$$

and apply the global sections functor Γ to them. The difficulty is that Γ is not right exact: given an exact sequence of $\mathcal{O}_X$-modules on a variety X

$$(*) \qquad\qquad 0 \longrightarrow \mathcal{F} \longrightarrow \mathcal{G} \xrightarrow{p} \mathcal{H} \longrightarrow 0$$

there is an exact sequence

$$(**) \qquad\qquad 0 \longrightarrow \Gamma(X, \mathcal{F}) \longrightarrow \Gamma(X, \mathcal{G}) \xrightarrow{\pi} \Gamma(X, \mathcal{H}),$$

but π is not generally surjective.

This situation often arises in algebraic geometry: we often need to calculate the dimension of a space of global sections $\Gamma(X, \mathcal{F})$ (which in projective space is always finite dimensional if $\mathcal{F}$ is coherent, cf. 4.6). For example:

1) If X is a variety, $\dim \Gamma(X, \mathcal{O}_X)$ is the number of connected components of X.

2) If C is a curve in $\mathbf{P}^3$, $\dim \Gamma(\mathbf{P}^3, \mathcal{J}_C(s))$ is the number of (independent) degree s surfaces containing C.

The method used to calculate these numbers is to insert the sheaves in question into some exact sequence as in the proof of Bézout's theorem, and we always come across the problem of the non-exactness of Γ. (In a sequence of type $(**)$ two dimensions are not enough to calculate the third.)

Cohomology was invented (partly) to get around this problem. We introduce new groups associated to $\mathcal{F}$, which we denote by $H^i(X, \mathcal{F})$ (or $H^i(\mathcal{F})$ if there is no ambiguity), defined for $i \geqslant 0$ and such that H^0 is simply Γ. Their most important property is that given an exact sequence of sheaves $(*)$ there is a long exact sequence

$$0 \longrightarrow H^0(\mathcal{F}) \longrightarrow H^0(\mathcal{G}) \xrightarrow{\pi} H^0(\mathcal{H}) \xrightarrow{\delta}$$
$$H^1(\mathcal{F}) \longrightarrow H^1(\mathcal{G}) \longrightarrow H^1(\mathcal{H}) \longrightarrow H^2(\mathcal{F}) \longrightarrow \cdots$$

which we hope will allow us to calculate the image of π. (In practice we can only do this if some of the spaces H^i are zero.)

The existence and uniqueness of these groups $H^i\mathcal{F}$ is a very general fact proved using the theory of derived functors, cf. [H] or [Tohoku]. We will construct them explicitly, using Čech cohomology (but will have to admit their uniqueness).

To understand the origin of this Čech cohomology we return to the example of the exact sequence of $\mathcal{O}_X$-modules $(*)$ and the sequence $(**)$ of global sections. Our aim is to characterise the image of π.

Consider $h \in \Gamma(\mathcal{H})$. A priori h is not in the image of π, but it is locally in the image of π because π is surjective. There is therefore an open cover (U_i) of X and sections $g_i \in \Gamma(U_i, \mathcal{G})$ such that $p(g_i) = h|_{U_i}$. The problem is whether or not the sections g_i can be glued together (after possibly changing each g_i by an $f_i \in \Gamma(U_i, \mathcal{F})$, which doesn't change $p(g_i)$). Over each of the open sets $U_{ij} = U_i \cap U_j$, we consider the sections $f_{ij} = g_i - g_j$. As g_i and g_j have the same image in $\mathcal{H}$, f_{ij} is a section of $\mathcal{F}$. What is more, over the open set $U_{ijk} = U_i \cap U_j \cap U_k$ we have $f_{ij} + f_{jk} = f_{ik}$: this is called the cocycle relation. We say that the sections f_{ij} form a cocycle.

The sections g_i can be glued together if and only if there are elements $f_i \in \Gamma(U_i, \mathcal{F})$ such that $f_i + g_i = f_j + g_j$ or $f_{ij} = f_j - f_i$ (NB: we are now talking about sections of $\mathcal{F}$). We then say that the family f_{ij} is a coboundary. (Of course, a coboundary is a cocycle.)

The obstruction to h being in the image of π is that a cocycle is not always a coboundary. We are therefore led to define $H^1(X, \mathcal{F})$ to be the quotient of the group of cocycles by the group of coboundaries, the map $\delta : H^0(X, \mathcal{H}) \to H^1(X, \mathcal{F})$ being the map which associates to h the class of the cocycle f_{ij} defined above. We then have $h \in \operatorname{Im} \pi \Leftrightarrow h \in \operatorname{Ker} \delta$.

1 Some homological algebra

Definition 1.1. *A* Complex *(of abelian groups, for example, but it could also be a complex of A-modules or $\mathcal{O}_X$-modules) is a sequence $A^\bullet$:*

$$\cdots \longrightarrow A^{i-1} \xrightarrow{d^i} A^i \xrightarrow{d^{i+1}} A^{i+1} \longrightarrow \cdots$$

of abelian groups A^i indexed by $\mathbf{Z}$ or $\mathbf{N}$ such that $d^{i+1} \circ d^i = 0$ for all i. The homomorphisms d^i are called differentials (or coboundaries). NB: this sequence is not exact: $\operatorname{Im} d^i$ is contained in $\operatorname{Ker} d^{i+1}$, but equality does not necessarily hold. The quotient group $\operatorname{Ker} d^{i+1} / \operatorname{Im} d^i$ is called the ith cohomology group of the complex and is denoted by $H^i(A^\bullet)$.

We can also define complexes whose differentials are degree-decreasing. The quotients of such a complex are called its homology. When dealing with cohomology (resp. homology) we generally write the terms of the complex with upper (resp. lower) indices.

Example 1.2. Over any open set U in $\mathbf{R}^n$ we have a complex of differential forms with the usual differential d (*cf.* [Go]). The kernel of d consists of closed forms, its image consists of exact forms and the vanishing of the cohomology groups encodes topological information on U. For example, if U is a star domain, Poincaré's lemma says that the cohomology groups H^i vanish. This is also the case if $n = 2$ and U is simply connected.

Definition 1.3. *Let $A^\bullet$ and $B^\bullet$ be complexes whose differentials are, respectively, d^i and δ^i. A morphism of complexes from $A^\bullet$ to $B^\bullet$ is given by the data of homomorphisms $f^i : A^i \to B^i$ such that the following diagram commutes:*

$$
\begin{array}{ccc}
A^i & \xrightarrow{\ d^{i+1}\ } & A^{i+1} \\
{\scriptstyle f^i}\big\downarrow & & \big\downarrow {\scriptstyle f^{i+1}} \\
B^i & \xrightarrow{\ \delta^{i+1}\ } & B^{i+1}
\end{array}
$$

We write $f : A^\bullet \to B^\bullet$. An exact sequence of complexes

$$0 \longrightarrow A^\bullet \xrightarrow{\ f\ } B^\bullet \xrightarrow{\ g\ } C^\bullet \longrightarrow 0$$

is the data of two morphisms of complexes f, g such that for all i, the sequence

$$0 \longrightarrow A^i \xrightarrow{\ f^i\ } B^i \xrightarrow{\ g^i\ } C^i \longrightarrow 0$$

is exact.

It follows immediately from the definition of a morphism that f^i sends the kernel and image of d^i to the kernel and image of δ^i and hence induces a morphism $\overline{f^i} : H^i(A^\bullet) \to H^i(B^\bullet)$.

We have the following fundamental proposition on exact sequences.

Proposition 1.4. *Let $0 \to A^\bullet \xrightarrow{\ f\ } B^\bullet \xrightarrow{\ g\ } C^\bullet \to 0$ be an exact sequence of complexes. There are homomorphisms, called connecting morphisms,*

$$\partial^n : H^n C^\bullet \longrightarrow H^{n+1} A^\bullet$$

such that there is a long exact sequence

$$\cdots H^n A^\bullet \xrightarrow{\ \overline{f^n}\ } H^n B^\bullet \xrightarrow{\ \overline{g^n}\ } H^n C^\bullet \xrightarrow{\ \partial^n\ } H^{n+1} A^\bullet \xrightarrow{\ \overline{f^{n+1}}\ } H^{n+1} B^\bullet \cdots$$

Proof.

1) We consider the following commutative diagram, whose lines are complexes and whose columns are exact sequences:

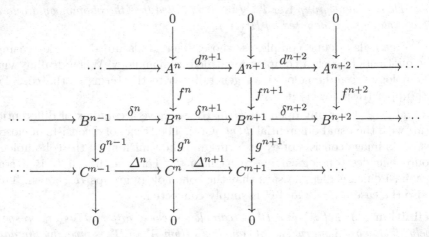

2) Possibly the hardest part of the solution is defining the homomorphism $\partial^n : H^n C^\bullet \to H^{n+1} A^\bullet$. By definition $H^n C^\bullet = \operatorname{Ker} \Delta^{n+1} / \operatorname{Im} \Delta^n$. Consider $\overline{x} \in H^n C^\bullet$: $\overline{x}$ is the class of some $x \in C^n$ such that $\Delta^{n+1}(x) = 0$.

Consider such an x. As g^n is surjective, $x = g^n(y)$, $y \in B^n$ and $\Delta^{n+1}(x) = \Delta^{n+1} g^n(y) = g^{n+1} \delta^{n+1}(y) = 0$, so $\delta^{n+1}(y) \in \operatorname{Im} f^{n+1}$: there is therefore a unique $z \in A^{n+1}$ such that $\delta^{n+1}(y) = f^{n+1}(z)$. We have $z \in \operatorname{Ker} d^{n+2}$. Indeed, $f^{n+2} d^{n+2}(z) = \delta^{n+2} f^{n+1}(z) = \delta^{n+2} \delta^{n+1}(y) = 0$ since $B^\bullet$ is a complex and the result follows since f^{n+2} is injective. Let $\overline{z}$ be the image of z in $H^{n+1} A^\bullet$.

We check first that $\overline{z}$ does not depend on the choice of the lifting y of x. Given another y' lifting x we have $y' = y + f^n(t)$ for some $t \in A^n$, and hence $\delta^{n+1}(y') = \delta^{n+1}(y) + f^{n+1} d^{n+1}(t)$. As f^{n+1} is injective, $z' = z + d^{n+1}(t)$, but then $\overline{z}' = \overline{z}$ in H^n.

We have therefore constructed a map $D : \operatorname{Ker} \Delta^{n+1} \to H^{n+1} A^\bullet$.

We now have to show that:

a) D is a homomorphism. (This is clear: we can lift $x + x'$ to $y + y'$.)

b) D vanishes on $\operatorname{Im} \Delta^n$. Indeed, if $x = \Delta^n(u)$, $u \in C^{n-1}$, then we lift u to $v \in B^{n-1}$ and we have $x = g^n \delta^n(v)$. We can therefore lift x to $y = \delta^n(v)$, but then $\delta^{n+1} \delta^n(v) = 0$ and hence $z = 0$.

We deduce that D factors through a map $\partial^n : H^n C^\bullet \to H^{n+1} A^\bullet$, given by $\partial^n(\overline{x}) = \overline{z}$ with the above notation.

3) It remains to prove that the sequence is exact. This is an easy diagram chase. We will prove exactness at $H^n C^\bullet$, the two other cases being left to the reader.

1) We have $\partial^n \overline{g^n} = 0$, so $\operatorname{Im} \overline{g^n} \subset \operatorname{Ker} \partial^n$. Indeed, consider $\overline{y} \in H^n B^\bullet$, coming from $y \in B^n$ such that $\delta^{n+1}(y) = 0$. We have $\overline{g^n}(\overline{y}) = \overline{g^n(y)} = \overline{x}$. By definition of ∂^n, we have $\partial^n(\overline{x}) = \overline{z}$, where $f^{n+1}(z) = \delta_{n+1}(y) = 0$, and since f^{n+1} is injective, $z = 0$.

2) Conversely, consider $\bar{x} \in \operatorname{Ker} \partial^n$. With the above notation $\bar{z} = 0$, *i.e.*, $z = d^{n+1}(t)$ with $t \in A^n$ and hence $f^{n+1}(z) = \delta^{n+1}(y) = \delta^{n+1}f^n(t)$, so that $y - f^n(t) \in \operatorname{Ker} \delta^{n+1}$, but then $x = g^n(y - f^n(t))$, so $\bar{x} \in \operatorname{Im} \bar{g}^n$.

2 Čech cohomology

a. Definition

Let X be a topological space and let $\mathcal{F}$ be a sheaf of abelian groups on X. Let $\mathcal{U} = (U_i)_{i \in I}$ be a finite open covering of X indexed by the ordered set $I = \{0, 1, \ldots, n\}$.

We denote the intersections $U_{ij} = U_i \cap U_j, \ldots, U_{i_0 \cdots i_p} = U_{i_0} \cap \cdots \cap U_{i_p}$. We now define a complex of abelian groups $C^\bullet(\mathcal{U}, \mathcal{F})$.

For all $0 \leqslant p \leqslant n$ we set

$$C^p(\mathcal{U}, \mathcal{F}) = \prod_{i_o < \cdots < i_p} \mathcal{F}(U_{i_0 \cdots i_p}).$$

If $p > n$, we set $C^p(\mathcal{U}, \mathcal{F}) = 0$.

An element of $C^p(\mathcal{U}, \mathcal{F})$ is called a cochain. It consists of the data of a section of $\mathcal{F}$, $s_{i_0 \cdots i_p}$, over each intersection $U_{i_0 \cdots i_p}$. For example, C^0 consists of sections s_i over U_i, C^1 of sections s_{ij} over $U_{ij}, \ldots$

We now define the differential $d^{p+1} : C^p \to C^{p+1}$ by the formula

$$(d^{p+1}s)_{i_0 \cdots i_{p+1}} = \sum_{k=0}^{p+1} (-1)^k s_{i_0 \cdots \widehat{i_k} \cdots i_{p+1}} |_{U_{i_0 \cdots i_{p+1}}},$$

where the symbol $\widehat{i_k}$ means that we leave out the index i_k.

Proposition-Definition 2.1. *We have $d^{p+1} \circ d^p = 0$ for all $p \geqslant 0$. The complex thus constructed is called the Čech complex of $\mathcal{F}$ relative to the open cover $\mathcal{U}$. Its cohomology groups are called the Čech cohomology groups of $\mathcal{F}$ relative to $\mathcal{U}$. They are denoted by $\check{H}^p(\mathcal{U}, \mathcal{F})$.*

Proof. Set $t = d^p(s)$. We calculate

$$d^{p+1}(t)_{i_0 \cdots i_{p+1}} = \sum_{k=0}^{p+1} (-1)^k t_{i_0 \cdots \widehat{i_k} \cdots i_{p+1}} |_{U_{i_0 \cdots i_{p+1}}}$$

$$= \sum_{k=0}^{p+1} (-1)^k \sum_{l=0, l \neq k}^{p+1} (-1)^{a(k,l)} s_{i_0 \cdots \widehat{i_k} \cdots \widehat{i_l} \cdots i_{p+1}} |_{U_{i_0 \cdots i_{p+1}}},$$

where the integer $a(k, l)$ is l if $l < k$ and $l + 1$ if $l > k$.

Consider the term $s_{i_0 \cdots \widehat{\alpha} \cdots \widehat{\beta} \cdots i_{p+1}}$, $\alpha < \beta$ in the total sum. It appears twice, once for $l = \alpha$, $k = \beta$ with the sign $(-1)^\alpha (-1)^{\beta+1}$, and once for $k = \alpha$, $l = \beta$ with the sign $(-1)^\alpha (-1)^\beta$. These two terms cancel each other and the result follows.

Example 2.2. If $s \in C^0$, then $s = (s_i)$, where $s_i \in \mathcal{F}(U_i)$ and ds is given by $(ds)_{ij} = (s_j - s_i)|_{U_{ij}}$. If $t \in C^1$, $t = (t_{ij}) \in \mathcal{F}(U_{ij})$, then its image under d is $(dt)_{ijk} = (t_{jk} - t_{ik} + t_{ij})|_{U_{ijk}}$. We see that in this example, $d \circ d$ does indeed vanish. We recover the description of cocycles and coboundaries given in the introduction.

We will now show that the cohomology groups we have defined have all the properties claimed for them in the introduction. The group H^0 is what we expect.

Proposition 2.3. *We have $\check{H}^0(\mathcal{U}, \mathcal{F}) = \Gamma(X, \mathcal{F})$.*

Proof. The group H^0 is simply the kernel of the first differential. It therefore consists of those $s_i \in \mathcal{F}(U_i)$ such that $ds = 0$, *i.e.*, such that $s_i = s_j$ over U_{ij}. As $\mathcal{F}$ is a sheaf the s_i can be glued together to form a global section $s \in \Gamma(X, \mathcal{F}) = \mathcal{F}(X)$.

Remarks 2.4.
1) Čech cohomology is functorial: given a homomorphism of sheaves of groups $u : \mathcal{F} \to \mathcal{G}$ there is an associated homomorphism of complexes from $C^p(\mathcal{U}, \mathcal{F})$ to $C^p(\mathcal{U}, \mathcal{G})$ and group homomorphisms $H^p(u) : \check{H}^p(\mathcal{U}, \mathcal{F}) \to \check{H}^p(\mathcal{U}, \mathcal{G})$.

2) It is immediate from the definition of Čech cohomology that it commutes with direct sums, *i.e.*,

$$\check{H}^p\left(X, \bigoplus_{i \in I} \mathcal{F}_i\right) = \bigoplus_{i \in I} \check{H}^p(X, \mathcal{F}_i).$$

3) When X is an algebraic variety over a field k and $\mathcal{F}$ a $\mathcal{O}_X$-module, the groups $\mathcal{F}(U_{i_0 \cdots i_p})$ have natural k-vector space structures and the same is true of the groups $C^p(\mathcal{U}, \mathcal{F})$. As the differentials are k-linear, all the sets $\check{H}^p(\mathcal{U}, \mathcal{F})$ are also k-vector spaces.

b. The cohomology groups H^i vanish in affine geometry

Theorem 2.5. *Let X be an affine variety, let $A = \Gamma(X)$ be its ring, let M be an A-module, let $\mathcal{F} = \widetilde{M}$ be the associated quasi-coherent sheaf and let $\mathcal{U} = (U_i)$ $(i = 0, \ldots, m)$ be an open cover of X by standard affine open sets. Then for all $p > 0$ we have $\check{H}^p(\mathcal{U}, \mathcal{F}) = 0$.*

Proof. We will basically only deal with a doubly special case: 1) we set $p = 1$ and 2) we assume A is an integral domain and M is torsion free. We will also give an idea of the general proof. It is not conceptually more difficult but is rather long.

We recall that M is torsion free if, for all $a \in A$ and $x \in M$ such that $ax = 0$, $a = 0$ or $x = 0$.

We set $U_i = D(f_i)$, with $f_i \in A$, $f_i \neq 0$.

Let $\alpha = (\alpha_{ij})$ be a 1-cocycle. A priori α_{ij} is only defined for $i < j$ and the cocycle relation

$$(*) \qquad\qquad \alpha_{jk} - \alpha_{ik} + \alpha_{ij} = 0$$

only holds for $i < j < k$. It will be convenient to extend this definition to all pairs i, j by setting $\alpha_{ii} = 0$ for all i and $\alpha_{ij} = -\alpha_{ji}$ if $i > j$. We can then check that the relation $(*)$ holds for all triplets i, j, k.

We have $\alpha_{ij} \in \Gamma(U_{ij}, \widetilde{M}) = M_{f_i f_j}$. We write $\alpha_{ij} = \beta_{ij}/f_i^n f_j^n$, where $\beta_{ij} \in M$. (We can use the same n for all pairs i, j because the open cover is finite.)

The relation $(*)$ can then be written over U_{ijk} as

$$\frac{\beta_{jk}}{f_j^n f_k^n} - \frac{\beta_{ik}}{f_i^n f_k^n} + \frac{\beta_{ij}}{f_i^n f_j^n} = 0$$

or, alternatively, $f_i^n \beta_{jk} - f_j^n \beta_{ik} + f_k^n \beta_{ij} = 0$. A priori this holds in $M_{f_i f_j f_k}$, but as M is torsion free, it also holds in M.

Alternatively, we can write this relation in the form

$$(**) \qquad\qquad f_k^n \alpha_{ij} = -\frac{\beta_{jk}}{f_j^n} + \frac{\beta_{ik}}{f_i^n},$$

which holds over the open set U_{ij}.

We note that this relation is not very far from what we seek: we want to show that a_{ij} is a coboundary, i.e., that it can be written as $\alpha_{ij} = \gamma_j - \gamma_i$. The relation $(**)$ says that α_{ij} is a coboundary after inverting f_k, i.e., on U_k. To glue all this together we will use a partition of unity.

Indeed, as we saw in Chapter III, 2.3, the fact that the sets $D(f_i) = D(f_i^n)$ cover X means there is a partition of unity: there are elements $a_k \in A$ such that $1 = \sum_{k=0}^{m} a_k f_k^n$. For all j we set

$$\gamma_j = -\sum_{k=0}^{m} a_k \frac{\beta_{jk}}{f_j^n} \in \Gamma(U_j, \widetilde{M}),$$

and over U_{ij} we have

$$\gamma_j - \gamma_i = \sum_{k=0}^{m} a_k \left(\frac{\beta_{ik}}{f_i^n} - \frac{\beta_{jk}}{f_j^n} \right) = \sum_{k=0}^{m} a_k f_k^n \alpha_{ij} = \alpha_{ij},$$

and hence α is indeed a coboundary.

The proof without assuming A is an integral domain and M is torsion free is very similar. The only thing we need to be careful of is that the relation $f_i^n \beta_{jk} - f_j^n \beta_{ik} + f_k^n \beta_{ij} = 0$ over U_{ijk}, (i.e., in $M_{f_i f_j f_k}$) means that there is an integer N (which can be taken to be the same for all triples) such that

$f_i^N f_j^N f_k^N (f_i^n \beta_{jk} - f_j^n \beta_{ik} + f_k^n \beta_{ij}) = 0$ in M. The rest of the proof is more or less the same. We have to use the exponent $n + N$ in the partition of unity and replace a_k by $a_k f_k^N$ in the expression of the sections g_j.

To deal with the group H^p (when A is an integral domain and M is torsion free), we set

$$\alpha_{i_0,\ldots,i_p} = \frac{\beta_{i_0,\ldots,i_p}}{f_{i_0}^n \cdots f_{i_p}^n},$$

where $\alpha_{i_0,\ldots,i_p}$ is our cocycle. There is a partition of unity $1 = \sum_{k=0}^m a_k f_k^n$. It is then enough to set

$$\gamma_{i_0,\ldots,i_{p-1}} = \sum_{k=0}^m a_k \frac{\beta_{k,i_0,\ldots,i_{p-1}}}{f_{i_0}^n \cdots f_{i_{p-1}}^n}$$

and we check (using the fact that $d\alpha = 0$) that $\alpha = d\gamma$.

Corollary 2.6. *Let X be an* affine *algebraic variety and let $0 \to \mathcal{F} \to \mathcal{G} \to \mathcal{H} \to 0$ be an exact sequence of* quasi-coherent *sheaves. There is then an exact sequence of global sections*

$$0 \longrightarrow \Gamma(X,\mathcal{F}) \longrightarrow \Gamma(X,\mathcal{G}) \longrightarrow \Gamma(X,\mathcal{H}) \longrightarrow 0.$$

Proof. By the calculations given in the introduction (using standard open sets, which is possible since these form a basis of open sets) this is exactly 2.5.

c. The exact sequence of Čech cohomology

We now need the concept of a separated variety (*cf.* Problem I, Part 4).

Definition 2.7. *Let X be an algebraic variety. We say that X is separated if the diagonal $\Delta = \{(x,y) \in X \times X \mid x = y\}$ is closed in the product $X \times X$. If X is separated and U and V are open affine sets of X, then $U \cap V$ is an open affine set of X.*

Proposition 2.8. *Affine, projective, quasi-affine and quasi-projective varieties are separated.*

We then have the following theorem.

Theorem 2.9. *Let X be a separated algebraic variety, let $\mathcal{U}$ be a finite affine cover of X and let $0 \to \mathcal{F} \to \mathcal{G} \to \mathcal{H} \to 0$ be an exact sequence of quasi-coherent sheaves. There is then a long exact sequence of Čech cohomology:*

$$0 \longrightarrow \check{H}^0(\mathcal{U},\mathcal{F}) \longrightarrow \check{H}^0(\mathcal{U},\mathcal{G}) \longrightarrow \check{H}^0(\mathcal{U},\mathcal{H}) \longrightarrow$$
$$\check{H}^1(\mathcal{U},\mathcal{F}) \longrightarrow \check{H}^1(\mathcal{U},\mathcal{G}) \longrightarrow \cdots$$

Proof. We consider the Čech complexes associated to $\mathcal{F}, \mathcal{G}, \mathcal{H}$. They form an exact sequence

$$0 \longrightarrow C^{\bullet}(\mathcal{U}, \mathcal{F}) \longrightarrow C^{\bullet}(\mathcal{U}, \mathcal{G}) \longrightarrow C^{\bullet}(\mathcal{U}, \mathcal{H}) \longrightarrow 0.$$

Indeed, as the sets U_i are affine and X is separated, the intersections $U_{i_0,\dots,i_p}$ are also affine. But as these sheaves are quasi-coherent we have by 2.5 an exact sequence

$$0 \longrightarrow \mathcal{F}(U_{i_0,\dots,i_p}) \longrightarrow \mathcal{G}(U_{i_0,\dots,i_p}) \longrightarrow \mathcal{H}(U_{i_0,\dots,i_p}) \longrightarrow 0.$$

We get the exact sequence of complexes by taking the finite product of these sequences for $i_0 < \cdots < i_p$ and the result follows by 1.4.

Remark 2.10. We can prove that under the hypotheses of 2.9 (*i.e.*, X separated, $\mathcal{F}$ quasi-coherent and $\mathcal{U}$ affine) Čech cohomology is *the* correct cohomology (which is unique if we require that $H^0 = \Gamma$, that the long exact sequence should exist and a certain vanishing theorem should hold). See [H], Chapter III, 4.5 for more details. In particular, and *we will use this fact without proof*, Čech cohomology does not depend on the choice of affine covering. (This can also be proved directly using spectral sequences, *cf.* [G], 5.9.2.) Henceforth, under the above hypotheses we will denote this ith cohomology group by $H^i(X, \mathcal{F})$, omitting the open cover and the sign $^{\vee}$.

Proposition 2.11. *Let X be a separated algebraic variety, let Y be a closed subvariety of X and let $j : Y \to X$ be the canonical injection. Let $\mathcal{F}$ be a quasi-coherent sheaf over Y. Then, for all i, $H^i(Y, \mathcal{F}) = H^i(X, j_*\mathcal{F})$.*

Proof. The hypotheses on X, Y and $\mathcal{F}$ allow us to calculate the cohomology groups using Čech cohomology. We take an affine cover U_i of X. The sets $Y \cap U_i$ form an affine cover of Y and we are done by the formula $\mathcal{F}(U \cap Y) = j_*\mathcal{F}(U)$, which is the definition of j_*.

Remark 2.12. As in Chapter III we will often identify $\mathcal{F}$ and its direct image $j_*\mathcal{F}$, and by abuse of notation we will denote the cohomology group $H^i(X, j_*\mathcal{F})$ by $H^i(X, \mathcal{F})$.

3 Vanishing theorems

The usefulness of cohomology (and particularly the long exact sequence) depends largely on our ability to prove that certain cohomology groups vanish. We already have Theorem 2.5 which says that on an affine algebraic variety there is no cohomology in degree > 0.

Here is another vanishing result linked to dimension.

Theorem 3.1. *Let V be a separated variety of dimension n and let $\mathcal{F}$ be a quasi-coherent sheaf. Then $H^i(V, \mathcal{F}) = 0$ for all $i > n$.*

Proof. We will prove this only for a projective algebraic variety. For the general case, see [H] Chapter III, 2.7 or [H] Exercise III, 4.8. In the projective case we have the following lemma.

Lemma 3.2. *Let V be a closed subvariety of $\mathbf{P}^N$ of dimension n. There is a linear subvariety $W \subset \mathbf{P}^N$, of codimension $n + 1$, such that $V \cap W = \varnothing$.*

Proof (of 3.2). We proceed by induction on N. This is clear for $N = 1$. To pass from $N - 1$ to N we prove that there is a hyperplane H in $\mathbf{P}^N$ which does not contain any component of V. The restriction of the said components to H are then empty or of dimension $\leqslant n - 1$, and we apply induction to H.

Set $E = k^{n+1}$. To prove the existence of H we note that the projective hyperplanes correspond to linear forms which do not vanish on E, two proportional forms giving rise to the same hyperplane. In other words, the space of hyperplanes is the projective space $\mathbf{P}(E^*)$. In terms of coordinates, the plane H defined by the equation $a_0 X_0 + \cdots + a_N X_N = 0$ corresponds to the point in $\mathbf{P}(E^*)$ with homogeneous coordinates $(a_0, \ldots, a_N)$.

If $V = V_1 \cup \cdots \cup V_r$, then we choose $x_i \in V_i$, $x_i = (x_{i0}, \ldots, x_{iN})$. In the space of hyperplanes, the set of hyperplanes not containing x_i is a non-empty open set Ω_i (defined by $\sum_{k=0}^{N} x_{ik} a_k \neq 0$). But then Ω, the intersection of the sets Ω_i, is non-empty and any $H \in \Omega$ will do.

We return to 3.1. Up to homography we may assume $W = V(X_0, \ldots, X_n)$. As $V \cap W$ is empty, $V \subset D^+(X_0) \cup \cdots \cup D^+(X_n)$ and V is therefore covered by the $n + 1$ open affine sets $V \cap D^+(X_i)$. But in the Čech complex associated to this covering all the groups C^i vanish for $i > n$ and a fortiori all the groups H^i vanish.

Remark 3.3. The above proof also shows that if $\mathcal{F}$ is a quasi-coherent sheaf on a variety V supported on a subvariety W of dimension d (*i.e.*, such that $\mathcal{F}(U) = 0$ for any open set U which does not meet W), then for all $p > d$, $H^p(V, \mathcal{F}) = 0$. In particular, this is the case if $\mathcal{F}$ is a sheaf of the form $j_*(\mathcal{G})$, where $\mathcal{G}$ is a sheaf over W and j is the canonical injection of W into V.

For example, if Z is a finite scheme contained in $\mathbf{P}^n$ (*cf.* Chapter VI), then $H^p(Z, \mathcal{O}_Z) = H^p(\mathbf{P}^n, j_* \mathcal{O}_Z) = 0$ for $p > 0$.

4 The cohomology of the sheaves $\mathcal{O}_{\mathbf{P}^n}(d)$

The cohomology of the sheaves $\mathcal{O}_{\mathbf{P}^n}(d)$, which play a fundamental role in algebraic geometry, is particularly simple.

Theorem 4.1. *Let n be an integer $\geqslant 1$ and consider $d \in \mathbf{Z}$. We denote by S_d the space of homogeneous polynomials of degree d in $n + 1$ variables. (Conventionally, if $d < 0$, $S_d = 0$.)*

a) We have $H^0(\mathbf{P}^n, \mathcal{O}_{\mathbf{P}^n}(d)) = S_d$ for all d.

b) We have $H^i(\mathbf{P}^n, \mathcal{O}_{\mathbf{P}^n}(d)) = 0$ for $0 < i < n$ and all d.

c) The vector space $H^n(\mathbf{P}^n, \mathcal{O}_{\mathbf{P}^n}(d))$ is isomorphic to the dual of the vector space $H^0(\mathbf{P}^n, \mathcal{O}_{\mathbf{P}^n}(-d - n - 1))$ for all $d \in \mathbf{Z}$.

Example 4.2. It is easy to remember the dimensions of these cohomology groups. Firstly, they all vanish except H^0 and H^n. Secondly, H^0 for given d is composed of homogeneous polynomials of degree d and is hence of dimension $\binom{n+d}{n}$. For $n = 3$, for example, the possible dimensions are $1, 4, 10, 20, 35, 56, \ldots$ Finally, the dimensions of the spaces H^n are the same but inverted (increasing as d tends to $-\infty$), and it is enough to remember that the first d for which $H^n(\mathbf{P}^n, \mathcal{O}_{\mathbf{P}^n}(d))$ does not vanish is $-n - 1$. The dimension of H^n for given d is $\binom{-d-1}{n}$. Note that the sheaves $\mathcal{O}_{\mathbf{P}^n}(-i)$ such that $1 \leqslant i \leqslant n$ have particularly simple cohomology groups!

Proof. Set $S = k[X_0, \ldots, X_n] = \bigoplus_{d \in \mathbf{N}} S_d$. We will prove the theorem for all integers $d \in \mathbf{Z}$ simultaneously using the sheaf $\mathcal{F} = \bigoplus_{d \in \mathbf{Z}} \mathcal{O}_{\mathbf{P}^n}(d)$. This sheaf is quasi-coherent (but not coherent) over $\mathbf{P}^n$ and is associated to the graded S-module $M = \bigoplus_{d \in \mathbf{Z}} S(d)$. As cohomology commutes with direct sums (*cf.* Remark 2.4.2), if we calculate the cohomology of $\mathcal{F}$ we get the cohomology of each of its factors.

Of course, we are going to calculate the cohomology of $\mathcal{F}$ using the standard open cover of $\mathbf{P}^n$ by open affine sets $U_i = D^+(X_i)$ for $i = 0, 1, \ldots, n$. We know (*cf.* Chapter III, 9) that $\Gamma(U_{i_0,\ldots,i_p}, \mathcal{O}_{\mathbf{P}^n}(d)) = S(d)_{(X_{i_0} \cdots X_{i_p})}$, the set of homogeneous degree d elements of the local ring $S_{X_{i_0} \cdots X_{i_p}}$, and we deduce from this the Čech complex of $\mathcal{O}_{\mathbf{P}^n}(d)$. The Čech complex of $\mathcal{F}$, which is the direct sum of the above, consists of the localised rings of S: we have $C^p = C^p(\mathcal{U}, \mathcal{F}) = \prod_{i_0 < \cdots < i_p} S_{X_{i_0} \cdots X_{i_p}}$, from which we obtain a complex

$$\prod_{i=0}^{n} S_{X_i} \longrightarrow \prod_{0 \leqslant i < j \leqslant n} S_{X_i X_j} \longrightarrow \cdots \longrightarrow \prod_{i=0}^{n} S_{X_0 \cdots \widehat{X_i} \cdots X_n} \xrightarrow{\delta_n} S_{X_0 X_1 \cdots X_n}.$$

We note that the group C^p with its natural decomposition as a direct sum is a graded S-module, and hence the same is true of $H^p(\mathbf{P}^n, \mathcal{F})$. The degree d homogeneous part of this group is simply $H^p(\mathbf{P}^n, \mathcal{O}_{\mathbf{P}^n}(d))$.

As the functor H^0 is equal to Γ (*cf.* 2.3) we have already calculated $H^0(\mathbf{P}^n, \mathcal{O}_{\mathbf{P}^n}(d))$ in Chapter III, 9.

To calculate H^n, we note that $S_{X_0 X_1 \cdots X_n}$ is an infinite dimensional space with a basis formed of all the monomials $X_0^{\alpha_0} \cdots X_n^{\alpha_n}$ with $\alpha_i \in \mathbf{Z}$. We then note that H^n is the cokernel of δ_n and that the image of this map consists of all fractions of the form

$$\sum_{i=0}^{n} (-1)^i \frac{F_i}{(X_0 \cdots \widehat{X_i} \cdots X_n)^r} = \sum_{i=0}^{n} (-1)^i \frac{X_i^r F_i}{(X_0 \cdots X_n)^r}.$$

We therefore have a basis of the image of δ_n formed of monomials $X_0^{\alpha_0} \cdots X_n^{\alpha_n}$ as above, such that at least one of the integer α_i is $\geqslant 0$. It follows that the cokernel $H^n(\mathbf{P}^n, \mathcal{F})$ has a basis consisting of the images of the monomials *all* of whose exponents are strictly negative, $H^n(\mathbf{P}^n, \mathcal{O}_{\mathbf{P}^n}(d))$ corresponding to monomials whose degree $\sum_{i=0}^{n} \alpha_i = d$. We note that this space is 0 for $d \geqslant -n$. For $d \leqslant -n - 1$ we have to count the degree d monomials in the variables X_i all of whose exponents are < 0. This is equivalent to counting the monomials of degree $-d - (n + 1)$ in variables $1/X_i$, and there are therefore $\binom{-d-1}{n}$ of them.

In the special case $d = -n - 1$ the space $H^n(\mathbf{P}^n, \mathcal{O}_{\mathbf{P}^n}(-n - 1))$ is of dimension 1 with a basis consisting of the image of the monomial $1/(X_0 \cdots X_n)$. Identifying this space with the base field we obtain a non-degenerate bilinear form

$$\varphi : H^n(\mathbf{P}^n, \mathcal{O}_{\mathbf{P}^n}(d)) \times H^0(\mathbf{P}^n, \mathcal{O}_{\mathbf{P}^n}(-d - n - 1)) \longrightarrow H^n(\mathbf{P}^n, \mathcal{O}_{\mathbf{P}^n}(-n - 1))$$

associating to the monomials $X_0^{\alpha_0} \cdots X_n^{\alpha_n}$ and $X_0^{\beta_0} \cdots X_n^{\beta_n}$, where $\sum_{i=0}^n \alpha_i = d$ and $\sum_{i=0}^n \beta_i = -d - n - 1$, the image of the monomial $X_0^{\alpha_0 + \beta_0} \cdots X_n^{\alpha_n + \beta_n}$ in $H^n(\mathbf{P}^n, \mathcal{O}_{\mathbf{P}^n}(-n - 1))$. As φ is non-degenerate (with a suitable choice of ordering on monomials its matrix is the identity), it induces an isomorphism between H^n and the dual of H^0.

It remains to prove b). We argue by induction on n. For $n = 1$ there is nothing to prove.

We consider the element X_0 in S and the complex $C^\bullet_{(X_0)}$ obtained by taking the homogeneous degree zero part of the localisation of the Čech complex of $\mathcal{F}$ with respect to X_0.

The proof of the following easy lemma (which follows from the fact that the localisation functor is exact) is left to the reader.

Lemma 4.3. *Let* $M^\bullet = M' \overset{v}{\longrightarrow} M \overset{u}{\longrightarrow} M''$ *be a complex of graded S-modules and let $H(M^\bullet)$ be the associated cohomology group (which is a graded S-module). Let $f \in S$ be a homogeneous element of degree $d > 0$ and $M^\bullet_{(f)} = M'_{(f)} \overset{v_{(f)}}{\longrightarrow} M_{(f)} \overset{u_{(f)}}{\longrightarrow} M''_{(f)}$ the complex obtained on taking the degree 0 elements of the localisation and let $H(M^\bullet_{(f)})$ be its cohomology group. Then $H(M^\bullet_{(f)}) = H(M^\bullet)_{(f)}$.*

The lemma shows that the cohomology groups of $C^\bullet_{(X_0)}$ can be obtained by localising the cohomology groups of $C^\bullet$: $H^p(C^\bullet_{(X_0)}) = H^p(C^\bullet)_{(X_0)}$.

Moreover, we know (*cf.* Chapter III, 8.5) that the operator $\flat$ gives us an isomorphism $S_{(X_0)} \simeq R = k[X_1, \ldots, X_n] = \Gamma(D^+(X_0), \mathcal{O}_{\mathbf{P}^n})$ and this isomorphism induces isomorphisms on the localisations: $(S_{X_{i_0} \cdots X_{i_p}})_{(X_0)} \simeq R_{X_{i_0} \cdots X_{i_p}}$ (replacing X_0 by 1 wherever necessary); in other words the complex $C^\bullet_{(X_0)}$ is isomorphic to the Čech complex $C'^\bullet$ of the sheaf $\mathcal{F}$ restricted to $D^+(X_0)$, with respect to the (affine) open cover of this open set by the sets $D^+(X_i) \cap D^+(X_0)$, for $i = 0, \ldots, n$.

As $D^+(X_0)$ is affine and $\mathcal{F}$ is quasi-coherent, the cohomology of $C'^\bullet$ vanishes for $p \geqslant 1$, as does that of $C^\bullet_{(X_0)}$, and we therefore have $H^p(\mathbf{P}^n, \mathcal{F})_{(X_0)} = 0$ for $p \geqslant 1$. This simply means that for all $x \in H^p(\mathbf{P}^n, \mathcal{F})$ (or $x \in H^p(\mathbf{P}^n, \mathcal{O}_{\mathbf{P}^n}(d))$) x vanishes on multiplication by a power of X_0.

To prove that $H^p(\mathbf{P}^n, \mathcal{O}_{\mathbf{P}^n}(d))$ vanishes for $p \geqslant 1$ it will therefore be enough to prove the following lemma.

Lemma 4.4. *Multiplication by X_0 induces an isomorphism*

$$H^p(\mathbf{P}^n, \mathcal{O}_{\mathbf{P}^n}(d)) \simeq H^p(\mathbf{P}^n, \mathcal{O}_{\mathbf{P}^n}(d+1)).$$

Proof (of 4.4). Consider the exact sequence of sheaves induced by multiplication by X_0 (*cf.* Chapter III, 10.a):

$$0 \longrightarrow \mathcal{O}_{\mathbf{P}^n}(d) \xrightarrow{X_0} \mathcal{O}_{\mathbf{P}^n}(d+1) \longrightarrow \mathcal{O}_H(d+1) \longrightarrow 0,$$

where H is the closed set $V(X_0)$, which is isomorphic to $\mathbf{P}^{n-1}$ (using the conventions of 2.12 above). We write the cohomology sequence associated to this exact sequence. It starts as follows;

$$0 \longrightarrow H^0(\mathcal{O}_{\mathbf{P}^n}(d)) \xrightarrow{X_0} H^0(\mathcal{O}_{\mathbf{P}^n}(d+1)) \xrightarrow{\pi} H^0(\mathcal{O}_H(d+1)),$$

and on calculating the dimension of the vector spaces appearing in this sequence we see that π is surjective. The long exact sequence ends with

$$H^{n-1}(\mathcal{O}_H(d+1)) \xrightarrow{i} H^n(\mathcal{O}_{\mathbf{P}^n}(d)) \xrightarrow{X_0} H^n(\mathcal{O}_{\mathbf{P}^n}(d+1)) \longrightarrow 0$$

(we have $H^n(\mathcal{O}_H(d+1)) = 0$ since H is of dimension $n-1$, *cf.* 3.1). Moreover, i is injective for dimension reasons. Since all the spaces $H^p(\mathcal{O}_H(d+1))$ vanish for $1 \leqslant p \leqslant n-2$ by the induction hypothesis, we obtain the required isomorphism for all p between 1 and $d-1$.

4.5. Application: back to Bézout. If $\mathcal{F}$ is a sheaf on X we denote by $h^i\mathcal{F}$ the dimension of the vector space $H^i(X, \mathcal{F})$.

We recall that our aim is to calculate $h^0\mathcal{O}_Z$, where Z is the finite scheme $V(F, G)$. We have the following two exact sequences:

$$0 \longrightarrow \mathcal{O}_{\mathbf{P}^2}(-s-t) \longrightarrow \mathcal{O}_{\mathbf{P}^2}(-s) \oplus \mathcal{O}_{\mathbf{P}^2}(-t) \longrightarrow \mathcal{J}_Z \longrightarrow 0,$$
$$0 \longrightarrow \mathcal{J}_Z \longrightarrow \mathcal{O}_{\mathbf{P}^2} \longrightarrow \mathcal{O}_Z \longrightarrow 0.$$

We write out the exact cohomology sequences (remembering that the numbers $h^1\mathcal{O}_{\mathbf{P}^2}(n)$ are zero). The second sequence gives us an equality $h^0\mathcal{O}_Z = 1 + h^1\mathcal{J}_Z$ (since $h^0\mathcal{J}_Z = 0$). This also gives us $h^2\mathcal{J}_Z = h^2\mathcal{O}_{\mathbf{P}^2} = 0$ (since $h^1\mathcal{O}_Z = h^2\mathcal{O}_Z = 0$ for dimension reasons). The first sequence then gives us

$$h^1\mathcal{J}_Z = h^2\mathcal{O}_{\mathbf{P}^2}(-s-t) - h^2\mathcal{O}_{\mathbf{P}^2}(-s) - h^2\mathcal{O}_{\mathbf{P}^2}(-t)$$
$$= \binom{s+t-1}{2} - \binom{s-1}{2} - \binom{t-1}{2} = st - 1,$$

and the result follows.

Another method for proving Bézout's theorem (which is essentially the one used in Chapter VI) is to calculate $h^0 \mathcal{O}_Z(d)$ using the above exact sequences (shifted by d). The key point in the proof is to show that $H^1 \mathcal{J}_Z(d)$ vanishes for d large enough (which is immediate using the resolution of $\mathcal{J}_Z$ and Theorem 4.1). We then calculate $h^0 \mathcal{J}_Z(d)$ and $h^0 \mathcal{O}_Z(d)$ using the exact cohomology sequences associated to the above exact sequences. The calculations are easy because both of the long sequences terminate at the H^0 terms.

The vanishing of the groups $H^1 \mathcal{J}_Z(d)$ for large enough d is a special case of a general theorem. It is a result due to J.-P. Serre, as is the finiteness of the dimensions of projective cohomology groups which we have already mentioned several times.

Theorem 4.6 (Serre). *Let X be a projective algebraic variety and let $\mathcal{F}$ be a coherent sheaf on X. Then*

1) For all $i \geqslant 0$, $H^i(X, \mathcal{F})$ is a finite-dimensional k-vector space.

2) There is an integer n_0 depending on $\mathcal{F}$ such that, for all $d \geqslant n_0$ and all $i > 0$, $H^i(X, \mathcal{F}(d)) = 0$.

Proof.

1) Assume that X is embedded in $\mathbf{P}^n$ and let i be the canonical injection. By 2.10, on replacing $\mathcal{F}$ by $i_* \mathcal{F}$ we can assume $X = \mathbf{P}^n$. We note that the sheaves $\mathcal{O}_{\mathbf{P}^n}(d)$ verify Properties 1 and 2 above.

2) We set $S = k[X_0, \dots, X_n]$. We know that $\mathcal{F} = \widetilde{M}$, where M is a graded S-module of finite type (*cf.* Chapter III, 9.8.2). If the generators $x_1, \dots, x_r$ of M are homogeneous of degrees $n_1, \dots, n_r$, then there is a surjection

$$p : L = \bigoplus_{i=1}^{r} S(-n_i) \longrightarrow M$$

associating the element x_i to the ith vector of the basis of L. Introducing the kernel N of p (which is also a graded S-module of finite type) and passing to sheaves, we have an exact sequence of coherent sheaves

$$0 \longrightarrow \mathcal{N} \longrightarrow \mathcal{L} \longrightarrow \mathcal{F} \longrightarrow 0,$$

where $\mathcal{L} = \bigoplus_{i=1}^{r} \mathcal{O}_{\mathbf{P}^n}(-n_i)$ is a sheaf whose cohomology satisfies 1) and 2) by 4.1.

3) We will now prove 1) by descending induction on i. For $i \geqslant n + 1$ we know that $H^i \mathcal{F} = 0$ for any coherent sheaf (*cf.* 3.1). Assume that the result is known for $i + 1$ and any coherent sheaf and let us pass to i. We have an exact sequence

$$\cdots \longrightarrow H^i \mathcal{L} \xrightarrow{f} H^i \mathcal{F} \xrightarrow{g} H^{i+1} \mathcal{N} \longrightarrow \cdots$$

and hence $h^i \mathcal{F} = \dim \operatorname{Im} f + \dim \operatorname{Im} g \leqslant h^i \mathcal{L} + h^{i+1} \mathcal{N}$: these last two quantities are finite. For $h^{i+1} \mathcal{H}$ this is our induction hypothesis and the other term is the special case dealt with in 4.1.

4) For 2) we also proceed by descending induction on i by proving the following property $P(i)$:

(P_i): *Let i be a strictly positive integer. For any coherent sheaf $\mathcal{F}$ there is an integer n_0 such that, for any $p \geqslant i$ and $d \geqslant n_0$, $h^p \mathcal{F}(d) = 0$.*

It is clear that (P_{n+1}) holds. Let i be an integer such that $1 \leqslant i \leqslant n$, assume that P_{i+1} is true and let us prove P_i. Let $\mathcal{F}$ be a coherent sheaf and let $\mathcal{L}$ and $\mathcal{P}$ be defined as in 2) above. There is an integer d_0 such that, for $d \geqslant d_0$ and $p \geqslant i+1$, $h^p \mathcal{N}(d) = h^p \mathcal{F}(d) = 0$. We take n_0 to be the sup of d_0 and the differences $n_i - n$ (where $i = 1, \ldots, r$). For $d \geqslant n_0$ there is an exact sequence:

$$\cdots \longrightarrow H^i \mathcal{L}(d) \overset{f}{\longrightarrow} H^i \mathcal{F}(d) \overset{g}{\longrightarrow} H^{i+1} \mathcal{N}(d) \longrightarrow \cdots$$

As $d - n_i \geqslant -n$ we have $h^i \mathcal{L}(d) = 0$ by 4.1. As $d \geqslant d_0$, we have $h^{i+1} \mathcal{N}(d) = 0$. It follows that $h^i \mathcal{F}(d) = 0$.

Exercises

1 In which we meet a non-affine variety

Let S be the open set $k^2 - \{(0,0)\}$ in affine space k^2 with its variety structure. Calculate the Čech cohomology group $\check{H}^1(\mathcal{U}, \mathcal{O}_S)$ relative to the covering of S by the open sets $D(X)$ and $D(Y)$ and prove that this group is non-trivial. Deduce that S is not an affine variety (*cf.* Exercise III, A.2).

2 A non-separated variety

1) Let X_1 and X_2 be two varieties and let U_i be an open set of X_i. We assume that the open sets U_1 and U_2 are isomorphic as varieties. Let $\varphi : U_1 \to U_2$ be an isomorphism. Prove that we can define a variety X (which we say is constructed by gluing X_1 and X_2 along U_1 and U_2) in the following way:
We take the *disjoint* union $Y = X_1 \cup X_2$, and we consider the equivalence relation $\mathcal{R}$ identifying U_1 and U_2 (*i.e.*, the only non-trivial equivalent pairs are the pairs $(x, \varphi(x))$ with $x \in U_1$). We denote the quotient space $Y/\mathcal{R}$ with its quotient topology by X. In other words, if we consider the natural (injective) maps $i_1 : X_1 \to X$ and $i_2 : X_2 \to X$, a subset $V \subset X$ is open if and only if $V_1 = i_1^{-1}(V)$ and $V_2 = i_2^{-1}(V)$ are open. The sheaf of rings $\mathcal{O}_X$ is then defined by

$$\mathcal{O}_X(V) = \{(s_1, s_2) \mid s_1 \in \mathcal{O}_{X_1}(V_1),\ s_2 \in \mathcal{O}_{X_1}(V_2) \text{ and } \varphi^*(s_2|_{V_2 \cap U_2}) = s_1|_{V_1 \cap U_1}\}.$$

2) With the above notation, we set $X_1 = X_2 = k^2$ and $U_1 = U_2 = k^2 - \{(0,0)\}$. Prove that the variety obtained by gluing is not separated. (Produce two open affine sets whose intersection is not affine.) Intuitively, this variety is simply the plane with a double origin.

The aim of the following exercises is to calculate some cohomology groups in projective space. If X is a projective variety, we denote by $h^i(X, \mathcal{F})$ (or $h^i(\mathcal{F})$ if there is no risk of confusion) the dimension of the k-vector space $H^i(X, \mathcal{F})$.

3 Hypersurfaces

Let $F \in S = k[X_0, \ldots, X_n]$ be a degree t homogeneous polynomial and let $Q = V(F)$ be the associated hypersurface. Assume $I(Q) = (F)$. There is an exact sequence $0 \to S(-t) \xrightarrow{j} S \to S/(F) \to 0$, where j denotes multiplication by F, and this gives us an exact sequence of sheaves

$$0 \longrightarrow \mathcal{O}_{\mathbf{P}^n}(-t) \longrightarrow \mathcal{O}_{\mathbf{P}^n} \longrightarrow \mathcal{O}_Q \longrightarrow 0.$$

Use this exact sequence to calculate $h^i(Q, \mathcal{O}_Q(d))$ for all $d \in \mathbf{Z}$.

Remark. This calculation is valid even if $I(Q) \neq (F)$, which corresponds to F with multiple factors. In this case, the calculated cohomology groups are those of the scheme Q, considered with multiplicities.

4 Complete intersections

We assume $n \geqslant 2$. We consider two polynomials $F, G \in S = k[X_0, \ldots, X_n]$, homogeneous of degrees s and t respectively, and without common factors. We set $Z = V(F, G)$ and assume $I(Z) = (F, G)$ (or simply work with schemes without knowing it!) and we then have (*cf.* Bézout's theorem) exact sequences

$$0 \longrightarrow \mathcal{J}_Z \longrightarrow \mathcal{O}_{\mathbf{P}^n} \longrightarrow \mathcal{O}_Z \longrightarrow 0$$

and
$$0 \longrightarrow \mathcal{O}_{\mathbf{P}^n}(-s-t) \longrightarrow \mathcal{O}_{\mathbf{P}^n}(-s) \oplus \mathcal{O}_{\mathbf{P}^n}(-t) \longrightarrow \mathcal{J}_Z \longrightarrow 0.$$

Calculate $h^i \mathcal{O}_Z(d)$ and $h^i \mathcal{J}_Z(d)$ for all $d \in \mathbf{Z}$. (Note that $\dim Z = n - 2$.)

5 The space cubic

We use the notations and results of Exercise 4 of Chapter II.
   We have an exact sequence

$$0 \longrightarrow R(-3)^2 \xrightarrow{u} R(-2)^3 \xrightarrow{v} I(C) \longrightarrow 0$$

which can be translated into sheaves as

$$0 \longrightarrow \mathcal{O}_{\mathbf{P}^3}(-3)^2 \longrightarrow \mathcal{O}_{\mathbf{P}^3}(-2)^3 \longrightarrow \mathcal{J}_C \longrightarrow 0.$$

Moreover, we have an exact sequence

$$0 \longrightarrow \mathcal{J}_C \longrightarrow \mathcal{O}_{\mathbf{P}^3} \longrightarrow \mathcal{O}_C \longrightarrow 0.$$

Calculate $h^i \mathcal{O}_C(d)$ and $h^i \mathcal{J}_C(d)$ for all $d \in \mathbf{Z}$. (Note that $h^i \mathcal{O}_C(d)$ vanishes for $i \geqslant 2$. Deduce the value of $h^3 \mathcal{J}_C(d)$, then that of $h^2 \mathcal{J}_C(d)$. The rest is easy.)

6 Two disjoint lines

We use the notations and results of Exercise 5 in Chapter II. We have an exact sequence

$$0 \longrightarrow R(-4) \longrightarrow R(-3)^4 \longrightarrow R(-2)^4 \longrightarrow I(C) \longrightarrow 0.$$

We deduce from this an exact sequence of sheaves

$$0 \longrightarrow \mathcal{O}_{\mathbf{P}^3}(-4) \longrightarrow \mathcal{O}_{\mathbf{P}^3}(-3)^4 \longrightarrow \mathcal{O}_{\mathbf{P}^3}(-2)^4 \longrightarrow \mathcal{J}_C \longrightarrow 0.$$

As always, our aim is to calculate $h^i \mathcal{O}_C(d)$ and $h^i \mathcal{J}_C(d)$. We start by noting that as C is a union of disjoint lines D_1 and D_2 there is an isomorphism $\mathcal{O}_C \simeq \mathcal{O}_{D_1} \oplus \mathcal{O}_{D_2}$, which allows us to calculate $h^i \mathcal{O}_C(d)$ directly.

VIII

Arithmetic genus of curves and the weak Riemann-Roch theorem

We work over an algebraically closed base field k.

0 Introduction: the Euler-Poincaré characteristic

If $\mathcal{F}$ is a coherent sheaf over a projective variety X, then we have seen that $H^i\mathcal{F} = H^i(X, \mathcal{F})$ is a finite-dimensional k-vector space. Our aim is to calculate its dimension $h^i\mathcal{F}$. We will see below that this situation often arises in practice. This is not an easy problem in general. There is, however, an invariant of $\mathcal{F}$ which is much easier to calculate than the numbers $h^i\mathcal{F}$, namely the Euler-Poincaré characteristic $\chi(\mathcal{F}) = \sum_{i \geqslant 0}(-1)^i h^i\mathcal{F}$. (This sum is finite because $h^i\mathcal{F} = 0$ for $i > \dim X$.)[1] Indeed, given an exact sequence of sheaves

$$0 \longrightarrow \mathcal{F}' \longrightarrow \mathcal{F} \longrightarrow \mathcal{F}'' \longrightarrow 0,$$

then $\chi(\mathcal{F}) = \chi(\mathcal{F}') + \chi(\mathcal{F}'')$, hence, if the Euler[2] characteristics of two of the sheaves are known, so is the third. Using the cohomology exact sequence, we reduce the proof of this additivity property to the following lemma, which the careful reader will prove by induction on n using the rank-kernel theorem.

Lemma 0.1. *Given an exact sequence of k-vector spaces*

$$0 \longrightarrow A_0 \longrightarrow A_1 \longrightarrow \cdots \longrightarrow A_n \longrightarrow 0,$$

we have $\sum_{i=0}^{n}(-1)^i \dim_k A_i = 0$.

In its weakest form, the famous Riemann-Roch theorem for curves simply calculates a certain Euler characteristic. (In its strongest form it also contains a duality theorem.) For a more general version of the Riemann-Roch theorem (which is still an Euler characteristic calculation, given in terms of Chern classes) see, for example, [H] Appendix A.4.1.

---

[1] The first example of this kind of alternating sum was doubtless the Euler formula $v - e + f = 2$ linking the number of vertices, edges and faces of a convex polyhedron.

[2] And Poincaré!

1 Degree and genus of projective curves, Riemann-Roch 1

a. Theory

Let $C \subset \mathbf{P}^N$ be an irreducible projective curve (*i.e.*, an algebraic variety of dimension 1). We set $S = k[X_0, \dots, X_n]$ and $A = \Gamma_h(C) = S/I(C)$. The ring A is a graded integral domain and the associated sheaf $\tilde{A}$ is simply $\mathcal{O}_C$. Our aim is to calculate the Euler characteristic $\chi(\mathcal{O}_C(n)) = h^0 \mathcal{O}_C(n) - h^1 \mathcal{O}_C(n)$ for all $n \in \mathbf{Z}$. (We note that the numbers h^i are zero for $i \geqslant 2$ because C is of dimension 1.) Our method is to study the intersection of the curve with a sufficiently general hyperplane: this is a standard technique in projective geometry.

Proposition 1.1. *Let H be a hyperplane not containing C. We denote its equation by h and the image in A of h is denoted by $\overline{h}$. The multiplication by $\overline{h}$ in A induces an exact sequence of graded S-modules*

$$(*) \qquad 0 \longrightarrow A(-1) \xrightarrow{\cdot \overline{h}} A \longrightarrow A/(\overline{h}) \longrightarrow 0.$$

Proof. The only thing we need to check is that multiplication by $\overline{h}$ is injective. As A is an integral domain, this follows from the fact that $C \not\subset H$ and hence $h \notin I(C)$, so $\overline{h} \neq 0$ in A.

We now consider $Z = C \cap H$. As H does not contain C, this intersection is finite. We endow it with its finite scheme structure (*cf.* Chapter VI, 1.1) by defining $\mathcal{O}_Z$ as the sheaf associated to the graded k-algebra $A/(\overline{h})$ (*cf.* Chapter VI, 1.6 and 2.5). We note that, as in the proof of Bézout's theorem (*cf.* Chapter VI, 1.6 and 2.5), there is an isomorphism $\mathcal{O}_Z \simeq \mathcal{O}_Z(n)$ for all n, obtained (for example) by multiplying by the linear form X_0 if the hyperplane $X_0 = 0$ does not meet Z.

At the level of sheaves, the sequence $(*)$ then gives us an exact sequence

$$0 \longrightarrow \mathcal{O}_C(-1) \xrightarrow{\cdot \overline{h}} \mathcal{O}_C \longrightarrow \mathcal{O}_Z \longrightarrow 0$$

or, alternatively, after shifting

$$0 \longrightarrow \mathcal{O}_C(n-1) \xrightarrow{\cdot \overline{h}} \mathcal{O}_C(n) \longrightarrow \mathcal{O}_Z(n) \longrightarrow 0.$$

Taking the Euler characteristics in this sequence we get

$$\chi(\mathcal{O}_C(n)) = \chi(\mathcal{O}_C(n-1)) + \chi(\mathcal{O}_Z(n)).$$

The last term is simply $h^0 \mathcal{O}_Z(n)$ (since $\dim Z = 0$) and it is also equal to $h^0 \mathcal{O}_Z$. We set $d = h^0 \mathcal{O}_Z = h^0 \mathcal{O}_{C \cap H}$. This is an integer $\geqslant 1$ (since Z is non-empty, *cf.* Chapter IV, 2.9), which is simply the number of points, counted with multiplicity, in $C \cap H$ (*cf.* Chapter VI, 1). We then have $\chi \mathcal{O}_C(n) = \chi \mathcal{O}_C(n-1) + d$ and by induction $\chi \mathcal{O}_C(n) = nd + \chi \mathcal{O}_C$.

It follows that $\chi \mathcal{O}_C(n)$ is a polynomial of degree 1 in n. (This result is already a sort of Riemann-Roch theorem.) Incidentally, this proof also shows that the number d of points in $C \cap H$ does not depend on H.

Definition 1.2. *Let $C \subset \mathbf{P}^N$ be an irreducible projective curve. The number d of intersection points of C with a hyperplane H not containing C, counted with multiplicity, is called the degree of C.*

We still have to calculate $\chi \mathcal{O}_C = h^0 \mathcal{O}_C - h^1 \mathcal{O}_C$. We have already established the following (*cf.* Chapter III, 8.8 or Problem II):

Lemma 1.3. *If X is an irreducible projective algebraic variety, then $H^0(X, \mathcal{O}_X) = k$ (the only global functions are the constants) and hence $h^0(X, \mathcal{O}_X) = 1$.*

Proof. The ring $R = H^0(X, \mathcal{O}_X) = \Gamma(X, \mathcal{O}_X)$ is an integral domain. Indeed, if $fg = 0$ in this ring, then $fg = 0$ still holds after restriction to an open affine set U. But U is irreducible and its ring is an integral domain by Chapter I, 3.2. It follows that without loss of generality $f|_U = 0$, and hence $f = 0$ by density.

We also know that R is finite dimensional over k (*cf.* Chapter VII, 4.6) and is hence a field (*cf.* Chapter IV, 4.5), which is algebraic over k, but as k is algebraically closed, $R = k$.

We still have to deal with the $h^1 \mathcal{O}_C$ term. Unable to calculate it, we will give it a name.

Definition 1.4. *Let C be an irreducible projective curve. We call the positive integer $g = h^1 \mathcal{O}_C$ the arithmetic genus of C. (We will sometimes denote this number by p_a, where a stands for arithmetic.)*

The following theorem summarises the above.

Theorem 1.5 (Riemann-Roch 1). *Let $C \subset \mathbf{P}^N$ be an irreducible projective curve of degree d and genus g. Then, for all $n \in \mathbf{Z}$,*

$$h^0 \mathcal{O}_C(n) - h^1 \mathcal{O}_C(n) = nd + 1 - g.$$

Moreover, for large n, $h^0 \mathcal{O}_C(n) = nd + 1 - g$.

Proof. This follows from the above, plus (for the statement for large n), Chapter VII, 4.6.

Remarks 1.6.

1) There is a fundamental difference between the two invariants d and g. The genus $g = h^1 \mathcal{O}_C$ depends only on C and not on the choice of embedding into $\mathbf{P}^N$. However, the degree d depends fundamentally on the choice of embedding. This is clear on examining the definition involving hyperplanes in $\mathbf{P}^N$: it is also clear on examining the Riemann-Roch theorem (which can be used as an alternative definition of degree) since the sheaves $\mathcal{O}_C(n)$ are only defined relative to an embedding in $\mathbf{P}^N$ (*cf.* Chapter III, 9.8.0). For example, a line (which is a degree 1 subvariety of $\mathbf{P}^3$) is isomorphic (*cf.* Chapter III, 11.6) to a conic or a space cubic, which are of degrees 2 and 3 respectively (*cf.* 1.7 and 1.12 below).

2) In 1.5, if C is smooth, then we can show that $h^1 \mathcal{O}_C(n)$ vanishes if $nd > 2g - 2$ (*cf.* 2.14 below).

3) If C is smooth, there are hyperplanes H such that $C \cap H$ consists of d distinct points. Indeed, this is the case for "general" H (*cf.* Exercise VIII, 1).

4) If C is an irreducible curve, then $h^0 \mathcal{O}_C(n) = 0$ for $n < 0$. To see this, consider the exact sequence

$$0 \longrightarrow H^0 \mathcal{O}_C(n-1) \longrightarrow H^0 \mathcal{O}_C(n) \longrightarrow H^0 \mathcal{O}_Z(n).$$

For $n = 0$, as $H^0 \mathcal{O}_C$ is a field and $H^0 \mathcal{O}_Z \neq 0$, the ring homomorphism $H^0 \mathcal{O}_C \to H^0 \mathcal{O}_Z$ is injective, so $H^0 \mathcal{O}_C(-1)$ vanishes. The vanishing of the other spaces $H^0 \mathcal{O}_C(n)$ for $n < 0$ follows by induction.

5) We can extend the above to more general curves—non-irreducible or non-connected curves, for example, or even non-reduced curves (schemes). If the curve C is not connected, $h^0 \mathcal{O}_C$ is no longer 1, but the number of connected components of C. To have a Riemann-Roch formula ($\chi \mathcal{O}_C(n) = nd + 1 - g$) we have to set $g = 1 - \chi(\mathcal{O}_C)$. Warning: g can then be negative. For example, if C is the union of two disjoint lines in $\mathbf{P}^3$, then $g = -1$.

b. Examples and applications

There are two natural questions at this point:

1) What is the use of these invariants?

2) How can they be calculated?

i) Use of d and g.

1) A first use for these invariants is the one that justified their introduction: calculating cohomology groups. We will give just one example: our aim is to calculate whether or not a given curve $C \subset \mathbf{P}^3$ lies on a surface of degree d or, alternatively, whether or not $h^0 \mathcal{J}_C(d) > 0$. The answer can often be calculated using Riemann-Roch (along with 1.6.2) and the following exact sequence:

$$0 \longrightarrow \mathcal{J}_C(d) \longrightarrow \mathcal{O}_{\mathbf{P}^3}(d) \longrightarrow \mathcal{O}_C(d) \longrightarrow 0.$$

Hence, if C is a smooth curve of degree 7 and genus 5, then $h^0\mathcal{O}_C(2) = 2 \cdot 7 + 1 - 5 = 10 = h^0\mathcal{O}_{\mathbf{P}^3}(2)$, so C is not a priori on a surface of degree 2 (it remains possible that this might be the case if $h^1\mathcal{J}_C(2) > 0$). On the other hand, as $h^0\mathcal{O}_C(3) = 3 \cdot 7 + 1 - 5 = 17 < h^0\mathcal{O}_{\mathbf{P}^3}(3) = 20$, any such curve lies on a surface of degree 3 (and we can even conclude that it lies on three independent such surfaces).

2) Another use of these invariants is in the classification of curves by genus, for example. We will see in the following chapter what happens when $g = 0$ (such curves are rational curves, if g is the geometric genus). Likewise, curves of genus 1 are elliptic curves. If the base field is $\mathbf{C}$, the genus has a simple topological interpretation. A smooth projective algebraic curve is also a differentiable variety of dimension 1 over $\mathbf{C}$ and dimension 2 over $\mathbf{R}$. It is therefore a compact orientable surface, and we know (*cf.* [Gr]) that it is homeomorphic to a sphere (for $g = 0$) or a torus ($g = 1$) or a torus with g holes (genus g).

Closer to home, if we consider curves in $\mathbf{P}^2$ or $\mathbf{P}^3$ with invariants d and g, then many natural questions arise. What degrees, genuses and pairs (d, g) are possible (for smooth curves, say)? If these questions are easy to solve in $\mathbf{P}^2$, the same is not true in $\mathbf{P}^3$ (especially for the last question, which was definitively solved in 1982 *cf.* [GP]). Finally, if d and g are fixed, we can study the family $H_{d,g}$ of curves of degree d and genus g. This family can be equipped with an algebraic variety structure (or, more exactly, a scheme structure) and we can ask whether it is smooth or irreducible and what its dimension is. This question is still essentially open in $\mathbf{P}^3$ (*cf.* for example [MDP]).

3) The reader may be wondering why the author has such an obvious penchant for curves in $\mathbf{P}^3$. This is not only because, apart from plane curves, they are the most easily accessible curves. Consider a curve C in $\mathbf{P}^N$ and a point P not contained in C. We can consider the projection π centred on P to a hyperplane H which does not contain P (the image of a point $Q \in C$ is the intersection of H and the line $\langle PQ \rangle$). Let C' be the image of C under π. This operation enables us to map C into H, but for C and C' to be isomorphic under π the map π must be injective on C, or, in other words, P must not be collinear with any two distinct points of C. (That is, P should not be on any secant line of C.)

It is easy to see that the union of all the secant lines of C is a three-dimensional subvariety of $\mathbf{P}^N$, so we can always find a suitable point P if $N \geqslant 4$. We can therefore prove (*cf.* [H] Chapter IV, 3.6) that any smooth projective curve is isomorphic to a curve in $\mathbf{P}^3$, and hence all smooth curves can be embedded in $\mathbf{P}^3$. On the other hand, we cannot generally project a curve in $\mathbf{P}^3$ isomorphically onto a plane curve: moreover, we will see below that the genuses of plane curves are very special (*cf.* Chapter IX, 2.9, however).

ii) Calculating d and g: resolutions. The basic idea used to calculate d and g (and more generally dimensions of cohomology groups) is the resolution. If $\mathcal{F}$

is a coherent sheaf on $\mathbf{P}^N$, then a resolution of $\mathcal{F}$ is an exact sequence

$$0 \longrightarrow \mathcal{L}_n \longrightarrow \cdots \longrightarrow \mathcal{L}_1 \longrightarrow \mathcal{L}_0 \longrightarrow \mathcal{F} \longrightarrow 0,$$

where the sheaves $\mathcal{L}_j$ are of the form

$$\mathcal{L} = \bigoplus_{i=1}^{r} \mathcal{O}_{\mathbf{P}^N}(n_i), \quad n_i \in \mathbf{Z}.$$

If $\mathcal{F} = \widetilde{M}$, where M is a graded S-module of finite type, then to do this it is enough to find a free graded resolution of M:

$$0 \longrightarrow L_n \longrightarrow \cdots \longrightarrow L_1 \longrightarrow L_0 \longrightarrow M \longrightarrow 0,$$

with L_j of the form

$$L = \bigoplus_{i=1}^{r} S(n_i), \quad n_i \in \mathbf{Z}.$$

We saw in the proof of Chapter VII, 4.6 how to construct the first term of such a resolution, $\bigoplus_{i=1}^{r} S(n_i) \xrightarrow{p} M \to 0$, using the generators of M. We continue by applying the same method to the kernel of p and so on. A theorem of Hilbert's (which goes under the pretty name of the "syzygies theorem" *cf.* Chapter X) says that this method stops after a finite number of steps, and hence every graded module of finite type has a resolution.

Example 1.7: plane curves.

Theorem 1.7. *Let $F \in k[X, Y, T]$ be an irreducible homogeneous polynomial of degree $d > 0$, and let $C = V_p(F) \subset \mathbf{P}^2$ be the projective curve defined by F. We have:*

  1) The degree of C is d.
  2) The arithmetic genus of C is

$$\frac{(d-1)(d-2)}{2}.$$

Proof.
  1) This is simply Bézout's theorem: a hyperplane is just a line, and the number of intersection points of C and D, counted with multiplicity, is equal to d.

  2) We use the usual exact sequence

$(*)$ $$0 \longrightarrow \mathcal{J}_C \longrightarrow \mathcal{O}_{\mathbf{P}^2} \longrightarrow \mathcal{O}_C \longrightarrow 0$$

which gives us $\chi(\mathcal{O}_C) = 1 - g = \chi(\mathcal{O}_{\mathbf{P}^2}) - \chi(\mathcal{J}_C) = 1 - \chi(\mathcal{J}_C)$. Since C is a hypersurface in $\mathbf{P}^2$, we know (*cf.* Chapter III, 10.a) that $\mathcal{J}_C$ is isomorphic to $\mathcal{O}_{\mathbf{P}^2}(-d)$ and hence $g = \chi(\mathcal{O}_{\mathbf{P}^2}(-d)) = h^2(\mathcal{O}_{\mathbf{P}^2}(-d)) = \binom{d-1}{2}$. QED.

Remark 1.9. This allows us to answer the question in i) above for plane curves: there are smooth curves of all degrees $d > 0$ (*cf.* Chapter V, 2.7.3), and their arithmetic genus depends only on their degree. We note that the possible genuses are rare: $0, 1, 3, 6, 10, 15, \ldots$. In particular, there are no plane curves of arithmetic genus 2 or 4.

Example 1.10: complete intersections in $\mathbf{P}^3$. Let $C = V_p(F, G) \subset \mathbf{P}^3$ be a complete intersection (*cf.* Chapter III, 10.b). We assume that $F, G \in k[X, Y, Z, T]$ are homogeneous polynomials of degrees s and t and we assume $I_p(C) = (F, G)$. (We then say that C is a scheme-theoretic complete intersection: its ideal is generated by two generators, which obviously implies that it is a set-theoretic complete intersection of the two corresponding surfaces, but the converse is false, *cf.* 1.13.) To calculate the genus $g = h^1 \mathcal{O}_C$ we once again use the exact sequence $(*)$. On expanding the long exact cohomology sequence of $(*)$, we note that $h^1 \mathcal{O}_C$ is equal to $h^2 \mathcal{J}_C$. But we have a resolution of $\mathcal{J}_C$ (*cf.* Chapter III, 10.1):

$$0 \longrightarrow \mathcal{O}_{\mathbf{P}^3}(-s-t) \longrightarrow \mathcal{O}_{\mathbf{P}^3}(-s) \oplus \mathcal{O}_{\mathbf{P}^3}(-t) \longrightarrow \mathcal{J}_C \longrightarrow 0$$

and we deduce an exact sequence

$$0 \longrightarrow H^2 \mathcal{J}_C \longrightarrow H^3 \mathcal{O}_{\mathbf{P}^3}(-s-t) \longrightarrow H^3 \mathcal{O}_{\mathbf{P}^3}(-s) \oplus H^3 \mathcal{O}_{\mathbf{P}^3}(-t) \longrightarrow 0;$$

indeed, the cohomology group H^2 of the sheaves $\mathcal{O}_{\mathbf{P}^3}(n)$ vanish, as does $H^3 \mathcal{J}_C$ (consider $(*)$). We have the formula:

$$g = \binom{s+t-1}{3} - \binom{s-1}{3} - \binom{t-1}{3}$$

(with the convention that the binomial coefficient $\binom{n}{p}$ vanishes if $n < p$) and we get $g = \frac{1}{2}st(s+t-4)+1$. To calculate the degree we could calculate in a similar way $\chi(\mathcal{O}_C(1)) = d+1-g$, but it is easier to intersect C with a general plane H (for example, $T = 0$). We are reduced to calculating the intersection in H of two curves with equations $F(X, Y, Z, 0)$ and $G(X, Y, Z, 0)$ of degrees s and t. Bézout's theorem tells us that there are st intersection points, so the degree of C is st.

Remarks 1.11.

1) There are smooth complete intersection curves for all s, t (*cf.* Exercise VIII, 2).

2) For $s = 2$ and $t = 3$ we obtain by this method smooth curves of genus 4 (which are therefore not isomorphic to plane curves, *cf.* i.3 above) but we do not get any curves of genus 2 (*cf.* Exam January 1992).

Example 1.12: The space cubic. We refer to Exercise II, 4 for the definition and properties of this curve. Intersecting with an arbitrary plane, we see that its degree is 3. For the genus, we use the method given above. This time we have the following resolution of $\mathcal{J}_C$:

$$0 \longrightarrow \mathcal{O}_{\mathbf{P}^3}(-3)^2 \longrightarrow \mathcal{O}_{\mathbf{P}^3}(-2)^3 \longrightarrow \mathcal{J}_C \longrightarrow 0$$

giving an exact sequence $0 \to H^2\mathcal{J}_C \to H^3\mathcal{O}_{\mathbf{P}^3}(-3)^2 \to \cdots$. As the H^3 term vanishes, g is zero. (For the degree we can alternatively, calculate $\chi(\mathcal{O}_C(1)) = 4$, which gives us $d = 3$.)

Remark 1.13. The space cubic is not a scheme-theoretic complete intersection. Indeed, we would otherwise have $3 = st$, so without loss of generality $s = 1$ and $t = 3$, but the genus would then be 1 and not 0. On the other hand, it is a *set-theoretic* complete intersection of two surfaces of degrees 2 and 3, $Z^2 - YT$ and $Y^3 - 2XYZ + X^2T = Y(Y^2 - XZ) + X(XT - YZ)$. What happens is that C is double in this intersection (which is a scheme of degree 6). Let us mention that the question whether or not any curve in $\mathbf{P}^3$ is a set-theoretic complete intersection is still open.

2 Divisors on a curve and Riemann-Roch 2

If C is a curve in $\mathbf{P}^N$ and H is a hyperplane which does not contain C, we have seen there is a exact sequence given by multiplication by H:

$$0 \longrightarrow \mathcal{O}_C(-1) \longrightarrow \mathcal{O}_C \longrightarrow \mathcal{O}_{C \cap H} \longrightarrow 0,$$

where $C \cap H$ is equipped with its finite scheme structure. We will generalise this situation to all finite subschemes of C by introducing divisors. Conversely, using these divisors we can reconstruct the sheaves $\mathcal{O}_C(n)$ associated to certain embeddings of C in projective space. This study of C using divisors, independently of a choice of projective embedding of C, is known as the study of the "abstract" curve C.

Before talking about divisors, we need to define rational functions.

a. Rational functions

Proposition-Definition 2.1. *Let X be an irreducible algebraic variety. We consider pairs (U, f), where U is a non-empty open set in X and $f \in \Gamma(U, \mathcal{O}_X)$, and we consider the relation on these pairs given by $(U, f) \sim (V, g) \Leftrightarrow f|_{U \cap V} = g|_{U \cap V}$. This is an equivalence relation. A rational function on X is an equivalence class for this relation.*

Proof. This is clearly an equivalence relation. (Transitivity follows from the irreducibility of X.)

Remark 2.2. A rational function is therefore a function which is defined only on a part of X. It has a largest possible domain of definition, namely the union of all the open sets U of all its representatives (U, f).

Proposition 2.3.
   1) Rational functions on X form a field, denoted by $K(X)$.
   2) For every open affine set U in X, $K(X)$ is the field of fractions of $\Gamma(U, \mathcal{O}_X)$ and is hence the field of rational functions of U as defined in Chapter I, 6.15.
   3) For every point $P \in X$, $K(X)$ is the field of fractions of the local ring $\mathcal{O}_{X,P}$.
   4) If X is projective, $K(X)$ is the subfield of $\mathrm{Fr}(\Gamma_h(X))$ consisting of degree 0 elements.

Proof.
   1) This is easy: after taking small enough open sets, we can always add, multiply and take inverses.

   2) Clearly, $\Gamma(U, \mathcal{O}_X)$ is contained in $K(X)$ by definition. $K(X)$ therefore also contains the field of fractions of this ring. Conversely, consider $f \in K(X)$, which is defined over an open set V. After restriction, we may assume that V is contained in U and even that V is a standard open affine set in U: $V = D_U(g)$, where $g \in \Gamma(U, \mathcal{O}_X)$. But $f = h/g^n$ is then contained in the fraction field of $\Gamma(U)$.

   3) We use 2) to reduce to the case where X is affine, and the result is then clear, since the local ring is a localisation of $\Gamma(V)$ and hence has the same field of fractions.

   4) Consider $f/g \in \mathrm{Fr}(\Gamma_h)$, homogeneous of degree 0. It defines a section of $\mathcal{O}_X$ over $D^+(g)$ and hence defines a rational function. The converse follows from the fact that the sets $D^+(f)$ form a basis of open sets.

Examples 2.4.
   i) *The projective line.* By 2.3.4 a rational function on $\mathbf{P}^1$ is of the form

$$f = \frac{P(X,T)}{Q(X,T)},$$

where P, Q are homogeneous of the same degree d and can be assumed to be coprime. Up to multiplication by a non-zero constant we have

$$P(X,T) = X^\alpha T^\gamma \prod_i (X - a_i T)^{\alpha_i} \quad \text{and} \quad Q(X,T) = X^\beta T^\delta \prod_i (X - b_i T)^{\beta_i},$$

where the elements a_i and b_j are non-zero and distinct, $\alpha_i, \beta_i > 0$ and $\alpha, \beta, \gamma, \delta \geqslant 0$, but $\alpha\beta = \gamma\delta = 0$.
   We study first the open set $T \neq 0$. We can set $t = 1$, and we see that $f_\flat$, which is a rational function in X, has a zero of order α_i at a_i, a pole of order β_i

at b_i and at 0 a zero of order $\alpha - \beta$ or a pole of order $\beta - \alpha$ according to whether α or β is larger.

If now we look at the open set $X \neq 0$ we find, besides the points $t = 1/a_i$ and $1/b_i$ already considered, a zero of order $\gamma - \delta$ or a pole of order $\delta - \gamma$ at the point $(1, 0)$ *i.e.*, at infinity. On summing the multiplicities of the zeros (respectively of the poles) we see that $\alpha + \gamma + \sum_i \alpha_i$ (resp. $\beta + \delta + \sum_i \beta_i$), and both of these quantities are equal to d: the number of zeros of a rational function on $\mathbf{P}^1$ is equal to the number of its poles. We will see a generalisation of this in 2.7.

ii) *The cuspidal cubic.* This is the curve $C = V(Y^2T - X^3) \subset \mathbf{P}^2$. Calculating $K(C)$ on the open set $t \neq 0$, we get a field generated over k by two elements x and y such that $y^2 = x^3$. But if we calculate over the open set U given by $y \neq 0$, we see that $\Gamma(U) = k[X, T]/(T - X^3) \simeq k[X]$, and hence $K(C)$ is a field of rational functions in one variable $k(X)$ and C is rational (*cf.* Chapter IX).

b. Divisors on curves

The definition of a divisor generalises the situation arising in Example i) above: we have a finite number of points to which are assigned positive or negative multiplicities (like the zeros and poles of f, respectively).

Definition 2.5. *Let C be an irreducible smooth projective curve. A divisor D on C is a formal sum*

$$D = \sum_{x \in C} n_x \, x,$$

where the $n_x \in \mathbf{Z}$ are almost all zero.[3] *The support of D is the set of $x \in C$ such that $n_x \neq 0$. We denote by $\mathrm{Div}(C)$ the set of all divisors on C. It is equipped with an obvious addition (adding the coefficients n_x), which turns it into an abelian group. A divisor D is said to be* positive *(or* effective*), and we write $D \geqslant 0$ if $\forall x \in C$, $n_x \geqslant 0$. Any divisor D can be written in the form $D = D_1 - D_2$, where the divisors D_i are positive with disjoint supports. The degree of a divisor D is the integer $\deg D = \sum_{x \in C} n_x$. The map $\deg : \mathrm{Div}\, C \to \mathbf{Z}$ is a surjective homomorphism.*

Example 2.6. This example generalises the example seen above for $\mathbf{P}^1$. Let $f \in K(C)$ be a non-zero rational function on C. We will define a divisor (called a *principal* divisor) $\mathrm{div}(f) = Z(f) - P(f)$, where $Z(f)$ (resp. $P(f)$) represents the zeros (resp. the poles) of f with multiplicities. More precisely, consider $P \in C$ and let $\mathcal{O}_P$ be the local ring of C at P. This is a discrete valuation ring (*cf.* Chapter V, 4.1 and Problem IV), and let v_P be the valuation in question. As f is contained in $K(C) = \mathrm{Fr}(\mathcal{O}_P)$ and is $\neq 0$, the valuation $v_P(f)$ is defined and is a (possibly negative) integer.

---

[3] In sophisticated language, a divisor on C is a member of the free $\mathbf{Z}$-module whose basis consists of the points in C.

Moreover, this integer vanishes for almost all P. Let U be an open affine set of C and write $f = g/h$ with $g, h \in \Gamma(U)$, $g, h \neq 0$. We have $v_P(f) > 0$ exactly when $g \in m_P$, i.e., when $g(P) = 0$ (and hence P is a *zero* of f, of order $v_P(f)$); we have $v_P(f) < 0$ exactly when $h \in m_P$, i.e., $h(P) = 0$ (and hence P is a *pole* of f, of order $-v_P(f)$). As $V(g)$ and $V(h)$ are finite, there are only a finite number of points of U such that $v_P(f) \neq 0$. And finally, $C - U$ is a proper closed set of C, and is hence finite. We note that $f \in H^0\mathcal{O}_C \Leftrightarrow \forall P \in C$, $v_P(f) \geqslant 0$.

We can now define the divisor $\mathrm{div}(f)$:

$$\mathrm{div}(f) = \sum_{P \in C} v_P(f)P.$$

Proposition 2.7. *Consider $f \in K(C)$, $f \neq 0$. We have $\deg \mathrm{div}(f) = 0$, or, in other words, the number of zeros and poles of f (counted with multiplicities) are the same. We have:*

$$f \in H^0\mathcal{O}_C \iff \mathrm{div}(f) \geqslant 0 \iff \mathrm{div}\, f = 0$$

Proof. For a proof of this result in full generality see [F] Chapter 8 Prop. 1 or [H] Chapter II, 6.10. We will only deal with the case where C is a smooth plane curve.

Let C be $V(F)$, where $F \in k[X, Y, T]$ is a homogeneous irreducible polynomial of degree d, and consider $u \in K(C)$. We know that u can be written in the form $u = \overline{G}/\overline{H}$, where $\overline{G}, \overline{H} \in \Gamma_h(C) = k[X, Y, T]/(F)$ are the images of homogeneous polynomials of the same degree n which are not multiples of F. Our aim is to calculate $v_P(u)$ for $u \in K(C)$. After homography we may assume the point P is in the open affine set $T \neq 0$ and the local ring of C at P is then $\mathcal{O}_P(C) = \mathcal{O}_P/(F_\flat)$, where $\mathcal{O}_P$ is the local ring of $\mathbf{P}^2$ (or k^2) in P and $u = \overline{G}_\flat/\overline{H}_\flat$. We then use the formula for the valuation given in the following lemma.

Lemma 2.8. *Let A be a k-algebra which is a discrete valuation ring with valuation v. We assume k is isomorphic to the residue field A/m. Then $v(a) = \dim_k(A/(a))$ for all $a \in A$.*

Proof (of the lemma).. If π is a uniformising parameter, we write $a = u\pi^n$, where $n = v(a)$ and it is easy to see that $(1, \pi, \ldots, \pi^{n-1})$ is a basis for $A/(a)$.

We now return to 2.7. We know that $v_P(u) = v_P(\overline{G}_\flat) - v_P(\overline{H}_\flat)$ and by the lemma $v_P(\overline{G}_\flat) = \dim \mathcal{O}_P/(F_\flat, G_\flat)$. We recognise this as being the intersection multiplicity $\mu_P(F_\flat, G_\flat) = \mu_P(F, G)$ of the curves F and G at P. It follows that

$$\deg \mathrm{div}(u) = \sum_{P \subset C} \mu_P(F, G) - \sum_{P \in C} \mu_P(F, H),$$

but, after Bézout's theorem, each of these sums is equal to nd, hence their difference is zero.

It remains to the prove the claim on functions without poles. Let $f \in K(C)$ be a non-zero function without poles (*i.e.*, such that $\mathrm{div}(f)$ is $\geqslant 0$) and consider $P \in C$. As $f \in \mathcal{O}_{C,P}$ there is a representative of f defined on an affine neighbourhood of P. By definition of a rational function, these representatives can be glued together and hence define a regular function. But f is then a constant function and $\mathrm{div}(f) = 0$.

Remark 2.9. The set $P(C)$ of principal divisors of C is a subgroup of $\mathrm{Div}\, C$. Two divisors which differ by a principal divisors are said to be equivalent. The quotient group $\mathrm{Div}\, C/P(C)$ is called the group of divisor classes of C or the Picard group of C. By 2.7, the degree homomorphism factorises through the Picard group. Its kernel is called the *Jacobian* of C and is denoted $\mathrm{Pic}^0 C$. It is a key invariant of the curve C.

c. The invertible sheaf associated to a divisor

Let C be an irreducible smooth projective curve. We will associate to every divisor D on C a sheaf on C denoted by $\mathcal{O}_C(D)$ (or sometimes $\mathcal{L}(D)$).

To understand where this sheaf comes from, let us start by considering a divisor $D = \sum n_P P \geqslant 0$. We can associate to D a closed finite subscheme of C, also denoted D, whose support consists of those points P such that $n_P > 0$ and which has the property that the multiplicity of D at P as a finite scheme is exactly n_P. To do this we take the local ring of the scheme D at P to be $\mathcal{O}_{C,P}/(\pi^{n_P})$ (*cf.* 2.8). We then have $n_P = \mu_P(D)$. Conversely, a finite subscheme X in C defines a divisor $\sum \mu_P(X)P$. We note that $h^0 \mathcal{O}_D = \chi \mathcal{O}_D = \sum_P \mu_P(D) = \sum_P n_P = \deg D$.

If D is a divisor $\geqslant 0$, then there is an exact sequence

$$0 \longrightarrow \mathcal{J}_{D/C} \longrightarrow \mathcal{O}_C \longrightarrow \mathcal{O}_D \longrightarrow 0,$$

where $\mathcal{J}_{D/C}$ is the ideal of functions which vanish on D (with prescribed multiplicities). We denote this sheaf by $\mathcal{O}_C(-D)$, and its sections over an open set U in C are simply the rational functions in C which are defined on U and which vanish at every point P with multiplicity at least n_P:

$$\Gamma(U, \mathcal{O}_C(-D)) = \{f \in K(C) \mid \forall P \in U, \ v_P(f) \geqslant n_P\}.$$

(The fact that the n_P are $\geqslant 0$ implies $f \in \Gamma(U, \mathcal{O}_C)$.)

Let D now be an arbitrary divisor. The sheaf $\mathcal{O}_C(D)$ is defined in a similar way, but the signs have changed:

$$\Gamma(U, \mathcal{O}_C(D)) = \{f \in K(C) \mid \forall P \in U, \ v_P(f) \geqslant -n_P\}.$$

NB: these functions are not defined everywhere on U since they can have poles at points where $n_P < 0$. However, at such P the order of the pole must be $\leqslant -n_P$. We are considering rational functions with poles and zeros whose orders are controlled by the divisor D.

Remarks 2.10.

0) If $D = 0$ it is easy to see that $\mathcal{O}_C(D) = \mathcal{O}_C$.

1) We will be especially interested in global sections of the sheaf $\mathcal{O}_C(D)$, which can be written in the following way:

$$H^0(C, \mathcal{O}_C(D)) = \Gamma(C, \mathcal{O}_C(D)) = \{f \in K(C) \mid \mathrm{div}(f) + D \geqslant 0\}.$$

(In particular, a divisor D is equivalent to a divisor $\geqslant 0$ if and only if the sheaf $\mathcal{O}_C(D)$ has non-zero sections.) The purpose of the Riemann-Roch theorem is to calculate the dimension $h^0(C, \mathcal{O}_C(D))$ of this space. We already know that if $\deg D < 0$, then $\Gamma(C, \mathcal{O}_C(D)) = 0$. Indeed, if $f \neq 0$, then $\deg \mathrm{div}(f) + \deg D = \deg D$ by 2.7 and this degree must be $\geqslant 0$, which is impossible.

2) If D and D' are equivalent, *i.e.*, if $D' = D + \mathrm{div}(g)$, then there is an isomorphism of associated sheaves $\mathcal{O}_C(D) \simeq \mathcal{O}_C(D')$ given by $f \mapsto f/g$.

3) Assume $C \subset \mathbf{P}^r$, let H be a hyperplane which does not contain C and let $D = C \cap H$ be the finite intersection subscheme (*cf.* 1). We can think of D as a positive divisor on C. Then the sheaf $\mathcal{O}_C(D)$ is simply $\mathcal{O}_C(1)$. Let U be an open set in C: a section of $\mathcal{O}_C(1)$ over U is a fraction $f = A/B$ such that $A, B \in \Gamma_h(C)$ are homogeneous, $\deg A = \deg B + 1$ and $B \neq 0$ over U (NB: this object is not a rational function). We associate to it the section $f/H = A/BH$ which is a rational function whose poles are the points of $C \cap H = D$ which are contained in U and which appear with the required multiplicity. This gives us a section of $\mathcal{O}_C(D)$.

Likewise, $\mathcal{O}_C(n) \simeq \mathcal{O}_C(nD)$. When C is embedded in $\mathbf{P}^r$, the sheaves $\mathcal{O}_C(n)$ are special cases of sheaves corresponding to divisors: they correspond to multiples of the hyperplane divisors $C \cap H$. It is clear that if H and H' are two hyperplanes not containing C, then their corresponding divisors are equivalent: they differ by a principal divisor $\mathrm{div}(H/H')$.

4) It is possible to see that $\mathcal{O}_C(D + D') \simeq \mathcal{O}_C(D) \otimes_{\mathcal{O}_C} \mathcal{O}_C(D')$ for all divisors D, D'. In particular, $\mathcal{O}_C \simeq \mathcal{O}_C(D) \otimes_{\mathcal{O}_C} \mathcal{O}_C(-D)$: $\mathcal{O}_C(-D)$ is the "inverse" of $\mathcal{O}_C(D)$. The Picard group of C is isomorphic via the map $D \mapsto \mathcal{O}_C(D)$ to the group of invertible (*i.e.*, locally free of rank 1, *cf.* Exercise III, A.7) sheaves over C with the tensor product law.

d. The Riemann-Roch theorem

We now try to calculate $h^0(C, \mathcal{O}_C(D))$. (See e. below for an application of this calculation.) We have the following crucial lemma.

Lemma 2.11.

1) Let D be a positive divisor (which we consider as a finite subscheme of C). There is an exact sequence

$$0 \longrightarrow \mathcal{O}_C(-D) \xrightarrow{i} \mathcal{O}_C \xrightarrow{p} \mathcal{O}_D \longrightarrow 0.$$

2) Let D be an arbitrary divisor, which we assume written in the form $D = D_1 + D_2$, *where* $D_1 \geqslant 0$. *There is an exact sequence*

$$0 \longrightarrow \mathcal{O}_C(D_2) \xrightarrow{\ i\ } \mathcal{O}_C(D) \xrightarrow{\ p\ } \mathcal{O}_{D_1} \longrightarrow 0.$$

Proof. We have already proved 1), which is also a special case of 2). To prove 2), let U be an open set in C. It is clear from the definition that $\Gamma(U, \mathcal{O}_C(D_2))$ is contained in $\Gamma(U, \mathcal{O}_C(D))$, which gives us the injective map i. To define p, recall that

$$\Gamma(U, \mathcal{O}_{D_1}) = \prod_{P \in D_1 \cap U} \mathcal{O}_{C,P}/\pi_P^{n_{1,P}},$$

where π_P is a uniformising parameter for $\mathcal{O}_{C,P}$ and the integers $n_{1,P}$ are the coefficients of D_1. Given $f \in \Gamma(U, \mathcal{O}_C(D))$, we associate to f the image of $\pi_P^{n_P} f$ in $\mathcal{O}_{C,P}/\pi_P^{n_{1,P}}$ at the point P (which is meaningful because $\pi_P^{n_P} f$ is in $\mathcal{O}_{C,P}$ since $v_P(f) + n_P$ is $\geqslant 0$ by definition). Let us check that p is a surjective sheaf morphism. This question is local so we can choose an open set U containing a unique point P in D_1. Consider $\overline{g} \in \mathcal{O}_{C,P}/\pi_P^{n_{1,P}}$. We can assume $g \in \Gamma(U, \mathcal{O}_C)$. Then $f = g/\pi_P^{n_P}$ is in $\Gamma(U, \mathcal{O}_C(D))$ and $p(f) = \overline{g}$.

Finally, we check that $\operatorname{Ker} p = \operatorname{Im} i$: if $f \in \Gamma(U, \mathcal{O}_C(D_2))$, then $v_P(f) \geqslant -n_{2,P}$ and hence $n_P + v_P(f) \geqslant n_{1,P}$, so the image of $\pi_P^{n_P} f$ in $\mathcal{O}_{C,P}/\pi_P^{n_{1,P}}$ vanishes. The converse is immediate.

There is an alternative proof of 2) using 1) applied to D_1 and tensoring by $\mathcal{O}_C(D)$.

We can now prove a second version of the Riemann-Roch theorem.

Theorem 2.12 (Riemann-Roch 2). *Let C be an irreducible smooth projective curve of genus g and let D be a divisor on C.*
   1) We have

$$\chi \mathcal{O}_C(D) = h^0 \mathcal{O}_C(D) - h^1 \mathcal{O}_C(D) = \deg D + 1 - g.$$

   2) There is an integer N such that if $\deg(D) \geqslant N$, then $h^1 \mathcal{O}_C(D) = 0$, and hence $h^0 \mathcal{O}_C(D) = \deg D + 1 - g$.

Proof.
   1) We set $D = D_1 - D_2$, where $D_i \geqslant 0$. By Lemma 2.11.2 $\chi \mathcal{O}_C(D) = \chi \mathcal{O}_C(-D_2) + \chi \mathcal{O}_{D_1}$, and 2.11.1 applied to D_2 gives us that $\chi \mathcal{O}_C(-D_2) = \chi \mathcal{O}_C - \chi \mathcal{O}_{D_2}$. By definition of the genus $\chi \mathcal{O}_C = 1 - g$. For the finite scheme D_i we know that $\chi \mathcal{O}_{D_i} = h^0 \mathcal{O}_{D_i} = \deg D_i$ (*cf.* c) above), and the formula follows.
   2) We can assume C is embedded in $\mathbf{P}^r$. If H is the hyperplane divisor (*cf.* 2.10.3), then $h^1 \mathcal{O}_C(nH) = h^1 \mathcal{O}_C(n) = 0$ for $n \geqslant n_0$ by Serre's theorem (*cf.* Chapter VII, 4.6). Let D be a divisor. By 1), we know that $h^0 \mathcal{O}_C(D - n_0 H) \geqslant \deg(D - n_0 H) + 1 - g$ and this is > 0 as soon as

$\deg(D) \geqslant N = n_0 \deg(H) + g$. If this holds and $f \in \Gamma(C, \mathcal{O}_C(D - n_0 H))$ does not vanish, then $D \geqslant n_0 H - \operatorname{div}(f)$ (cf. 2.10.1), or, alternatively, $D = D_1 + (n_0 H - \operatorname{div}(f))$, where $D_1 \geqslant 0$. Writing the long exact sequence associated to 2.11.2 we get an exact sequence

$$\cdots \longrightarrow H^1 \mathcal{O}_C(n_0 H - \operatorname{div}(f)) \longrightarrow H^1 \mathcal{O}_C(D) \longrightarrow H^1 \mathcal{O}_{D_1} \longrightarrow 0.$$

As $n_0 H - \operatorname{div}(f)$ is equivalent to $n_0 H$, the associated sheaves are the same (cf. 2.10.2) and hence their cohomology groups H^1 vanish. As D_1 is finite, $H^1 \mathcal{O}_{D_1} = 0$, and hence $H^1 \mathcal{O}_C(D) = 0$.

Of course, 2.10.3 implies that the first version of the Riemann-Roch theorem is a special case of the second version.

The third version of Riemann-Roch is in fact a duality theorem. We quote it without proof.

Theorem 2.13 (Riemann-Roch 3). *Let C be a smooth irreducible projective curve of genus g. There is a positive divisor K on C, called the canonical divisor, such that for any divisor D on C the vector space $H^1(C, \mathcal{O}_C(D))$ is isomorphic to the dual of the vector space $H^0(C, \mathcal{O}_C(K - D))$. Moreover, K has degree $2g - 2$. In particular,*

$$h^0 \mathcal{O}_C(D) = \deg D + 1 - g + h^0 \mathcal{O}_C(K - D).$$

Proof. See [F] Chapter 8 §8 or [H] Chapter IV, 1.3. The canonical divisor is of the form $\operatorname{div}(\omega)$, where ω is not a function but a differential form.

The following corollary gives us an explicit integer N which works in 2.12.2.

Corollary 2.14. *With the above notation, if $\deg D > 2g - 2$, then $h^1 \mathcal{O}_C(D) = 0$, and hence $h^0 \mathcal{O}_C(D) = \deg D + 1 - g$.*

Proof. Indeed, we then have $\deg(K - D) < 0$ and hence (cf. 2.10.1) $h^0 \mathcal{O}_C(K - D) = 0$.

e. An application

Proposition 2.15. *Let C be an irreducible smooth projective curve. The following are equivalent:*

1) C is isomorphic to $\mathbf{P}^1$.

2) C is of genus 0.

3) There is a point $P \in C$ such that $h^0 \mathcal{O}_C(P) \geqslant 2$.

4) There are two distinct points $P, Q \in C$ such that the divisors P and Q are equivalent.

5) The fraction field $K(C)$ is isomorphic to the field of rational fractions in one variable $k(T)$.

Proof. The fact that $1 \Rightarrow 2$ follows from Chapter VII, 4.1 and the definition of the genus. For $2 \Rightarrow 3$ we take an arbitrary P and the result follows by Riemann-Roch (2.12). Let us show that $3 \Rightarrow 4$. Consider a non-constant $f \in H^0\mathcal{O}_C(P)$. The function f necessarily has a pole (since otherwise it would be in $H^0\mathcal{O}_C$, hence constant) and this pole must be a simple pole at P; hence f must have a unique simple zero Q, so $\mathrm{div}(f) = (Q) - (P)$ and $(P) \sim (Q)$.

We now prove that $4 \Rightarrow 5$. Consider $f \in K(C)$ such that $\mathrm{div}(f) = (Q) - (P)$. We then have $f \in H^0\mathcal{O}_C(P)$ and f non-constant (which, incidentally, proves 3).

Let us prove that $1, f$ is a basis for $H^0\mathcal{O}_C(P)$. Let π be a uniformising parameter for $\mathcal{O}_{C,P}$. We have $f = u/\pi$ with $u \in \mathcal{O}_{C,P}$ invertible, so $u(P) \neq 0$ and we can assume $u(P) = 1$. Consider $g \in H^0\mathcal{O}_C(P)$. At P, $g = v\pi^{-1}$, where $v \in \mathcal{O}_{C,P}$. We set $\lambda = v(P)$ and consider $h = g - \lambda f$. We have $h = (v - \lambda u)/\pi$, where $(v - \lambda u)(P) = 0$, so $h \in \mathcal{O}_{C,P}$ and as h has no other pole, it is an element of $H^0\mathcal{O}_C$. The function h is therefore a constant μ, so $g = \lambda f + \mu$. It follows that $h^0\mathcal{O}_C(P) = 2$ and by Riemann-Roch $h^1\mathcal{O}_C(P) = g$.

We now consider $H^0\mathcal{O}_C(nP)$. This space obviously contains the functions $1, f, f^2, \ldots, f^n$. These functions are independent (if they were linked by a linear relationship, we could multiply it by f^{-n} and evaluate at P). It follows that $h^0\mathcal{O}_C(nP) \geqslant n+1$. By Riemann-Roch $h^0\mathcal{O}_C(nP) = n+1-g+h^1\mathcal{O}_C(nP)$.

Consider the exact sequence (*cf.* 2.11)

$$0 \longrightarrow \mathcal{O}_C((n-1)P) \longrightarrow \mathcal{O}_C(nP) \longrightarrow \mathcal{O}_P \longrightarrow 0.$$

From the long exact sequence it follows that $h^1\mathcal{O}_C(nP) \leqslant h^1\mathcal{O}_C((n-1)P)$, and by induction $h^1\mathcal{O}_C(nP) \leqslant h^1\mathcal{O}_C(P) = g$. In other words, $h^0\mathcal{O}_C(nP) \leqslant n+1$ and equality holds. (We note that this implies $h^1\mathcal{O}_C(nP) = g$ for all n, so $g = 0$ by 2.12.2: this proves 2.) It follows that $1, f, \ldots, f^n$ is a basis for $H^0\mathcal{O}_C(nP)$.

Let us now consider $K(C)$. This contains the element f, which is transcendental over k (since otherwise it would be a constant) and hence, since C is a curve, f is a transcendental basis for $K(C)$ over k. Consider $u \in K(C)$. We will show that u is contained in the subfield $k(f)$. We will have proved that $K(C) = k(f)$, which is exactly 5.

To do this, note that u is algebraic over $k(f)$: on replacing u by $ua(f)$, where a is a polynomial in f, we can even assume that u is integral over the ring $k[f]$. We then have

$$(*) \qquad u^n + a_{n-1}(f)u^{n-1} + \cdots + a_0(f) = 0,$$

with $a_i(f) \in k[f]$. It follows that u has no poles in any point Q different from P. Indeed, as $f \in \mathcal{O}_Q(C)$, u is integral over $\mathcal{O}_Q(C)$, and as this ring is a discrete valuation ring, and is hence integrally closed, u is contained in $\mathcal{O}_Q(C)$. But this implies that u is contained in $H^0\mathcal{O}_C(nP)$ for some n, and is hence a polynomial in f. QED.

It remains to prove that $5 \Rightarrow 1$, which will be done in Chapter IX (*cf.* Chapter IX, 2.5.1).

See Exercises VIII, 4 for similar results on elliptic curves (*i.e.*, of genus 1).

Exercises

In what follows we work over an algebraically closed base field k of characteristic zero.

0 Back to theory

Let $I \subset R = k[X_0,\ldots,X_n]$ be a homogeneous ideal and assume that $Z = V(I)$ is finite. Let h be a non-zero linear form such that $V(h) \cap Z = \varnothing$. We are going to give an algebraic proof of the existence of the multiplication by h isomorphism $\mathcal{O}_Z(-1) \simeq \mathcal{O}_Z$, denoted μ_h.

a) Prove that, for large n, $\mu_h : (R/I)_{n-1} \to (R/I)_n$ is surjective. (Use the Nullstellensatz to prove that the ideal $I + (h)$ contains a power of the irrelevant ideal.)

b) Assume that I is saturated (*cf.* Chapter X, 1.1). Prove that μ_h is injective (use the Nullstellensatz again). Use this to complete the proof of the theorem.

1 General hyperplane sections of a curve

Let C be an irreducible smooth curve in $\mathbf{P}^N$ of degree d.

Prove that for a general hyperplane H the scheme $Z = C \cap H$ contains d distinct points. (Useful hints are to be found in Problem VI, Theorem 3 and Summary 4.5). Prove that this result still holds if C is not assumed to be smooth.

2 Existence of smooth complete intersections

We work in $\mathbf{P}^3$, but this exercise can be easily generalised to arbitrary dimensions.

We fix two positive integers s and t. Our aim is to prove that if F and G are sufficiently general homogeneous polynomials in X,Y,Z,T of degrees s and t, then the complete intersection curve $C = V(F,G)$ is smooth and irreducible.

We consider the spaces $L_s = H^0(\mathbf{P}^3, \mathcal{O}_\mathbf{P}(s))$ and $L_t = H^0(\mathbf{P}^3, \mathcal{O}_\mathbf{P}(t))$ and the projective (why?) variety $V = \mathbf{P}^3 \times \mathbf{P}(L_s) \times \mathbf{P}(L_t)$, whose points are triples (P,F,G), given by coordinates $P = (x,y,z,t)$, $F = (a_{i,j,k,l})$ and $G = (b_{i,j,k,l})$, such that

$$F = \sum_{i+j+k+l=s} a_{i,j,k,l} X^i Y^j Z^k T^l \quad \text{and} \quad G = \sum_{i+j+k+l=t} b_{i,j,k,l} X^i Y^j Z^k T^l.$$

We set

$$M = \{(P,F,G) \in V \mid F(P) = G(P) = 0\}.$$

We consider the projections $\pi_1 : M \to \mathbf{P}^3$ and $\pi_2 : M \to \mathbf{P}(L_s) \times \mathbf{P}(L_t)$. What are the fibres of π_2?

1) Prove that M is a closed subvariety of V (and is hence a projective variety). (Give explicit equations for M in open affine sets of the form $x \neq 0$, $a_i \neq 0$, $b_j \neq 0$.)
2) Prove that the fibres of π_1 are irreducible projective varieties of constant dimension. Deduce that M is irreducible and of codimension 2 in V.
3) Prove that M is smooth. (Calculate the tangent space $T_{(P,F,G)}(M)$: it is the kernel of a matrix with 2 lines and many columns and we have to show that its rank is 2.)
4) Applying the generic smoothness theorem (cf. Problem VI) to π_2, prove that for (F,G) in a non-empty open set of the product $\mathbf{P}(L_s) \times \mathbf{P}(L_t)$, $V(F,G)$ is a smooth curve.
   Prove that $V(F,G)$ is then irreducible (prove first that it is connected and then use Chapter V, 3.6).
5) Study the following example: $F = X^2 + Y^2 + Z^2 + T^2$, $G = X^3 + Y^3 + Z^3$.

3 Some degree and genus calculations

Determine the degrees and genuses of the curves in $\mathbf{P}^3$ whose ideal I_C has the following graded resolution:

$$0 \longrightarrow R(-a-1)^a \longrightarrow R(-a)^{a+1} \longrightarrow I_C \longrightarrow 0, \quad \text{where } a \in \mathbf{N}^*,$$

$$0 \longrightarrow R(-s-t-1)^2 \longrightarrow R(-s-t-2) \oplus R(-t) \oplus R(-s) \longrightarrow I_C \longrightarrow 0,$$

where $s, t \in \mathbf{N}^*$,

$$0 \longrightarrow R(-5) \longrightarrow R(-4)^4 \longrightarrow R(-2) \oplus R(-3)^3 \longrightarrow I_C \longrightarrow 0.$$

4 Elliptic curves

The aim of this exercise is to prove that any curve of genus 1 is isomorphic to a plane cubic.

We assume that the base field k is of characteristic different from 2. Let C be a smooth irreducible projective curve of genus 1, and set $P_0 \in C$ and $C' = C - \{P_0\}$.

1) Prove that for all $n \in \mathbf{N}^*$, $h^0 \mathcal{O}_C(nP_0) = n$.
2) Prove that we can find $x, y \in K(C)$ such that $1, x$ (resp. $1, x, y$) are a basis of $H^0 \mathcal{O}_C(2P_0)$ (resp. of $H^0 \mathcal{O}_C(3P_0)$) over k.
3) Prove that the quantities $1, x, y, x^2, xy, y^2, x^3$ are linearly dependent over k. Let $P(x,y)$ be the dependence relation. Prove that the coefficients of y^2 and x^3 in P are non-zero.
4) Prove that up to change of basis we can assume that $P(x,y)$ is of the form $y^2 - x(x-1)(x-\lambda)$, where $\lambda \neq 0, 1$. (Start by eliminating the terms in y and xy by completing the square and then use an affine transformation of k to get a cubic polynomial with roots at $0, 1$ and λ.)
5) Consider the map $\varphi : C' \to k^2$ sending P to $x(P), y(P)$. Prove that φ is an isomorphism from C' to the affine curve whose equation is $y^2 = x(x-1)(x-\lambda)$.
6) Prove using Chapter IX, 2.4 that C is isomorphic to the plane curve whose equation is $Y^2 T = X(X-T)(X-\lambda T)$.

For more information on the extensive and beautiful theory of elliptic curves, cf. [H] Chapter II, 6.10.2 and Chapter IV, 4 (and references therein) or [F] Chapter V, 6 and VIII.

IX

Rational maps, geometric genus and rational curves

We work over an algebraically closed base field k.

0 Introduction

We saw in the book's introduction how useful it can be to have rational parameterisations of curves (notably for resolving Diophantine equations or calculating primitives). We then say the curve is rational. The aim of this chapter is to give a method for calculating whether or not a curve is rational. We will prove that this is equivalent to the (geometric) genus of the curve being zero and we will give methods for calculating this geometric genus.

1 Rational maps

These are generalisations of the rational functions seen in Chapter VIII (*cf.* also Problem V).

Proposition-Definition 1.1. *Let X and Y be irreducible algebraic varieties. We consider pairs (U, φ), where U is a non-empty open set of X and $\varphi : U \to Y$ is a morphism and we consider the relation on these pairs given by $(U, \varphi) \sim (V, \psi) \Leftrightarrow \varphi|_{U \cap V} = \psi|_{U \cap V}$. This is an equivalence relations on such pairs. A rational map from X to Y is an equivalence class for this relation. By abuse of notation, it is also denoted $\varphi : X \to Y$.*

Remark 1.2. As for rational functions, rational morphisms are morphisms which are not defined everywhere. There is a largest possible open set U on which φ is defined and which is called the domain of definition of φ.

Examples 1.3.

i) A morphism is a rational map.

ii) A rational function on X is a rational map from X to k.

iii) The parameterisation of the affine plane curve $C = V(X^3 - Y^3 - XY)$ obtained by intersecting with the line $y = tx$ is a rational map φ from k to C defined on k minus the cube roots of unity by

$$\varphi(t) = \left(\frac{t}{1 - t^3}, \frac{t^2}{1 - t^3} \right).$$

Proposition-Definition 1.4.

a) Let $\varphi : X \to Y$ be a rational map. We say that φ is dominant *if the image of φ is dense in Y. (It is possible to prove that this does not depend on the choice of element representing φ.)*

b) Let $X \xrightarrow{\varphi} Y \xrightarrow{\psi} Z$ be dominant rational maps. Then the composition $\psi\varphi$ is a rational map defined as follows. Consider (U, φ) and (V, ψ) representing φ and ψ. The inverse image $\varphi^{-1}(V)$ is a non-empty open set in U. Replacing U by $\varphi^{-1}(V)$ we can assume $\varphi(U) \subset V$. Then $\psi\varphi$ is well defined and defines a dominant rational map. (We check that the composition does not depend on the choice of representations of φ and ψ.)

c) We say that a dominant rational map $\varphi X \to Y$ is birational *if there is a dominant rational map $\psi : Y \to X$ such that $\psi\varphi = \mathrm{Id}_X$ and $\varphi\psi = \mathrm{Id}_Y$. (These equalities are equalities of rational maps, i.e., they hold on some non-empty open subset).*

d) Two irreducible algebraic varieties X and Y are said to be birationally equivalent *if there is a birational map $\varphi : X \to Y$. This means that X and Y have isomorphic non-empty open sets. X and Y are therefore of the same dimension. If C is a curve, C is said to be* rational *if it is birationally equivalent to $\mathbf{P}^1$ (or k, which is obviously the same thing): this means that C has a rational parameterisation which is an isomorphism on some open set.*

Proof. The proofs of the above statements are immediate. The claim concerning the dimensions follows from the fact that a non-empty open subset has the same dimension as the ambient variety.

Example 1.5.

i) Of course, the inclusion of an open subvariety U in X is birational: from the birational point of view, we can always restrict ourselves to affine varieties.

ii) All the examples of rational parameterisations of the form $y = tx$ are birational since there is an inverse rational map $(x, y) \mapsto t = y/x$ (for example: the curves of equations $X^2 + Y^2 - Y, X^3 - Y^3 - XY, Y^2 - X^3, \ldots$ are rational).

iii) For a trickier example, consider the tricuspidal quartic, *cf.* Exercise VI, 4.

The following theorem is the analogue for rational maps of Chapter I, 6.13.

Theorem 1.6. *There is an equivalence of categories $X \mapsto K(X)$ between irreducible algebraic varieties with dominant rational maps on one hand and finite-type extensions K of k with field homomorphisms which are trivial on k on the other.*

Proof.

1) The functor is obtained as in the affine case: given a rational map $\varphi :$ $X \to Y$ we obtain a field homomorphism $\varphi^* : K(Y) \to K(X)$ by associating to the rational function f on Y the composition $f\varphi \in K(X)$. (Here we have used the fact that f is dominant.) We hence obtain a contravariant functor $X \mapsto K(X)$ and our aim is to show that this functor is fully faithful and essentially surjective (*cf.* Chapter I).

2) The functor is faithful. Let φ and ψ be two rational applications from X to Y such that $\varphi^* = \psi^*$. We take open affine sets U and V in X and Y such that φ and ψ define morphisms from U to V (which is possible on restricting U). This yields ring homomorphisms φ^* and ψ^* from $\Gamma(V)$ to $\Gamma(U)$ which have the same extensions to the fraction fields. These homomorphisms are therefore the same and we are done by Chapter I, 6.7 (faithfulness for regular maps.)

3) The functor is fully faithful. Let $\theta : K(Y) \to K(X)$ be a homomorphism of fields which restricts to the identity on k. Choose open affine subsets U, V of X, Y. Their algebras $\Gamma(U)$ and $\Gamma(V)$ have generators $\xi_1, \ldots, \xi_n$ and $\eta_1, \ldots, \eta_m$ respectively. The images of the elements η_i under θ are contained in some local ring $\Gamma(U)_f$, so θ induces an algebra homomorphism from $\Gamma(V)$ to $\Gamma(U)_f$. By Chapter I, 6.7 this homomorphism comes from a morphism $\varphi : D_U(f) \to V$, *i.e.*, a rational map from X to Y, and it follows that $\theta = \varphi^*$.

4) It remains to prove that the functor is essentially surjective. Let K be a finite-type extension of k, let $\xi_1, \ldots, \xi_n$ be generators of K and let A be the k-algebra generated by the elements ξ_i. The ring A is then both an integral domain and a finite-type k-algebra, and hence A is isomorphic to an algebra $\Gamma(X)$ for some affine irreducible X, and as $K = \mathrm{Fr}(A)$, $K \simeq K(X)$.

Corollary 1.7. *Two irreducible algebraic varieties are birationally equivalent if and only if their functions fields are isomorphic. In particular, a curve is rational if and only if its field of functions is isomorphic to the field of rational fractions in one variable, $k(T)$.*

Example 1.8. This gives us a new proof of the fact that $V(Y^2 - X^3)$ is rational: on considering the open set $Y \neq 0$ in projective space we see that $K(C) = k(X)$ (*cf.* Chapter VIII, 2.4)

Proposition 1.9. *Let X be an irreducible algebraic variety of dimension n. Then X is birationally equivalent to a hypersurface in k^{n+1} (or $\mathbf{P}^{n+1}$). In particular, every curve is birationally equivalent to a plane curve.*

Proof. See Problem V; in characteristic 0 it is enough to take a transcendence basis $\xi_1, \ldots, \xi_n$ of $K(X)$ over k and note that by the primitive element theorem $K(X)$ is an extension of $k(\xi_1, \ldots, \xi_n)$ generated by a unique element.

2 Curves

We now come to the heart of the problem. Our aim is to study what birational equivalence says about curves, particularly whether or not a given irreducible curve is birationally equivalent to a "nicer" curve. What do we mean by "nice"? There are three fairly natural criteria: being smooth (a nice curve should not have singular points), being projective (a nice curve should be "complete," *cf.* Problem II), and being plane (a "nice" curve should be embedded in a small projective space), and there are many natural questions to be asked, depending on whether we want one, two or all three criteria to be satisfied. We start by summarising known results: proofs of most of them will be given in the coming paragraphs.

We note that the answer to the question of whether every curve is birationally equivalent to a nice curve is yes if we only require that one of our criteria be satisfied.

1) As the singular locus S of C is finite (*cf.* Problem V), C is birationally equivalent to its open smooth locus $C - S$.

2) Given an arbitrary C, take an affine open set U in C, embed U into k^n and consider the closure $\overline{U}$ of U in $\mathbf{P}^n$. Then C is birationally equivalent to the projective curve $\overline{U}$.

3) The "plane curve" criterion was dealt with in 1.9.

Things become more complicated with two criteria.

1) Plane and projective is easy: we consider an equivalent affine plane curve and take its closure.

2) For plane and smooth the answer is yes if we consider any open set of a projective plane curve to be a plane curve. A more delicate (but probably not very interesting) question is whether or not there exists a smooth plane curve which is closed in k^2 and which is birationally equivalent to the given curve C (if C is a plane curve of degree 5 with three non-collinear double points, for example). The author does not know the answer to this question.

3) For smooth and projective the answer is yes, but the proof is difficult enough to make it a theorem (*cf.* 5.11).

Theorem 2.1 (Desingularisation). *Any irreducible curve is birationally equivalent to a smooth projective curve. More precisely, if C is an irreducible projective curve, then there is an irreducible smooth projective curve X and a morphism $\pi : X \to C$ which is finite (and hence surjective, cf. 3.4) and birational. We say that X is the desingularisation or normalisation of C.*

Remark 2.2. The analogous theorem for surfaces is true in all characteristics (Abhyankar). In arbitrary dimension the result has only officially been proved in characteristic 0 (Hironaka), but it seems that a proof of the general case has been announced.

And finally, if we want all three criteria to be satisfied, the answer to the question is generally no (*cf.* 2.6).

Having determined whether there is a nice birational model, we can ask whether it is unique. The answer is yes for the smooth projective model.

Theorem 2.3. *Let C, C' be two irreducible smooth projective curves. Assume that C and C' are birationally equivalent. Then they are isomorphic.*

Proof. The proof relies on the following lemma.

Lemma 2.4. *Let C be a smooth irreducible curve, consider $P \in C$ and let $\varphi : C - \{P\} \to X$ be a morphism towards a projective variety X. Then there is a unique morphism $\overline{\varphi} : C \to X$ extending φ.*

Proof.

1) Uniqueness. We reduce to the case of an affine C, and uniqueness follows immediately from the irreducibility of C.

2) Existence. As the variety X is embedded in $\mathbf{P}^n$, we can assume $X = \mathbf{P}^n$. We note that if U is an open set in C containing P and we know how to extend $\varphi|_{U-\{P\}}$, then we are done (we simply glue this extension to φ on $U - \{P\}$). We can therefore replace C by any open set $U \subset C$ containing P. We can thus suppose that C is affine with associated ring A. Moreover, let U_0 be the open set $x_0 \neq 0$ in $\mathbf{P}^n$. We can assume that the image of φ meets U_0, and after shrinking C we can assume it is contained in U_0. Let π be a uniformising parameter for C at P (we know that C is smooth).

After again shrinking C if necessary we can assume that π comes from a function f which is regular on C and has no other zeros than P on C. We have therefore reduced to the case where $C - \{P\}$ is the open affine set $D(f)$ in C. As the image of φ is contained in U_0, $\varphi = (1, h_1, \ldots, h_n)$, with $h_i \in A_f$. After shrinking C for the last time we can assume that the elements h_i are of the form $h_i = u_i f^{-\alpha_i}$ with $u_i \in A$, $u_i(P) \neq 0$ and $\alpha_i \in \mathbf{Z}$.

If all the integers α_i are $\leqslant 0$, then φ can be extended in the obvious way. Otherwise, let $\alpha = \alpha_k$ be the largest of the integers α_i: as the image of φ is in $\mathbf{P}^n$, we can write $\varphi = (f^\alpha, u_1 f^{\alpha-\alpha_1}, \ldots, u_k, \cdots)$, or, alternatively, as u_k is non-zero in a neighbourhood of P, $\varphi = (f^\alpha/u_k, u_1 f^{\alpha-\alpha_1}/u_k, \ldots, 1, \ldots)$, so φ can indeed be extended to P.

We now return to the proof of Theorem 2.3. We have a morphism $\varphi : U \to C'$ defined on an open set of C or, in other words, defined on C minus a certain number of points. As C is smooth and C' is projective, we see by Lemma 2.4 that φ can be extended to C. Likewise, we extend the (rational) inverse of φ, *i.e.*, ψ, to the whole of C'. We are therefore done by comparing $\varphi\psi$ (resp. $\psi\varphi$) and the identity on C' (resp. C), according to uniqueness.

Remarks 2.5.

1) Theorem 2.3 allows us to finish the proof of Chapter VIII, 2.15 ($5 \Rightarrow 1$): as $K(C)$ is isomorphic to $k(T)$, C is birationally equivalent to $\mathbf{P}^1$ (*cf.* 1.7); as C and $\mathbf{P}^1$ are irreducible, smooth and projective, they are isomorphic by 2.3.

2) Theorem 2.3 allows us to talk about *the* normalisation of C: this normalisation is unique up to isomorphism. We call it the projective model of C or of $K(C)$. As far as birational geometry goes, we can restrict ourselves to studying smooth projective curves. We can hence define the geometric genus of C as follows.

Definition 2.6. *Let C be a curve and X its normalisation. We call the arithmetic genus of X the* geometric *genus of C.*

We note that for a smooth projective curve the two genuses coincide. For the rest of this chapter we will denote the geometric genus of C by $g(C)$ and the arithmetic genus of C by $p_a(C)$.

Proposition 2.7. *If two curves are birationally equivalent, then they have the same geometric genus. The converse is true if $g = 0$: a curve is rational if and only if its geometric genus is zero.*

Proof. The first statement is trivial. The second follows from Chapter VIII, 2.15, *cf.* Remark 2.5.1.

Remarks 2.8.

1) If $g \geqslant 1$, then the irreducible smooth curves of genus g are not all isomorphic. It is possible to show that the set of such curves (up to isomorphism) forms a variety of dimension 1 (resp. $3g - 3$) if $g = 1$ (resp. $g > 1$).

2) We can now show that the answer to the question of whether a triply nice (projective plane and smooth) curve birational to a given curve necessarily exists is no. Indeed, a smooth plane curve has genus equal to $0, 1, 3, 6, \ldots$, but not to 2 or 4. However, there are curves of genus 2 or 4 (*cf.* Exam 1992 and Exercise VIII, 2) and indeed of any genus (*cf.* [H] Chapter III, Exercise 5.6). (On the other hand, we can show that any curve is isomorphic to a curve in $\mathbf{P}^3$ (*cf.* Chapter VIII, 2.i.3 or [H] Chapter IV, 3.6).)

The birational-classification-of-curves problem is therefore directly linked to the problem of calculating genuses. This is not an easy problem, especially for curves in $\mathbf{P}^3$. One way of advancing is to use another type of curve which represents an acceptable compromise, namely projective plane curves with only ordinary singularities (*cf.* Chapter V, 4). This is justified by the following theorem, whose proof is contained in Problem VIII.

Theorem 2.9. *Let C be an irreducible curve. Then C is birationally equivalent to a projective plane curve with only ordinary singular points (i.e., with distinct tangents).*

And finally, we know how to calculate the genus of such curves (which we call "ordinary" curves).

Theorem 2.10. *Let C be an irreducible projective plane curve with only ordinary singularities of degree d. We denote by μ_P the multiplicity of C at P (cf. Chapter V, 4.4). The following formula holds:*

$$g(C) = \frac{(d-1)(d-2)}{2} - \sum_{P \in C} \frac{\mu_P(\mu_P - 1)}{2}.$$

In fact, the method for calculating the genus suggested by 2.9 and 2.10 is not very satisfactory in practice and we will give another one which is much more efficient (*cf.* 5.12 and 5.15).

3 Normalisation: the algebraic method

We will now tackle the problem of desingularising curves (Theorem 2.1), initially using an essentially algebraic method. We will need several auxiliary results whose statements are simple but whose proofs are not always easy: we will mainly give only sketch proofs or references for these results.

a. Some preliminaries

a.1. Some results on finite morphisms.

Definition 3.1. *Let $\varphi : X \to Y$ be a dominant morphism of irreducible algebraic varieties. We say that φ is* affine *if it satisfies one of the following equivalent properties.*

1) For any open affine subset of Y, U, $\varphi^{-1}(U)$ is an open affine subset of X.

2) There is a covering of Y by open affine subsets U_i $(i = 1, \ldots, n)$ such that $\varphi^{-1}(U_i)$ is an open affine subset of X for every i.

For the equivalence of the above two properties, which is not obvious, see [M] Chapter II, 7.5 and Chapter III, 1.5.

Definition 3.2. *Let $\varphi : X \to Y$ be a dominant morphism of irreducible algebraic varieties. We say that φ is* finite *if it is affine and for every open affine set U in Y the ring morphism $\varphi^* : \Gamma(U) \to \Gamma(\varphi^{-1}(U))$ is integral (and hence finite). It is enough to check this property on some open affine covering of Y.*

To show that it is enough to check this property on an open covering we reduce to the affine case and then prove that $\Gamma(X)$ is a $\Gamma(Y)$-module of finite type; this follows from the following lemma.

Lemma 3.3. *Let A be a ring, let M be an A-module and let $f_1, \ldots, f_n \in A$ be elements which generate the unit ideal. We assume that for all i the localised module M_{f_i} is of finite type over A_{f_i}. Then M is of finite type over A.*

Proof. Exercise. (Consider the submodule M' of M generated by the generators of all the localised rings and the conductor

$$I = \{a \in A \mid aM \subset M'\}.)$$

Proposition 3.4. *Let* $\varphi : X \to Y$ *be a finite morphism.*
  1) We have $\dim X = \dim Y$.
  2) The morphism φ *is surjective and its fibres are finite.*
  3) The morphism φ *is closed (i.e., transforms closed sets into closed sets).*
  4) If Y *is a complete variety (cf. Problem II), then so is* X.
  5) If Y *is a separated variety (cf. Problem I), then so is* X.

Proof.
  1) We restrict ourselves to the affine case and the result is clear by the transcendence degree characterisation of dimension.

  2) We may assume that X and Y are affine with rings B and A. Surjectivity then follows by Chapter IV, 4.2. If $y \in Y$ corresponds to the maximal ideal m of A, then the points of the corresponding fibre correspond to maximal ideals of the finite k-algebra B/mB and hence there are a finite number of them.

  3) Closedness is a local property, so we may restrict to the affine case. If F is a closed subset of X, then consider the restricted morphism $\varphi : F \to \overline{\varphi(F)}$. It is again finite, and is hence surjective, so $\varphi(F) = \overline{\varphi(F)}$.

  4) It will be enough to prove that φ is proper, *i.e.*, that if Z is a variety, $\varphi \times \mathrm{Id}_Z$ is closed. By 2) it will be enough to show that this morphism is finite. We reduce to the case where X, Y and Z are affine with rings A, B and C, and it remains to prove that if A is integral over B, then $A \otimes_k C$ is integral over $B \otimes_k C$, which is immediate.

  5) We have to prove that the diagonal in $X \times X$ is closed, or, alternatively, that its complement is open. Consider $(x, x') \in X \times X$ such that $x \neq x'$. If $\varphi(x) \neq \varphi(x')$, then we are done because Y is separated. If $\varphi(x) = \varphi(x')$, then since φ is affine, x and x' are contained in the same open affine set and we are done because any affine variety is separated.

Proposition 3.5. *Let* $\varphi : X \to Y$ *be a finite morphism. Assume that* Y *is a projective curve and* X *is smooth. Then* X *is a projective curve.*

Proof. See [H] Chapter I, 6.8 for the details of the argument. The idea is the following: we note that X is a complete curve so it would be enough to find an embedding of X in a projective variety (*cf.* Problem II). We cover X by open affine sets $U_1, \ldots, U_n$. We embed each U_i in a projective space and we denote by Y_i its closure in this space: Y_i is a projective curve. There is a morphism $\varphi_i : U_i \to Y_i$ which extends to a morphism on X (again denoted by φ_i) by 2.4. A priori, this morphism is not an embedding, but we can take the diagonal embedding $\varphi = (\varphi_1, \ldots, \varphi_n)$ into the product of the curves Y_i. We then prove (using the smoothness of X) that φ is an embedding and we are done.

Remark 3.6. The above claim is actually true even for non-smooth X (*cf.* [H] Chapter III, Exercise 5.7.d). We will assume this to be the case throughout the following discussion, especially in Paragraph 5. See Appendix 6 for a proof of the theorems in Paragraph 5 which does not use Proposition 3.5.

Proposition 3.7. *Let X, Y be separated varieties, $\varphi : X \to Y$ an affine morphism and $\mathcal{F}$ a quasi-coherent sheaf over X. For all integers i*

$$H^i(X, \mathcal{F}) \simeq H^i(Y, \varphi_* \mathcal{F}).$$

Proof. We note first that $\varphi_* \mathcal{F}$ is quasi-coherent. For $i = 0$ the statement of 3.7 is simply the definition of the direct image. For $i > 0$ take a finite affine covering $V_1, \ldots, V_n$ of Y and consider the covering $U_i = \varphi^{-1}(V_i)$ of X (which is also affine) and calculate the associated Čech cohomologies. The respective complexes are obtained by taking products of spaces of sections $\Gamma(V_{i_0, \ldots, i_p}, \varphi_* \mathcal{F})$ and $\Gamma(U_{i_0, \ldots, i_p}, \mathcal{F})$, but these spaces are equal by definition of φ_*, so the Čech complexes are identical.

a.2 Some results on gluing varieties. Let X_1 and X_2 be two varieties and let U_i be an open subset of X_i. We assume that the open sets U_1 and U_2 are isomorphic as varieties. Let $\varphi : U_1 \to U_2$ be such an isomorphism. We define a variety X (obtained by gluing X_1 and X_2 together along U_1 and U_2) in the following way.

We take the *disjoint* union $Y = X_1 \cup X_2$ and we consider the equivalence relation $\mathcal{R}$ which identifies U_1 and U_2 (*i.e.*, the only non-trivially equivalent pairs are the pairs $(x, \varphi(x))$, where $x \in U_1$) and we denote the quotient set $Y/\mathcal{R}$ with its quotient topology by X. This means that, considering the natural (injective) maps $i_1 : X_1 \to X$ and $i_2 : X_2 \to X$, a subset $V \subset X$ is open if and only if $V_1 = i_1^{-1}(V)$ and $V_2 = i_2^{-1}(V)$ are open. The sheaf of rings $\mathcal{O}_X$ is then defined by

$$\mathcal{O}_X(V) = \big\{ (s_1, s_2) \mid s_1 \in \mathcal{O}_{X_1}(V_1), \; s_2 \in \mathcal{O}_{X_1}(V_2)$$
$$\text{and } \varphi^*(s_2|_{V_2 \cap U_2}) = s_1|_{V_1 \cap U_1} \big\}.$$

(This is an exercise, *cf.* Exercise VII, 2.)

b. Normalisation

We now sketch a first (brutal but efficient) method for desingularising curves. If C is an irreducible curve, we know that C is smooth if and only if its local rings are discrete valuation rings (*cf.* Chapter V, 4.1). If, moreover, C is affine, this simply means that $\Gamma(C)$ is integrally closed (*cf.* Chapter V, 4.2).

We therefore start with an affine curve C. Let $K(C)$ be the field of rational functions on C. If C is not smooth, the ring $A = \Gamma(C)$ is not integrally closed. Let A' be the integral closure of A in $K(C)$. The ring A' is integrally closed and is an A-module of finite type (*cf.* Summary 1.7) and is hence a k-algebra of finite type. There is therefore an affine algebraic variety C' such that $\Gamma(C') = A'$ and a morphism $\varphi : C' \to C$ corresponding to the inclusion of A in A' and which is therefore a finite morphism. It follows that C' is an irreducible smooth curve and that φ is finite and birational. The map φ is hence an isomorphism away from the singular points of C (cf. midterm 1991 II). We have therefore found a desingularisation of C (in the sense given in 2.1).

In the general case, we cover the curve C with affine open sets $U_1, \ldots, U_n$ such that each U_i contains at most one singular point of C. We set $A_i = \Gamma(U_i) \subset K(C)$ and we consider the integral closure A_i' of A_i and the corresponding affine curve C_i. There is a morphism $\varphi_i = C_i \to U_i$ which is finite and birational, and which is an isomorphism except possibly at one point. We glue C_i and C_j together along the open sets $\varphi_i^{-1}(U_i \cap U_j)$ and $\varphi_j^{-1}(U_i \cap U_j)$. We then iterate this method until we get a map $\varphi : C' \to C$ which is finite and birational and has the property that C' is smooth and irreducible, so we have desingularised C. By 3.5, if C is projective the same is true of C'.

This method, which is pleasingly simple, has a disadvantage: we have little control over the morphism φ. For genus calculations in particular, we need detailed information on the local behaviour of φ. This is what we will now obtain using blow-ups.

4 Affine blow-ups

a. Introduction

In this course we will study the blow-up of a point in the plane only. For generalisations see (for example) [H] Chapter II, 7 (for the general case) and Chapter V, 3 (for the blow-up of a point on a surface).

The blow-up is a method for desingularising a plane curve C at a point. As the problem is local, we can restrict to the case of an affine curve $C \subset k^2$. We assume that C is singular at a point P and, after change of coordinates, we can assume $P = (0,0)$. The idea of the blow-up for a double ordinary point $(C = V(X^3 + Y^2 - X^2)$, for example) is to separate the two branches of the curve passing through P by replacing P by two points, corresponding to the two tangents of C at P (cf. Figure 1). To do this we have to pass into three-dimensional space and modify the plane by replacing the point P with the set of all the tangent lines passing through P. The algebraic translation of this principle is easy: the lines passing through P are the lines $y = tx$ and we consider the affine algebraic set

$$B = \{(x, y, t) \in k^3 \mid y = tx\}.$$

This is an irreducible surface in k^3 admitting a morphism $\pi : B \to k^2$ which associates (x, y) to (x, y, t) and whose fibres $\pi^{-1}(x, y)$ are as follows.

1) If $x \neq 0$, the fibre contains a unique point $(x, y, y/x)$.

2) If $x = 0$ and $y \neq 0$, the fibre is empty.

3) If $x = y = 0$ (*i.e.*, over P), the fibre is the line L (called the *exceptional* line) consisting of points $(0, 0, t)$ with $t \in k$.

We say that (B, π) is the *blow-up* of the plane at the point P.

We note that if B' (resp. U) is the open subset of B (resp. k^2) given by $x \neq 0$, then π induces an isomorphism of B' and U whose inverse is given by $(x, y) \mapsto (x, y, y/x)$.

The inverse image of $C = V(X^3 + Y^2 - X^2)$ under π is the set of points (x, y, t) such that $y = tx$ and $x^3 + y^2 - x^2 = 0$, or, alternatively, $y = tx$ and $x^2(x + t^2 - 1) = 0$. We see that $\pi^{-1}(C)$ is reducible and can be decomposed into the line L defined by $x = y = 0$ and a curve C' whose equations are $y = tx$ and $x + t^2 - 1 = 0$. This curve is called the *strict transform* of C and is smooth. To see this it is enough to project B onto the (x, t) plane by $\pi' : B \to k^2$ (this is an isomorphism whose inverse is given by $(x, t) \mapsto (x, xt, t)$): C' is isomorphic under this projection to the plane curve $C'' = V(X + T^2 - 1)$, which is smooth (it is a parabola). The curve C' (or C'') is indeed a desingularisation of C as $\pi : C' \to C$ is birational. We note that there are two points of C' over P corresponding to the two tangent lines of C at P: $(0, 0, \pm 1)$.

In fact, in what follows, we will forget the surface B and study directly the transformation $\psi = \pi \pi'^{-1}$ from the (x, t) plane to the (x, y) plane.

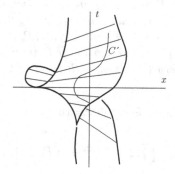

Fig. 1. The blow-up of the affine plane

b. Modification of the plane

We consider the morphism $\psi : k^2 \to k^2$, which associates $(x, y) = (x, xt)$ to (x, t). This is a birational map whose image is $U \cup \{P\}$, where U is the open set $x \neq 0$ and P is the point $(0, 0)$. Over U the inverse of ψ is given by $(x, y) \mapsto (x, y/x)$, which is an isomorphism of U on the open set $\psi^{-1}(U)$ (also defined by $x \neq 0$). The fibre $L = \psi^{-1}(P)$ is the (exceptional) line whose equation is $x = 0$.

Let C be a curve in the (x,y)-plane, set $C_0 = C \cap U$, $C_0' = \psi^{-1}(C_0)$ (which is isomorphic to C_0 via ψ) and let C' be the closure of C_0'.

Proposition 4.1. *Assume* $C = V(F)$, *where* $F \in k[X,Y]$ *is irreducible. We write* $F = F_r + \cdots + F_n$, *where* $r = \mu_P(C) \leqslant n = \deg F$ *and* F_i *is homogeneous of degree* i. *Then* $C' = V(F')$, *where*

$$F'(X,T) = F_r(1,T) + X F_{r+1}(1,T) + \cdots + X^{n-r} F_n(1,T)$$

and $\psi|_{C'}$ *is a birational morphism from* C' *to* C *inducing an isomorphism from* C_0' *to* C_0.

Proof. We have $\psi^{-1}(C) = V(F^\psi)$, where $F^\psi(X,T) = F(X,XT) = X^r F_r(1,T) + \cdots + X^n F_n(1,T) = X^r F'(X,T)$. We write this equation in the form $F(X,Y) = X^r F'(X,Y/X)$, and it follows that F' is irreducible. Consider $(x,t) \in k^2$ such that $x \neq 0$: (x,t) is in C_0' if and only if $F(x,xt) = 0$, or, alternatively, $F'(x,t) = 0$, since $x \neq 0$. In other words, $C_0' \subset V(F')$ and hence $C' \subset V(F')$, but as F' is irreducible, equality holds. The rest of the proposition is obvious.

Remark 4.2. If $r = 0$, i.e., if the point P is not in C, then $F'(X,T) = F(X,XT)$ and $C_0' = V(F')$ is closed in k^2.

Proposition 4.3. *Assume that the line* $X = 0$ *is not tangent to* C *at* P *(we can always reduce to this case by an affine isomorphism). We can then write* $F_r = \prod_{i=1}^{s}(Y - \lambda_i X)^{r_i}$, *where the* $\lambda_1, \ldots, \lambda_s \in k$ *are distinct and* $r = \sum_{i=1}^{s} r_i = r$. *We then have* $\psi^{-1}(P) = \{P_1, \ldots, P_s\}$, *where* $P_i = (0, \lambda_i)$ *and* $1 \leqslant \mu_{P_i}(C') \leqslant \mu_{P_i}(C', L) = r_i$. *Moreover, if* P *is an ordinary multiple point (i.e.,* $r_i = 1$ *for all* i*) every* P_i *is a smooth point of* C' *and the line* $X = 0$ *is not tangent to* C' *at* P_i.

Proof. From the definition of F it follows that

$$F'(X,T) = \prod_{i=1}^{s}(T - \lambda_i)^{r_i} + X F_{r+1}(1,T) + \cdots$$

We now determine the points over P: these are the points $(0,t)$ such that $F'(0,t) = 0$, i.e., the points $P_i = (0, \lambda_i)$. We compare the multiplicity of C' at P_i and the intersection multiplicity of C' and L at this point: $\mu_{P_i}(C') \leqslant \mu_{P_i}(C' \cap L)$ (*cf.* Problem VII, Axiom 5), but $\mu_{P_i}(C' \cap L) = \mu_{P_i}(F', X) = \mu_{P_i}(\prod_{i=1}^{s}(T - \lambda_i)^{r_i}, X) = r_i$ (*cf.* Problem VII, loc. cit.). If $r_i = 1$, the point P_i is therefore smooth, and as the intersection multiplicity of C' and the line $X = 0$ at this point is 1, this line is not tangent to C (*cf.* Problem VII, loc. cit.).

Remark 4.4. The example $F(X,Y) = Y^2 - X^3$ shows that 4.3 is false if $r_i \geqslant 2$: the line $X = 0$ is tangent to $F'(X,T) = T^2 - X$ at the point $(0,0)$.

Summarising, the situation is as follows. 1) We have a transformation $\psi : C' \to C$ which is an isomorphism over the open set $x \neq 0$. 2) If P is an ordinary multiple point of order r of C, then it is replaced in C' by r smooth points.

Be careful, however: using this method we lose control of the points of C such that $x = 0$, which can produce singular points of the projective curve $V(F'^{\sharp})$ at infinity. For example, if we start with $F = X^4 + Y^3 + Y^2 - X^2$ and we blow up the origin (which is the only singular point of C, including infinity), we get a curve of equation $X^2 + T^3 X + T^2 - 1$ which has no singular points in affine space but has one singular point at infinity, $(1,0,0)$. To get around this problem we will have to use projective blow-ups and gluings (*cf.* Theorem 5.7).

In any case, we will need a better understanding of the local structure of blow-ups, which is the object of the next paragraph.

c. Local structure of a blow-up

We use the notations of the above section. The point $P = (0,0)$ is assumed of multiplicity $r \geqslant 1$ in C. Once again, we assume that the line $V(X)$ (*i.e.*, $X = 0$) is not tangent to C at P.

We denote by x, y (resp. x, t) the coordinate functions on k^2. The fields of rational functions are $K(C) = k(x,y) = K(C') = k(x,t)$, and the elements x, y (resp. x, t) are related by an equation $F(x,y) = 0$ (resp. $F'(x,t) = 0$). Moreover, $y = tx$. We start by showing that the element t satisfies an algebraic equation with coefficients in the ring $k[x,y]$ generated by x and y whose dominant coefficient is $\neq 0$ at P.

Lemma 4.5. *With the above notation t satisfies an equation of the form $h_r(y)t^r + \sum_{k=0}^{r-1} h_k(x,y)t^k = 0$, where the coefficients h_k are polynomials. Moreover, if we set $h = h_r(y)$, then $h(P) \neq 0$, and if $D_C(h)$ is the affine open subset of C defined by $h \neq 0$, then $D_C(h) \cap V(X) = \{P\}$.*

Proof. Set $F(X,Y) = \sum_{d=r}^{n} \sum_{j=0}^{d} a_{d-j,j} X^{d-j} Y^j$. We have $F(x,y) = 0$ and $F'(x,t) = \sum_{d=r}^{n} x^{d-r} \sum_{j=0}^{d} a_{d-j,j} t^j = 0$. The monomials $x^{d-r}t^j$ can be reduced using the equation $y = tx$.

1) If $j \leqslant d - r$, then $x^{d-r}t^j = x^{d-r-j}y^j$, which does not contain t.

2) If $j > d - r$, then $x^{d-r}t^j = y^{d-r}t^{j-d+r}$ and the degree in t of this monomial is $\leqslant r$ and equal to r for $j = d$.

We therefore have an equation of the desired form with $h = h_r(y) = \sum_{d=r}^{n} a_{0,d} y^{d-r}$. We note that $h(P) = a_{0,r} = 1$ (as $V(X)$ is not tangent to C at P, $F_r = \prod_{i=1}^{s} (Y - \lambda_i X)^{r_i}$). Moreover, the formula $F(0,y) = h(y)y^r$ shows that the line $V(X)$ does not meet the open set $D_C(h)$ other than at P.

We then have the following corollary.

Corollary 4.6. *We use the above notation and set $W = D_C(h)$ and $V = \psi^{-1}(W) = D'_C(h)$. Then W and V are affine varieties, W contains P and $B = \Gamma(V) = k[x, t]_h$ is finite over $A = \Gamma(W) = k[x, y]_h$. The morphism $\psi : V \to W$ is finite and birational and is an isomorphism outside of $\psi^{-1}(P)$.*

We now calculate the dimension of the quotient space B/A.

Lemma 4.7 (Gorenstein's lemma). *With the above notation and denoting by m the ideal $[x, y]$ in A the following hold.*

1) The ideal $x^{r-1}B \in B$ is contained in A and is equal to

$$m^{r-1} = (x^{r-1}, x^{r-2}y, \ldots, y^{r-1}).$$

2) We have $\dim_k B/x^{r-1}B = r(r-1)$.
3) We have $\dim_k A/m^{r-1} = r(r-1)/2$, and hence $\dim_k B/A = r(r-1)/2$.

Proof.

1) The A-module B is generated by $1, t, \ldots, t^{r-1}$, but $x^{r-1}t^j = x^{r-1-j}x^j t^j = x^{r-1-j}y^j \in m^{r-1}$, so $x^{r-1}B \subset m^{r-1}$. As m^{r-1} is generated by the monomials $x^{r-1-j}y^j$, the converse is clear.

2) We calculate the quotient ring $k[x, t]/(x)$. This is the quotient:

$$k[X, T]/(F'(X, T), X) = k[T]/(F'(0, T)) = k[T] \Big/ \prod_1^s (T - \lambda_i)^{r_i},$$

and as $F'(0, T)$ is unitary of degree r, this quotient is of dimension r over k. Moreover the image of h in this ring is equal to the image of $h_r(tx) = 1 + a_{0,1}tx + \cdots + a_{0,n-r}(tx)^{n-r}$, and as $x = 0$ in the quotient, this image is equal to 1 and is hence invertible, so the quotient ring is isomorphic to B/xB. It follows that $\dim_k B/xB = r$. Continuing, we consider the filtration of B by the ideals $x^i B$: $0 \subset x^{r-1}B \subset \cdots xB \subset B$. We have $\dim_k B/x^{r-1}B = \sum_{i=0}^{r-2} \dim_k x^i B/x^{i+1}B$. There is an isomorphism $B/xB \to x^i B/x^{i+1}B$ given by multiplication by x^i (which is injective because B is an integral domain), and hence $\dim_k x^i B/x^{i+1}B = \dim_k B/xB = r$, so $\dim_k B/x^{r-1}B = r(r-1)$.

3) Denoting the ideal (X, Y) of $k[X, Y]$ by m_P, the quotient ring A/m^{r-1} is simply $(k[X, Y]/(m_P^{r-1}, F))_h$, but as $F \in m_P^{r-1}$ and $h_r \notin m_P^{r-1}$, this is just $k[X, Y]/m_P^{r-1}$, which is of dimension $r(r-1)/2$ as we saw in Chapter V, 4.6. (A basis of this space is given by the elements $1, x, y, x^2, xy, y^2, \ldots, x^{r-2}, x^{r-3}y, \ldots, y^{r-2}$.)

Remark 4.8. A priori, the open set W may contain singular points $P_1, \ldots, P_r$ other than P: however, we can eliminate them by replacing W with a smaller open affine set. Indeed, if $P_i = (\alpha_i, \beta_i)$ is such a point, then $\alpha_i \neq 0$ because W only meets $V(X)$ in P. We set $\alpha = \alpha_i$. We localise A and B along $g = x - \alpha$. The open sets obtained are again affine and the restricted morphism is again finite. The only thing we need to check is that the dimension of B/A has not changed. To do this we need the following lemma.

Lemma 4.9. *With the above notation the multiplication by $g = x - \alpha$ induces an automorphism of B/A.*

Proof. As B/A is finite dimensional over k it will be enough to prove injectivity. Consider $b \in B$ such that $(x - \alpha)b \in A$. It will be enough to show that b is contained in A. We know that $x^{r-1}b \in A$. Let i be the smallest integer such that $x^i b \in A$. If $i > 0$, then $x^{i-1}(x - \alpha)b \in A$, so $x^{i-1}\alpha b \in A$, and as $\alpha \in k^*$, $x^{i-1}b \in A$, which is a contradiction, so $i = 0$, and we are done.

Returning to 4.8 we have an isomorphism $B_g/A_g \simeq (B/A)_g$ and as multiplication by g is an automorphism of B/A, we also have $B/A \simeq (B/A)_g$, so the dimension of the quotient has not changed.

We can now define what we mean by a standard local blow-up.

Definition 4.10. *A standard local blow-up with centre P is a morphism $\psi :$ $V \to W$ satisfying the following conditions:*

1) V and W are isomorphic to affine open subsets of irreducible plane curves,

2) P is a point of W with multiplicity $\mu_P \geqslant 1$ and $W - \{P\}$ is smooth,

3) ψ is finite and birational and is an isomorphism away from the fibre $\psi^{-1}(P)$,

4) If P is an ordinary singular point of W, V is smooth,

5) If $\psi^* : \Gamma(W) \to \Gamma(V)$ is the (injective) homomorphism induced by the morphism ψ, then $\dim_k \Gamma(V)/\Gamma(W) = \mu_P(\mu_P - 1)/2$.

Remark 4.11. In particular, the last condition of 4.10 shows that if P is a smooth point of W (*i.e.*, $\mu_P = 1$) then the morphism ψ is an isomorphism.

The following theorem summarises all the results of this section.

Theorem 4.12. *Let C be a variety which is isomorphic to an open affine subset of an irreducible plane curve and consider $P \in C$. There is an open affine set W in C containing P and a standard local blow-up $\psi : V \to W$.*

Remark 4.13. In the local blow-up of the variety W we replace the ring $A = \Gamma(W)$ by $B = \Gamma(V)$, which is integral over A and has the same fraction field, and hence is contained in the integral closure A' of A. When the singularity at P is not ordinary the ring B is not generally equal to A' and hence V is not the normalisation of W, but Property 5) says that its "distance" from the normalisation has decreased. To obtain the normalisation it is enough to repeat this operation a finite number of times (*cf.* 5.11).

5 Global blow-ups

a. Definition

Not having defined blow-ups of points of a surface (other than the plane) we will have to limit ourselves to rather special curves throughout the following discussion.

Definition 5.1. *Let X be an irreducible curve. We say that X is locally planar if for any $P \in X$ there is an open affine subset of X containing P which is isomorphic to an open affine subset of a plane curve.*

Remark 5.2. We will see further on (*cf.* 5.13) that every irreducible smooth curve is locally planar but (*cf.* 5.14) there are curves in $\mathbf{P}^3$ which are not locally planar.

Definition 5.3. *Let X and C be irreducible curves and let $\pi : X \to C$ be a morphism. We say that π is a global blow-up if there is an affine open cover $U_1, \ldots, U_n$ of C such that each U_i is isomorphic to an open affine set of a plane curve and the restriction $\pi : \pi^{-1}(U_i) \to U_i$ is either a standard local blow-up with centre $P_i \in U_i$ or an isomorphism.*

Remarks 5.4.
1) It follows from the definition of local blow-ups that $\pi^{-1}(U_i)$ is also isomorphic to an open affine subset of a planar curve: in a global blow-up the curves X and C are locally planar.

2) In a global blow-up all the singular points of X necessarily lie over singular points of C and their multiplicities are equal to or less than the multiplicities of the points of C.

Proposition 5.5. *Let $\pi : X \to C$ be a global blow-up. Then π is a finite birational morphism, which is an isomorphism away from the blow-up centres. (In particular, this is the case away from the singular points of C.) If C is projective, then so is X.*

Proof. This follows immediately from the definitions, except for the last part which, follows from Proposition 3.5 when X is smooth and from the result quoted in Remark 3.6 otherwise (*cf.* also [H] Chapter II, 7.16). ∎

b. Existence of a desingularisation of an ordinary curve

Definition 5.6. *An* ordinary curve *is a projective irreducible plane curve having only ordinary multiple point singularities.*

For an ordinary curve desingularisation by blow-up is easy.

Theorem 5.7. *Let C be an ordinary curve. There is a global blow-up $\pi : X \to C$ whose blow-up centres are the singular points of C such that X is smooth and projective.*

Proof. We will give two proofs of 5.7, first by gluing and then by projective blow-up.

Proof of 5.7 by gluing. In fact, by 4.10.4, Theorem 5.7 follows from the following slightly more general statement.

Proposition 5.8. *Let C be a projective curve which is locally planar. There is a global blow-up $\pi : X \to C$, whose blow-up centres are the singular points of C such that X is projective and locally planar.*

Proof (of 5.8). We proceed by gluing as in 3.2 above. We isolate each singularity P_i of C in an open affine subset U_i which is isomorphic to an open subset of an affine plane curve. We blow up U_i at P_i and we obtain a variety V_i. After shrinking U_i and V_i we may assume that $\pi_i : V_i \to U_i$ is a standard local blow-up. We then glue together all the sets V_i and the open set $C - \{P_1, \ldots, P_r\}$ to form a variety X which is the global blow-up we seek. The fact that X is projective follows from Proposition 5.5 (which depends on Remark 3.6). When C is ordinary, X is smooth, and we are done by Proposition 3.5.

Proof of 5.7 by projective blow-up. We start by blowing up a single point. The method is similar to that used in the affine case, but we have to pay a bit more attention to the denominators. Indeed, we have seen that the problem with the transformation $y = tx$ (or, alternatively, $t = y/x$) is that we lose the points $x = 0$ which correspond to infinite t. The idea is to replace the point t of the affine line by the point (u, v) of the projective line, with $t = u/v$ when $v \neq 0$. This is equivalent to replacing the surface B given by $y = tx$ by a surface of equation $vy = ux$: more precisely, we consider the product variety $\mathbf{P}^2 \times \mathbf{P}^1$ (which is a projective variety via the Segre embedding into $\mathbf{P}^5$) and the closed subvariety[1] $B \subset \mathbf{P}^2 \times \mathbf{P}^1$ consisting of points $(x, y, z; u, v)$ such that $vy - ux = 0$. It can be checked that this is an irreducible smooth projective surface.

There is a morphism $\pi : B \to \mathbf{P}^2$, which is the restriction of the first projection from $\mathbf{P}^2 \times \mathbf{P}^1$ onto $\mathbf{P}^2$. Let P be the point $(0, 0, 1)$ in $\mathbf{P}^2$. The fibre of π over a point $Q = (x, y, z) \neq P$ is a single point $(x, y, z; y, x)$ and the fibre over P is the exceptional line E formed of points $(0, 0, 1; u, v)$ with $(u, v) \in \mathbf{P}^1$. The morphism π is the projective blow-up of the point P of $\mathbf{P}^2$. It is surjective and induces an isomorphism of $V = B - E$ and $U = \mathbf{P}^2 - \{P\}$.

The link with affine blow-ups can be obtained on considering the open sets Ω in $\mathbf{P}^2$ defined by $z \neq 0$ (or, alternatively, $z = 1$) or, more precisely, the open set Ω_0 in Ω defined by $x, z \neq 0$. The inverse image of Ω_0 under π is contained in the open set $z \neq 0$, $v \neq 0$ of B, and on setting $z = v = 1$ we obtain an affine blow-up. (NB: if we consider points (x, y) such that $x = 0$ and $y \neq 0$, then the preimage of $(0, y, 1)$ is $(0, y, 1; 1, 0)$, which is "at infinity" in $\mathbf{P}^1$.)

---

[1] In a product space $\mathbf{P}^{n_1} \times \cdots \times \mathbf{P}^{n_r}$ the closed subsets are defined by polynomials which are homogeneous with respect to each set of variables corresponding to a projective space (*cf.* Problem 1).

Let C be a curve in $\mathbf{P}^2$, set $P = (0,0,1)$ and assume (at least to begin with) that P is the unique singular point of C and that P is ordinary. Set $C_0 = C \cap U$ and let C' be the closure of $\pi^{-1}(C_0)$. Then C' is a projective curve, the restriction of π to C' is a finite birational map which is an isomorphism except at P (so C' is non-singular away from the fibre over P). Moreover, if C is not tangent to the line $X = 0$ at P (which we may assume to be the case after homography), then π is isomorphic in a neighbourhood of P to an affine blow-up. In particular, since P is ordinary, C' is non-singular over P, so C' is non-singular everywhere and we have desingularised C.

Let us now deal with the general case of an ordinary curve C with several singular points $P_1, \ldots, P_n \in \mathbf{P}^2$. We may assume these points have homogeneous coordinates $P_i = (a_i, b_i, 1)$. We consider the product of the plane $\mathbf{P}^2$ with n copies of the line $\mathbf{P}^1$ with (partially homogeneous) coordinates $(x, y, z; u_1, v_1; \cdots ; u_n, v_n)$. In this product we consider the closed subvariety B defined by the n equations $u_i(x - a_i z) - v_i(y - b_i z)$ for $i = 1, \ldots, n$. We check that B is an irreducible surface and the projection of B onto $\mathbf{P}^2$ is an isomorphism except at the points P_i. At P_i the fibre E_i is a line (called the exceptional line) the coordinates of whose points are given by $x = a_i, y = b_i, z = 1$, $u_j = b_j - b_i, v_j = a_j - a_i$ if $j \neq i$ and arbitrary u_i, v_i. If U (resp. V) is the open subset of $\mathbf{P}^2$ (resp. B), which is the complement of the points P_i (resp. the lines E_i), then π induces an isomorphism of V onto U and there is a neighbourhood of P_i in which π is isomorphic to an affine blow-up. If we set $C_0 = C \cap U$ and $C' = \overline{\pi^{-1}(C_0)}$, then we can prove as above that C' is the desingularisation of C we seek.

c. Genus of ordinary curves

The following theorem allows us to calculate the change in the arithmetic genus of a curve under a global blow-up.

Theorem 5.9. *Let C and X be irreducible projective curves and let $\pi : X \to C$ be a global blow-up. Then*

$$p_a(X) = p_a(C) - \sum \mu_P(\mu_P - 1)/2,$$

the sum being taken over all blow-up centres.

(Recall that p_a denotes the arithmetic genus and μ_P the multiplicity of P in C.)

Proof. We recall that $\chi(\mathcal{O}_X) = 1 - p_a(X)$ and $\chi(\mathcal{O}_C) = 1 - p_a(C)$. Moreover, 3.7 implies $\chi(\mathcal{O}_X) = \chi(\pi_*\mathcal{O}_X)$. But by definition of a morphism there is a sheaf morphism $i : \mathcal{O}_C \to \pi_*\mathcal{O}_X$. This morphism is injective because π is dominant. Let T be the cokernel of i, so there is an exact sequence

$$(*) \qquad\qquad 0 \longrightarrow \mathcal{O}_C \xrightarrow{\ i\ } \pi_*\mathcal{O}_X \xrightarrow{\ p\ } T \longrightarrow 0,$$

and hence $\chi(\pi_* \mathcal{O}_X) = \chi(\mathcal{O}_C) + \chi(\mathcal{T})$. As π is an isomorphism away from the blow-up centres $P_1, \ldots, P_r$, the sheaf $\mathcal{T}$ is supported at $\{P_1, \ldots, P_r\}$, and it follows from Chapter VII, 3.3 that its only cohomology group is H^0. Moreover, if we cover C with open affine sets W_i, $i = 1, \ldots, r$ corresponding to local blow-ups $V_i \to W_i$ with centre P_i, there is an isomorphism:

$$H^0(C, \mathcal{T}) \longrightarrow \prod_{i=1}^r H^0(W_i, \mathcal{T})$$

(gluing being automatically possible because $\mathcal{T}$ vanishes on the intersections). Since W_i is affine, over each W_i there is an exact sequence $0 \to \Gamma(W_i) \to \Gamma(V_i) \to H^0(W_i, \mathcal{T}) \to 0$ deduced from $(*)$, so $h^0(W_i, \mathcal{T}) = \dim_k(\Gamma(V_i)/\Gamma(W_i)) = \mu_{P_i}(\mu_{P_i} - 1)/2$ by 4.10.5. We have therefore proved that

$$\chi(\mathcal{T}) = h^0 \mathcal{T} = \sum_{i=1}^r \mu_{P_i}(\mu_{P_i} - 1)/2,$$

and the theorem follows immediately.

Corollary 5.10. *Let C be an ordinary curve of degree d. We denote by μ_P the multiplicity of C at P (cf. Chapter V, 4.4). The geometric genus of C is given by the formula*

$$g(C) = \frac{(d-1)(d-2)}{2} - \sum_{P \in C} \frac{\mu_P(\mu_P - 1)}{2}.$$

Proof. This follows from 5.7 and 5.9.

d. General desingularisations

Corollary 5.11 (General desingularisations). *Let C be a locally planar projective irreducible curve. There is a sequence of global blow-ups*

$$X = C_n \xrightarrow{\pi_n} C_{n-1} \longrightarrow \cdots \longrightarrow C_1 \xrightarrow{\pi_1} C_0 = C$$

such that X is smooth, irreducible and projective.

Proof. If C is not smooth we blow up its singular points to obtain C_1, which is locally planar and projective with $0 \leqslant p_a(C_1) < p_a(C)$ (cf. 5.9). If C_1 is not smooth, we repeat the procedure and as the arithmetic genus of an irreducible curve is $\geqslant 0$, this process terminates after at most p_a blow-ups.

Corollary 5.12 (calculating the geometric genus). *Let C be an irreducible projective plane curve of degree d. Let*

$$X = C_n \xrightarrow{\pi_n} C_{n-1} \longrightarrow \cdots \longrightarrow C_1 \xrightarrow{\pi_1} C_0 = C$$

be a sequence of blow-ups as above with X smooth and projective. Then

$$g(C) = \frac{(d-1)(d-2)}{2} - \sum \frac{\mu_P(\mu_P - 1)}{2},$$

where the sum is taken over all the points P of all the curves $C = C_0, C_1, \ldots, C_n$ (we say that the sum is taken over all the "infinitesimal neighbours" of C).

Considering only the singular points of C we get

$$g(C) \leqslant \frac{(d-1)(d-2)}{2} - \sum_{P \in C} \frac{\mu_P(\mu_P - 1)}{2}.$$

Proof. This is clear by induction on n using 5.9.

Corollary 5.13. *Any smooth irreducible curve is locally planar.*

Proof.

1) If C is projective, consider a curve Γ which is planar and projective and birationally equivalent to C. There is then a sequence of global blow-ups which desingularises Γ to X. As the fact of being locally planar is preserved under blow-up, X is locally planar. But X and C are then projective, smooth and birationally equivalent, so they are isomorphic, and we are done.

2) If C is not projective, we can assume C is affine. We embed C in a projective curve $\overline{C}$ and we construct a desingularisation X of $\overline{C}$ which is smooth and projective. There is therefore a finite birational map $\pi : X \to \overline{C}$ which is an isomorphism over the smooth locus of $\overline{C}$ and is hence an isomorphism over C. As X is locally planar by 1), we are done.

Remark 5.14. A singular curve C, on the other hand, is not generally locally planar. Indeed, the tangent space at a point of a plane curve is of dimension $\leqslant 2$. Consider the curve C in k^3 whose equations are $X^2 - Y^3 = Y^2 - Z^3 = 0$. It is easy to see that C is irreducible and $I(C) = (X^2 - Y^3, Y^2 - Z^3)$ (use the parameterisation $x = t^9, y = t^6, z = t^4$.) But the tangent space to C at $(0,0,0)$ is of dimension 3 because the Jacobian matrix vanishes, so C is not locally planar.

5.15. Algorithm for calculating the geometric genus of a plane curve. Let C be an irreducible projective plane curve of degree d. (We know that any curve is birationally equivalent to such a curve, *cf.* 1.9). The method for calculating the geometric genus of C is as follows.

0) Calculate $p_a(C) = (d-1)(d-2)/2$.

1) Determine the singular points of C.

2) For each singular point P of C, do a local blow-up of centre P to obtain points $P_1, \ldots, P_r$ (the infinitesimal neighbours of P). If these points are smooth (which happens if P is ordinary), we are done. Otherwise, we do local blow-ups centred at each of the singular points P_i. We carry on until all

the infinitesimal neighbours of P are smooth. Theorem 5.11 guarantees that this process stops after a finite number of steps. We note that this stage of the calculation is *local* (we work only with P and its successive fibres.)

3) Calculate the multiplicity of each singular infinitesimal neighbour of C and apply 5.12.

Example 5.16. 1) By 5.12 a cubic with a singular point is of genus 0 and hence is rational. Likewise, a quartic with a triple point or three double points is of genus 0. This is the case of the trefoil defined by $F(X,Y,T) = (X^2 + Y^2)^2 + 3X^2YT - Y^3T$ (which has a triple point at the origin) or the tricuspidal quartic defined by $F(X,Y,T) = Y^2T^2 + T^2X^2 + X^2Y^2 - 2XYT(X+Y+T)$ (three cusp points at $(0,0,1)$, $(0,1,0)$ and $(1,0,0)$) or the regular trefoil given by

$$F(X,Y,T) = 4(X^2 + Y^2)^2 - 4X(X^2 - 3Y^2)T - 27(X^2 + Y^2)T^2 + 27T^4$$

(which has three double ordinary points). The curious reader will find more beautiful curves in the special edition number 8, July 1976, of the magazine *Revue du Palais de la Découverte*. To get an explicit parameterisation of these quartics, we intersect the curve with a varying line passing through the triple point or a pencil of conics passing through the three double points (plus a fixed fourth point, *cf.* Exercises VI, 4).

2) Consider the curve C in $\mathbf{P}^3$ given by the equations

$$XT - YZ, \ X^2Z + Y^3 + YZT, \ XZ^2 + Y^2T + ZT^2.$$

It is easy to show that C is an irreducible smooth curve (*cf.* the problem on the January 1992 exam paper). To calculate its genus we project C into the $[x,y,t]$ plane. We obtain a curve C_0 whose equation is $F(X,Y,T) = Y^4 + X^3T + XYT^2$, which is birationally equivalent to C and has a unique double ordinary point at the origin, and which therefore has geometric genus 2 by 5.10. The initial curve therefore also has genus 2.

3) Consider C the quadrifoil whose equation is $(X^2+Y^2)^3 - 4X^2Y^2T^2 = 0$. Its arithmetic genus is 10 and it has a quadruple point P at the origin of the x,y affine plane, $(0,0,1)$, with two double tangents, plus two cusp points at infinity, the cyclic points $(1,i,0)$ and $(1,-i,0)$ (the easiest way to see this is to perform the homography given by $U = X + iY$ and $V = X - iY$, which transforms the initial equation into $4U^3V^3 + (U^2 - V^2)^2T^2 = 0$). The genus of C is therefore $\leqslant 2$. To calculate this genus we blow up the origin. NB: as the two axes are tangent to C at P, it is preferable to use the second form of the equation. On setting $u = vw$ the blow-up gives a curve having equation $4w^3v^2 + (1 - w)^2(1 + w)^2$, with two points over P, $(0,1)$ and $(0,-1)$ which are ordinary double points. These two points are therefore to be counted as infinitesimal neighbours with multiplicity 2, so the genus of C is 0.

It is easy to find a parameterisation of C using polar coordinates and the rational parameterisations of the functions sine and cosine.

4) We finish with the example of the plane curve C defined by the equation $F(X, Y, T) = (X^2 - YT)^2 + Y^3(Y - T)$. This curve has arithmetic genus 3 and a unique double point (a cusp) at $P = (0, 0, 1)$. Its geometric genus is therefore $\leqslant 2$. We will show that it is in fact zero. We work in the affine (x, y) plane: in this plane the equation of C is $x^4 + y^4 - 2x^2 y - y^3 + y^2$. We perform a blow-up $y = tx$. We obtain a plane curve whose equation is $t^4 x^2 - t^3 x + (x - t)^2$. Over P there is a unique point $P_1 = (0, 0)$, which is again a cusp. We perform a blow-up $x = zt$, which gives us a curve $t^4 z^2 - t^2 z + (z - 1)^2$. Over P_1 there is a unique point P_2 whose coordinates are $z = 1$, $t = 0$.

We study the nature of this point by setting $u = z - 1$. We obtain the equation $t^4 u^2 + 2t^4 u + t^4 - t^2 u + t^2 - u^2$, which proves that P_2 is a double ordinary point. It follows that blowing up P_2 yields two smooth points and completes the desingularisation of C. There is therefore a sequence of blow-ups $C_3 \to C_2 \to C_1 \to C$, where C_3 is smooth, and there are three infinitesimal neighbours which are double points, $P \in C$, $P_1 \in C_1$ and $P_2 \in C_2$, and hence the arithmetic genus of C_3 is the genus of C minus 3 which is zero.

To find an explicit parameterisation of C we use the osculating conics of C at P, i.e., the polynomials G of degree 2 such that $\mu_P(F, G) \geqslant 7$. It is easy to prove that these are conics with affine equations of the form $G_\lambda(x, y) = x^2 + xy - y + \lambda y^2$, with $\lambda \in k$. These conics form a pencil, i.e., a linear family of dimension 1 in the projective space of conics. If we intersect F and G_λ, then the resulting scheme contains P with multiplicity 7 plus another point in C whose coordinates depend linearly on λ. More precisely, $x(1 - 2\lambda) = y(\lambda^2 - \lambda + 1)$, which gives us a parameterisation

$$x = \frac{(1 - 2\lambda)(\lambda^2 - \lambda + 1)}{\lambda^4 + 2\lambda^2 - 4\lambda + 2}, \qquad y = \frac{(1 - 2\lambda)^2}{\lambda^4 + 2\lambda^2 - 4\lambda + 2}.$$

6 Appendix: review of the above proofs

In the preceding paragraph, we have freely used the fact (which was quoted without proof, cf. 3.6) that any curve, even singular, which is finite over a projective curve is projective (or, more generally, any complete curve is projective). We now sketch how to avoid using this result.

We note first that if C is a complete irreducible curve, then $H^0(C, \mathcal{O}_C) = k$ (cf. Problem II, 3.b). We can therefore reformulate 3.9 as follows.

Theorem 6.1. *Let C be a locally planar irreducible separated complete curve such that $h^1 \mathcal{O}_C = p_a(C)$ is finite and let $\pi : X \to C$ be a global blow-up. Then X is a locally planar irreducible separated complete curve, $h^1 \mathcal{O}_X = p_a(X)$ is finite and*

$$p_a(X) = p_a(C) - \sum \mu_P(\mu_P - 1)/2,$$

where the sum is taken over the centres of all the blow-ups.

Proof. We apply 3.4 to establish the separatedness and completeness of X: the rest of the proof is identical to that of 5.9. The finiteness of $h^1 \mathcal{O}_X = h^1 \pi_* \mathcal{O}_X$ comes from the long exact sequence associated to the exact sequence $(*)$ in 5.9. In fact, the finiteness of cohomology holds for complete varieties (*cf.* [EGA], III).

The rest of Paragraph 5 is unchanged. In 5.11 we can no longer assert a priori that the intermediate curves C_i are projective, but X is indeed projective (since it is smooth).

X

Liaison of space curves

We assume that the field k is algebraically closed. We will use the following notation:

1) If $\mathcal{F}$ is a coherent sheaf on $\mathbf{P}^n$, we set

$$H_*^i \mathcal{F} = \bigoplus_{k \in \mathbf{Z}} H^i \mathcal{F}(k)$$

for any $i \in \mathbf{N}$;

2) By a *split* sheaf we will mean a sheaf over $\mathbf{P}^n$ of the form

$$\mathcal{F} = \bigoplus_{i=1}^{r} \mathcal{O}_{\mathbf{P}^n}(-n_i),$$

where the numbers n_i are integers. If $\mathcal{F}$ is such a sheaf, then $H_*^i \mathcal{F} = 0$ for any i such that $1 \leqslant i \leqslant n-1$ (*cf.* Chapter VII, 4.1).

0 Introduction

In this section, a space curve means a curve in $\mathbf{P}^3$ (*cf.* below for more details). A space curve C is said to a be a scheme-theoretic complete intersection if its ideal $I(C)$ is generated by two generators (*cf.* Chapter III, 10.b). We saw (*cf.* Chapter VIII, 1.13) that space curves are not generally scheme-theoretic complete intersections. For example, the space curve C defined by the ideal $I = (XT-YZ, Y^2-XZ, Z^2-YT)$ (*cf.* Exercise II, 4) is not a scheme-theoretic complete intersection: the ideal I cannot be generated by only two generators. However, this curve is close to being a complete intersection in the following sense.

If we consider the two quadrics whose equations are $XT-YZ = Y^2-XZ$, for example, their intersection is the union of the cubic C and a line $D = V(X, Y)$. We then say that C and D are *linked* by the given quadrics. The curve C, whilst not a complete intersection, is linked to a complete intersection. In what follows, we will study this liaison operation and characterise the curves which, like C, are linked (possibly in several steps) to complete intersections. We will characterise such curves both in cohomological terms and in terms of resolutions.

1 Ideals and resolutions

In this paragraph R denotes the polynomial ring $k[X_0, \ldots, X_n]$.

a. Subschemes of $\mathbf{P}^n$

Even if we are mainly interested in varieties (particularly smooth curves), using liaison requires us to work with schemes. A smooth curve can be linked to a singular or even non-reduced curve (*cf.* Examples 2.7). We refer the reader to the appendix on schemes and the references its contains for more details of this concept. We recall simply that if I is a homogeneous ideal of R and $S = R/I$ is the quotient ring, then we define the closed subscheme $X = \text{Proj}(S)$ in $\mathbf{P}^n = \mathbf{P}_k^n$ to be the ringed space $(\underline{X}, \mathcal{O}_X)$ whose underlying topological space[1] $\underline{X}$ is the closed subspace $V(I)$ in $\mathbf{P}^n$ with its Zariski topology and whose sheaf of rings is given over a basis of standard open sets $D^+(f)$ of $\underline{X}$ by setting $\Gamma(D^+(f), \mathcal{O}_X) = S_{(f)}$ (*cf.* Chapter III, 8.1). If i denotes the inclusion of X in $\mathbf{P}^n$, then the sheaf $i_* \mathcal{O}_X$ is simply $\widetilde{S}$. If I is not a radical ideal, then these rings are not necessarily reduced, contrary to the rings of the variety $V(I)$ (for which, *cf.* Chapter III, 8.a, we take I to be the ideal $I(\underline{X})$ of all the polynomial functions which vanish on $\underline{X}$).

Let $\mathcal{J} = \widetilde{I}$ be the sheaf associated to the ideal I. This is a sheaf of ideals on $\mathcal{O}_{\mathbf{P}^n}$ and by definition of the sheaf of rings $\mathcal{O}_X$ we have the following exact sequence:
$$0 \longrightarrow \mathcal{J} \longrightarrow \mathcal{O}_{\mathbf{P}^n} \longrightarrow \mathcal{O}_X \longrightarrow 0,$$
so writing $\mathcal{J} = \mathcal{J}_X$ is compatible with Chapter III, 6.10. We say that $\mathcal{J}_X$ is the sheaf of ideals defining the scheme $X = \text{Proj}(R/I)$.

It follows from the above that the ideal I (which we say is *a* defining ideal for X) determines X entirely. Conversely, X entirely determines *the* sheaf $\mathcal{J}_X = \widetilde{I}$ but does not uniquely determine the ideal I (*cf.* Chapter III, 9.8.3). We will need more details of what's going on. We start with a definition.

Definition 1.1. *Let I be a homogeneous ideal of R. We set*
$$\text{sat}(I) = \{f \in R \mid \exists N \in \mathbf{N} \quad \forall i = 0, \ldots, n \quad X_i^N f \in I\}.$$

$\text{sat}(I)$ is then a homogeneous ideal containing I, which is called the saturation *of I. We say that I is* saturated *if it is equal to $\text{sat}(I)$.*

---

[1] In fact, by abuse of notation, we will often use the same notation for the scheme X and the underlying topological space $\underline{X}$.

Examples 1.2.
1) The saturation of the ideal (X^2, XY, XZ, XT) in $k[X, Y, Z, T]$ is the ideal (X).

2) If I is a radical homogeneous ideal of $k[X_0, \dots, X_n]$ (*i.e.*, equal to its radical), then $\mathrm{sat}(I) = I$, unless I is the irrelevant ideal $m = (X_0, \dots, X_n)$ in which case $\mathrm{sat}(I) = R$. This is clear if $I = R$ or m. Otherwise, consider $f \in \mathrm{sat}(I)$ homogeneous of degree > 0 (if f is a constant, then we are in the above case). There is an N such that, for all i, $X_i^N f \in I$. But as f is homogeneous of degree > 0, we can write it as a linear combination $f = \sum_{i=0}^n a_i X_i$ and using the $(n+1)$-omial formula we see that $f^k \in I$ for $k \geqslant nN + N - n$, and hence, as I is radical, $f \in I$.

We have the following proposition.

Proposition 1.3.
1) Let I be a homogeneous ideal of R and $J = \mathrm{sat}(I)$. Then $\mathrm{Proj}(R/I) \simeq \mathrm{Proj}(R/J)$.
2) Set $X = \mathrm{Proj}(R/I)$. We have

$$\mathrm{sat}(I) = H^0_*(\mathcal{J}_X) = \Gamma_*(\mathcal{J}_X) = \bigoplus_{d \in \mathbf{N}} \Gamma(\mathbf{P}^n, \mathcal{J}_X(d)).$$

This ideal depends only on X (and not on the choice of defining ideal I of X). We call it the saturated ideal of X and we denote it by I_X. It is the largest ideal which defines X.

Proof.
1) We have already proved the set-theoretic equality $V(I) = V(J)$. As $I \subset J$, we know that $V(J) \subset V(I)$. Conversely, if $P = (x_0, \dots, x_n) \in V(I)$, then one of the coordinates of P is non-zero: for example $x_0 \neq 0$. Consider $f \in V(J)$. There is an $N \in \mathbf{N}$ such that $X_0^N f \in I$, and hence $x_0^N f(P) = 0$, so $f(P) = 0$ and $P \in V(J)$.

To prove equality of sheaves we consider the canonical surjection $R/I \to R/J$ given by $\overline{a} \mapsto \widehat{a}$. This induces a surjection of local rings $(R/I)_{(f)} \to (R/J)_{(f)}$ which associates $\widehat{a}/f^r$ to $\overline{a}/f^r$ for homogeneous a, f with $\deg f > 0$ and $\deg a = r \deg f$. It will be enough to show that this map is also injective. But if $\widehat{a}/f^r = 0$, then $f^m a$ is in J. There is therefore an integer N such that $X_0^N f^m a, \dots, X_n^N f^m a \in I$ and for k large enough (*cf.* Example 1.2.2) $f^k a \in I$, so $\overline{a}/f^r = 0$ is in $(R/I)_{(f)}$.

2) We consider the natural homogeneous degree 0 map $I \to \Gamma_*(\mathcal{J}_X)$ (*cf.* Chapter III, 9.8.3). This map is injective because it is induced by the map from R to $\Gamma_*(\mathcal{O}_{\mathbf{P}^n})$ which was proved in Chapter III, 9.9 to be an isomorphism. It remains to show that $\Gamma_*(\mathcal{J}_X)$ is indeed the saturation of I. Consider $f \in \Gamma(\mathbf{P}^n, \mathcal{J}_X(d))$. The restriction of f to $D^+(X_0)$ is in $I_{(X_0)}$, so $f = f_0/X_0^N$ with $f_0 \in I$. We therefore have $X_0^N f \in I$, and the same holds for the other variables, so f is in the saturation of I.

Remark 1.4. In what follows we will systematically use the saturated ideal I_X as defining ideal of X. We note that this ideal is determined by the sheaf $\mathcal{J}_X$, which therefore also determines X. For varieties, the ideal $I(X)$ of polynomial functions which vanish on X is automatically saturated and is hence equal to I_X (*cf.* 1.2.2). In the non-reduced case, on the other hand, we generally have $I_V \subset I(V)$ but $I_V \neq I(V)$.

b. Resolutions

Notations 1.5. Let E and F be graded free R-modules

$$E = \bigoplus_{j=1}^{s} R(-m_j), \qquad F = \bigoplus_{i=1}^{r} R(-n_i),$$

where the numbers m_j and the n_i are integers such that $n_1 \leqslant \cdots \leqslant n_r$ and $m_1 \leqslant \cdots \leqslant m_s$. The degree of E (resp. F) is defined to be the integer $\sum_j -m_j$ (resp. $\sum_i -n_i$). Let $u : E \to F$ be a graded R-linear homomorphism of degree zero (*i.e.*, sending an element of degree n to an element of degree n). We denote by e_j $(j = 1, \ldots, s)$ and ε_i $(i = 1, \ldots, r)$ respectively the vectors of the canonical bases of E and F. The homomorphism u is given in these bases by a matrix $A = (a_{ij})$ of size $r \times s$ whose coefficient of index i, j is a homogeneous polynomial of degree $m_j - n_i$. In particular, this polynomial is zero if $m_j < n_i$ and is a constant (*i.e.*, an element of k) if $m_j = n_i$.

We have seen (*cf.* Exercise III, B.2) that every graded module of finite type M over a graded Noetherian ring S has a *free graded resolution* (we will say a resolution for short):

$$\cdots \longrightarrow L_d \longrightarrow L_{d-1} \longrightarrow \cdots \longrightarrow L_0 \longrightarrow M \longrightarrow 0,$$

where the modules L_i are free graded modules, that is, modules of the form $\bigoplus_{i=1}^{r} S(n_i)$, and the maps are homogeneous of degree 0.

In general this resolution is not finite, but if R is a polynomial algebra, every graded module of finite type has a *finite* resolution. More precisely:

Theorem-Definition 1.6 (Hilbert's syzygies). *If $R = k[X_0, \ldots, X_n]$ and M is a graded R-module of finite type, then M has a resolution of the form*

$$0 \longrightarrow L_d \longrightarrow L_{d-1} \longrightarrow \cdots \longrightarrow L_0 \longrightarrow M \longrightarrow 0,$$

with $d \leqslant n + 1$. If L_d is non-zero, the resolution is said to be of length d. *The smallest integer d such that M has a resolution of length d is called the projective dimension of M and is written $\mathrm{dp}(M)$. If $M = 0$, then $\mathrm{dp}(M) = -1$ by convention.*

Proof. See Problem XI or [Pes] § 13.

Definition 1.7. *With the notations of 1.5 a homomorphism $u : E \to F$ is said to be* minimal *if its constant coefficients are zero. A resolution is said to be* minimal *if all the homomorphisms which compose it are minimal.*

Proposition 1.8. *Every graded R-module of finite type has a finite minimal resolution. Moreover, this resolution is unique up to isomorphism, its length is equal to the projective dimension of M and the rank of L_0 is the minimal number of generators of M.*

Proof. Here we will prove the existence of the minimal resolution: for its uniqueness and the other claims of 1.8, see Problem IX. Let M be a graded R-module of finite type. We know that it has a finite resolution. If this resolution is not minimal, then one of the maps of the resolution has a constant coefficient $\lambda \in k^*$ in its matrix. After change of notation, let $u : E \to F$ be the map in question. We will prove that we can replace the free modules E and F by modules E' and F', which are free of rank one less than E and F respectively, and u by u', without the coefficient λ, such that the kernel and cokernel of u and u' are the same. This means that we can replace u by u' in the resolution without changing the other terms and the theorem follows easily by induction on the number of constant coefficients.

The situation is therefore as follows: we have a homomorphism $u : E = R(-a) \oplus E' \to F = R(-a) \oplus F'$ whose block matrix is of the form

$$\begin{pmatrix} \lambda & v \\ w & u' \end{pmatrix},$$

with $\lambda \in k^*$. Let e_1 be the first vector of the basis of E. We note that after performing a change of basis $\varepsilon_1' = u(e_1)$, $\varepsilon_i' = \varepsilon_i$ for $i > 1$ in F we can assume $w = 0$ and $\lambda = 1$. We then easily check that we have the following commutative diagram:

$$
\begin{array}{ccccccc}
\operatorname{Ker} u' & \longrightarrow & E' & \xrightarrow{u'} & F' & \longrightarrow & \operatorname{Coker} u' \\
\downarrow{\scriptstyle i_0} & & \downarrow{\scriptstyle i} & & \downarrow{\scriptstyle j} & & \downarrow{\scriptstyle \varphi} \\
\operatorname{Ker} u & \longrightarrow & R(-a) \oplus E' & \xrightarrow{u} & R(-a) \oplus F' & \longrightarrow & \operatorname{Coker} u
\end{array}
$$

in which we have set $i(y) = (-v(y), y)$ and $j(z) = (0, z)$. The maps i_0 and φ are induced by i and j respectively, and it is easy to see that these are indeed isomorphisms. We say we have "simplified" the resolution by $R(-a)$.

Corollary 1.9. *Let I be a homogeneous ideal of $k[X_0, \ldots, X_n]$.*
  1) The ideal I is of projective dimension $\leqslant n$.
  2) The ideal I is saturated if and only if it is of projective dimension $\leqslant n - 1$.
  3) Set $d = \mathrm{dp}(I)$. Let Z be the subscheme defined by I. If $d = 0$, Z is empty or a hypersurface, if $d \geqslant 1$, then $\dim(Z) \geqslant n - d - 1$.

Proof.

1) The case $I = R$ is trivial. Let $0 \to L_d \xrightarrow{u_d} L_{d-1} \to \cdots \to L_0 \to I \to 0$ be the minimal resolution of $I \neq R$. The exact sequence $0 \to I \to R \to R/I \to 0$ then gives a minimal resolution of R/I:

$$0 \longrightarrow L_d \xrightarrow{u_d} L_{d-1} \longrightarrow \cdots \longrightarrow L_0 \longrightarrow R \longrightarrow R/I \longrightarrow 0.$$

As R/I has projective dimension $\leqslant n+1$ (*cf.* 1.6), it follows that I has projective dimension $\leqslant n$.

2) Let $\mathcal{J} = \widetilde{I}$ be the sheaf associated to I and consider the exact sequence of sheaves obtained from the above minimal resolution of I:

$$0 \longrightarrow \mathcal{L}_d \xrightarrow{\widetilde{u}_d} \mathcal{L}_{d-1} \longrightarrow \cdots \xrightarrow{\widetilde{u}_1} \mathcal{L}_0 \longrightarrow \mathcal{J} \longrightarrow 0.$$

We introduce the modules E_i which are defined to be the cokernels of the morphisms u_i (or the kernels of the morphisms u_{i-1} if you prefer) and the associated sheaves $\mathcal{E}_i$. We therefore have exact sequences:

(1) $0 \longrightarrow \mathcal{E}_1 \longrightarrow \mathcal{L}_0 \longrightarrow \mathcal{J} \longrightarrow 0,$

(2) $0 \longrightarrow \mathcal{E}_{i+1} \longrightarrow \mathcal{L}_i \longrightarrow \mathcal{E}_i \longrightarrow 0,$ for $1 \leqslant i \leqslant d-2,$

(3) $0 \longrightarrow \mathcal{L}_d \longrightarrow \mathcal{L}_{d-1} \longrightarrow \mathcal{E}_{d-1} \longrightarrow 0.$

Consider the natural map $j : I \to H^0_* \mathcal{J}$. We saw in the proof of 1.3 that this map is injective and the hypothesis that I is saturated means that it is also surjective. We consider the following diagram:

$$
\begin{array}{ccccccccc}
0 & \longrightarrow & E_1 & \longrightarrow & L_0 & \longrightarrow & I & \longrightarrow & 0 \\
& & \downarrow & & \downarrow{\scriptstyle\varphi} & & \downarrow{\scriptstyle j} & & \\
0 & \longrightarrow & H^0_* \mathcal{E}_1 & \longrightarrow & H^0_* \mathcal{L}_0 & \xrightarrow{\ g\ } & H^0_* \mathcal{J} & \longrightarrow & H^1_* \mathcal{E}_1 \longrightarrow 0
\end{array}
$$

obtained from the resolution of I, the long exact sequence associated to the exact sequence (1) and Chapter III, 9.8. As φ is an isomorphism (*cf.* Chapter III, 9.9), we see that j is surjective if and only if g is, that is to say, if and only if $H^1_* \mathcal{E}_1$ vanishes.[2] By induction using the long exact sequence associated to the exact sequence (2) we also obtain isomorphisms $H^1_* \mathcal{E}_1 \simeq H^i_* \mathcal{E}_i = 0$ for $1 \leqslant i \leqslant d-1$. And finally, I is saturated if and only if $H^{d-1}_* \mathcal{E}_{d-1}$ vanishes, or, alternatively, if the map $H^d_* \mathcal{L}_d \to H^d_* \mathcal{L}_{d-1}$ arising from the exact sequence (3) is injective. If I is of projective dimension $d \leqslant n-1$, then $d-1 \leqslant n-2$ and the space H^d in question vanish because the modules L_i are free, so the sheaves $\mathcal{L}_i$ are split, and the result follows by Chapter VII, 4.1. Conversely, let us assume that I is saturated and argue by contradiction, assuming that $d = n$ and $L_n \neq 0$. We are done by the following lemma.

---

[2] The careful reader will doubtless have noticed that the above only really applies for $n \geqslant 2$. Indeed, for $n = 1$ the term $H^1_* \mathcal{L}_0$ has no reason to vanish. However, the proof of the case $n = 1$ is very similar and we leave it as an exercise for the reader, as a reward for his or her astuteness.

Lemma 1.10. *Let $u : E \to F$ be a minimal graded homomorphism between two free modules, let $\tilde{u} : \mathcal{E} \to \mathcal{F}$ be the map induced by u on the associated sheaves and let $h(u) : H^n_* \mathcal{E} \to H^n_* \mathcal{F}$ be the map induced by $\tilde{u}$ on cohomology groups. If $h(u)$ is injective, then E is trivial.*

Proof (of 1.10). Assume $E \neq 0$. We use the notations of 1.5. We consider the restriction v of u to $R(-m_j)$. This can be written as $v = v_1 + \cdots + v_r$, with v_i taking values in $R(-n_i)$. The elements v_i are therefore homogeneous polynomials of degree $m_j - n_i$. It will be enough to show that $h(v)$ is non-injective or, alternatively, that its homogeneous part $h(v, m_j - n - 1) = \oplus h(v_i, m_j - n - 1)$

$$h(v, m_j - n - 1) : H^n \mathcal{O}_{\mathbf{P}^n}(-n-1) \longrightarrow \bigoplus_{i=1}^{r} H^n \mathcal{O}_{\mathbf{P}^n}(-n_i + m_j - n - 1)$$

is not injective. As the group $H^n \mathcal{O}_{\mathbf{P}^n}(-n-1)$ does not vanish, this will follow if we can show that $h(v, m_j - n - 1) = 0$. But as u is minimal, v_i is trivial if $n_i \geqslant m_j$, and hence $h(v_i, m_j - n - 1) = 0$ in this case. On the other hand, if $n_i < m_j$ the group $H^n \mathcal{O}_{\mathbf{P}^n}(-n_i + m_j - n - 1)$ vanishes, and, once again, $h(v_i, m_j - n - 1) = 0$. QED.

3) It remains to prove the final claim of 1.9. We use the above notation. If $d = 0$, I is isomorphic to L_0 and is hence a free R-module which is only possible for an ideal if it is principal (since two elements f, g in I are always linked by the relation $fg - gf = 0$). We therefore have $I = (f)$, where f is homogeneous of degree s and $Z = V(I)$ is empty if $s = 0$ and is a hypersurface if $s > 0$. Assume now that $d \geqslant 1$. After replacing I by its saturation we can assume $I = I_Z$. Lemma 1.10 shows that $H^n_* \mathcal{E}_{d-1} \neq 0$, and it follows recursively that $H^{n-i}_* \mathcal{E}_{d-i} \neq 0$ for $i = 1, \ldots, d-1$, so $H^{n-d}_* \mathcal{J}_Z \neq 0$, and finally $H^{n-d-1}_* \mathcal{O}_Z \neq 0$. It follows that Z is of dimension at least $n - d - 1$ by Chapter VII, 3.3.

2 ACM curves

Henceforth we will work in $\mathbf{P}^3$. We set $R = k[X, Y, Z, T]$.

Definition 2.1. *By a* space curve *we mean a closed subscheme of $\mathbf{P}^3$ of dimension 1.*[3]

We saw in 1.9 that the saturated ideal of a curve C has projective dimension $\leqslant 2$. We now study curves whose ideal is of projective dimension 1. Of course, there are many others (*cf.* Exercise II, 5 or 3.d below).

---

[3] This definition is not reasonable in general. There are good reasons, some of which will appear below (*cf.* 3.6), for imposing extra conditions on space curves, notably that the curves should be equidimensional and have no embedded components. However, for the ACM curves we will be dealing with these conditions are automatically satisfied.

Proposition-Definition 2.2. *Let C be a space curve, I_C its saturated ideal and $\mathcal{J}_C = \tilde{I}_C$ the sheaf of ideals defining C. The following conditions are equivalent:*

  i) $H^1 \mathcal{J}_C(n) = 0$ *for all $n \in \mathbf{Z}$.*
  ii) The ideal I_C has projective dimension 1.
  We then say that C is arithmetically Cohen-Macaulay *(ACM for short).*

Proof. Assume first that I_C has a free resolution of length 1: $0 \to E \to F \to I_C \to 0$. There is an exact sequence of sheaves $0 \to \mathcal{E} \to \mathcal{F} \to \mathcal{J}_C \to 0$, and passing to cohomology we deduce the existence of an exact sequence $\cdots \to H^1_* \mathcal{F} \to H^1_* \mathcal{J}_C \to H^2_* \mathcal{E} \to \cdots$. But as the sheaves $\mathcal{E}$ and $\mathcal{F}$ are split (see the notations at the start of this chapter), their cohomology groups H^1 and H^2 vanish, so the same holds for the cohomology groups $H^1 \mathcal{J}_C(n)$.

Conversely, we note that $\mathrm{dp}(I_C) \geqslant 1$ by 1.9.3.

Moreover, we have seen (*cf.* 1.9) that I_C has a minimal resolution of length at most 2: $0 \to L_2 \to L_1 \to L_0 \to I \to 0$. We pass to sheaves and introduce the cokernel $0 \to \mathcal{L}_2 \to \mathcal{L}_1 \to \mathcal{E} \to 0$. We then have $H^1_* \mathcal{J}_C = H^2_* \mathcal{E} = 0$ and we are done by Lemma 1.10.

The simplest ACM curves are the scheme-theoretic complete intersections (*cf.* Chapter III, 10.2).

Definition 2.3. *A curve C in $\mathbf{P}^3$ is said to be a* scheme-theoretic complete intersection *if its saturated ideal I_C is generated by two generators (which necessarily have no common factors).*

It follows from Chapter III, 10.1 that such a curve is ACM. The Peskine-Szpiro theorem, which we will see below, studies the converse of this statement.

The following proposition gathers together some important properties of ACM curves.

Proposition 2.4. *Let C be an ACM curve, d its degree, g its arithmetic genus and $0 \to E \xrightarrow{u} F \xrightarrow{p} I_C \to 0$ the minimal resolution of its saturated ideal.[4] With the notations of 1.5 the following hold.*

  0) C is connected (and hence has no isolated points).
  1) We have $s = r - 1$ and $\sum_{j=1}^{r-1} m_j = \sum_{i=1}^{r} n_i$ (i.e., $\deg E = \deg F$).
  2) We have $2d = \sum_{j=1}^{r-1} m_j^2 - \sum_{i=1}^{r} n_i^2$.
  3) We have

$$g = \sum_{j=1}^{r-1} \binom{m_j - 1}{3} - \sum_{i=1}^{r} \binom{n_i - 1}{3}$$

$$= 1 - 2d + \frac{1}{6}\left(\sum_{j=1}^{r-1} m_j^3 - \sum_{i=1}^{r} n_i^3\right).$$

---

[4] We will sometimes write a resolution (resp. a minimal resolution) of the curve C to mean a resolution (resp. a minimal resolution) of its saturated ideal I_C.

4) We have $n_1 < m_1$, and n_1 is the smallest degree of a surface containing C:

$$n_1 = s_0 = \inf\{n \in \mathbf{N} \mid h^0 \mathcal{J}_C(n) > 0\}.$$

In particular, $n_1 > 0$.

5) We have $n_r < m_{r-1}$ and $m_{r-1} = e+4$, where e is the index of speciality of C:

$$e = \sup\{n \in \mathbf{Z} \mid h^1 \mathcal{O}_C(n) > 0\}.$$

Proof. 0) follows from Chapter VIII, 1.6.5. We recall (*cf.* Chapter VII, 4.1) that the Euler characteristic of the sheaf $\mathcal{O}_{\mathbf{P}}(n)$ is $\chi\mathcal{O}_{\mathbf{P}}(n) = B(n)$, where B is the polynomial given by

$$B(n) = \frac{(n+3)(n+2)(n+1)}{6} = \begin{cases} \binom{n+3}{3} & \text{if } n \geqslant 0, \\ -\binom{-n-1}{3} & \text{if } n < 0. \end{cases}$$

1) 2) and 3) then follow by calculating the Euler characteristic $\chi\mathcal{J}_C(n)$ in two different ways, firstly using Riemann-Roch, which gives us

$$\chi\mathcal{J}_C(n) = \chi\mathcal{O}_{\mathbf{P}}(n) - \chi\mathcal{O}_C(n) = B(n) - nd - 1 + g,$$

for all $n \in \mathbf{Z}$, and secondly using the exact sequence of sheaves associated to the resolution of I_C

$$0 \longrightarrow \mathcal{E} \longrightarrow \mathcal{F} \longrightarrow \mathcal{J}_C \longrightarrow 0,$$

which gives us

$$\chi\mathcal{J}_C(n) = \chi\mathcal{F}(n) - \chi\mathcal{E}(n) = \sum_{i=1}^{r} B(n - n_i) - \sum_{j=1}^{s} B(n - m_j).$$

It is then enough to identify the terms of degrees $3, 2, 1, 0$ to obtain the four formulae we seek.

To prove 4), assume $m_1 \leqslant n_1$ and hence $m_1 \leqslant n_i$ for all i. The coefficient a_{1i} of the matrix u then vanishes for all i (it is of degree $\leqslant 0$ and the resolution is minimal). But this contradicts the injectivity of u. It is easy to check that $s_0 = n_1$.

The proof of 5) is a little more subtle. If $m_{r-1} \leqslant n_r$, then $m_j \leqslant n_r$ for all j and hence $a_{rj} = 0$ for all j. The canonical basis for F being denoted by $\varepsilon_1, \ldots, \varepsilon_r$, this means that the image of u is contained in the submodule generated by $\varepsilon_1, \ldots, \varepsilon_{r-1}$. Let f be the image of ε_r in I_C under p and let $g \in I_C$ be arbitrary. We will show that g is contained in the ideal (f) and hence $I_C = (f)$, which gives a contradiction (C would then be a surface).

We have $g = p\left(\sum_{i=1}^{r} g_i\varepsilon_i\right)$ with $g_i \in R$. We write the obvious relationship $fg - gf = 0$ in the form $fp\left(\sum_{i=1}^{r} g_i\varepsilon_i\right) - gp(\varepsilon_r) = 0$ or, alternatively, $p\left(f\sum_{i=1}^{r} g_i\varepsilon_i - g\varepsilon_r\right) = 0$. As the kernel of p is simply the image of u and this latter is contained in the submodule generated by $\varepsilon_1, \ldots, \varepsilon_{r-1}$, the coefficient of ε_r in this sum is zero, so $g = fg_r \in (f)$. QED.

To determine the speciality index e we note first that $h^1 \mathcal{O}_C(n) = h^2 \mathcal{J}_C(n)$ (this follows from the long exact sequence associated to the exact sequence $0 \to \mathcal{J}_C \to \mathcal{O}_{\mathbf{P}} \to \mathcal{O}_C \to 0$). Using the resolution, we have

$$0 \longrightarrow H^2 \mathcal{J}_C(n) \longrightarrow H^3 \mathcal{E}(n) \longrightarrow H^3 \mathcal{F}(n) \longrightarrow \cdots$$

If $n > m_{r-1} - 4$, then $H^3 \mathcal{E}(n) = 0$ and hence $H^2 \mathcal{J}_C(n) = 0$, so $e \leqslant m_{r-1} - 4$. If $n = m_{r-1} - 4$, then $H^3 \mathcal{E}(n) \neq 0$ and $H^3 \mathcal{F}(n) = 0$, so $H^3 \mathcal{J}_C(n) \neq 0$ and e is therefore equal to $m_{r-1} - 4$.

Example 2.5. The curve C whose minimal resolution is given by

$$0 \longrightarrow R(-4)^3 \longrightarrow R(-3)^4 \longrightarrow I_C \longrightarrow 0$$

is of degree 6, genus 3 and satisfies $s_0 = 3$ and $e = 0$.

Resolutions and minors: two lemmas. The following two lemmas will be particularly useful for dealing with ACM curves. Their proof is essentially an application of the expansion of a determinant along a row, which here reveals its full power.

Reminders and notation 2.6. We use the same notation as in 1.5, with $R = k[X, Y, Z, T]$. Let E and F be graded free R-modules,

$$E = \bigoplus_{j=1}^{s} R(-m_j), \qquad F = \bigoplus_{i=1}^{r} R(-n_i).$$

We assume $r \geqslant 2$, $s = r - 1$ and $\deg E = -\sum_{j=1}^{r-1} m_j = \deg F = -\sum_{i=1}^{r} n_i$ (*cf.* 2.4). Let $u : E \to F$ be a graded R-linear homomorphism of degree zero and let $A = (a_{ij})$ be its $(r \times (r-1))$ matrix in the canonical basis. For $i = 1, \ldots, r$ we let A_i be the matrix obtained from A by removing the line of index i. The $(r-1)$-minors of A are, up to sign, the determinants of the matrices A_i: more precisely the ith minor, φ_i, is given by

$$\varphi_i = (-1)^{i+1} \det A_i.$$

It is easy to check that φ_i is a homogeneous polynomial of degree n_i. If B is the matrix obtained by adding a column of the form $(a_1, \ldots, a_r)$ to the left of A, then

$$\det B = \sum_{i=1}^{r} a_i \varphi_i.$$

Let $i_1, i_2 \in [1, r]$ be two distinct integers and let j be an integer satisfying the inequalities $1 \leqslant j \leqslant r - 1$. We denote by $A_{i_1, i_2; j}$ the matrix obtained on removing from A the lines of index i_1 and i_2 and the column of index j. The corresponding $r - 2$nd minor is then defined by

$$\Delta_{i_1, i_2; j} = \begin{cases} (-1)^{i_1 + i_2 + j} \det A_{i_1, i_2; j} & \text{if } i_1 < i_2, \\ (-1)^{i_1 + i_2 + j + 1} \det A_{i_1, i_2; j} & \text{if } i_1 > i_2. \end{cases}$$

We extend this definition to the case $i_1 = i_2$ by setting $\Delta_{i,i;j} = 0$. The formula for expanding a determinant along column applied to A_i gives us

(1)
$$\sum_{k\neq i} a_{kj}\Delta_{i,k;l} = \sum_{k=1}^{r} a_{kj}\Delta_{i,k;l} = \delta_{jl}\varphi_i$$

whenever $1 \leqslant i \leqslant r$ and $1 \leqslant j,l \leqslant r-1$. Here δ_{ij} denotes the Kronecker delta symbol. We can interpret this formula in matrix terms by defining Δ_i to be the $(r-1) \times r$ matrix whose term of index (l,k) is $\Delta_{i,k;l}$. We then have $\Delta_i A = \varphi_i I_{r-1}$, where I_{r-1} is the identity matrix.

On the other hand, the formula for expanding a determinant along a row applied to A_{i_1} and A_{i_2} gives us

(2)
$$\sum_{j=1}^{r-1} a_{kj}\Delta_{i_1,i_2;j} = \delta_{k,i_2}\varphi_{i_1} - \delta_{k,i_1}\varphi_{i_2}$$

for all $i_1, i_2, k \in [1,r]$. In matrix terms, if Δ is the matrix $r^2 \times (r-1)$ with general term $\Delta_{i_1,i_2;j}$, then this equation means that the product matrix $\Delta^t A$ of size $r^2 \times r$ has the general term $\delta_{k,i_2}\varphi_{i_1} - \delta_{k,i_1}\varphi_{i_2}$.

Lemma 2.7. *With the notation of 2.6 the following hold.*

1) u is injective if and only if all the $(r-1)$-minors $\varphi_1, \ldots, \varphi_r$ of A are not all zero.

2) If the minors φ_i are not all zero, then the following are equivalent:

i) The polynomials φ_i have no common factor.

ii) Let J be the ideal generated by the polynomials φ_i. We have an exact sequence

$$0 \longrightarrow E \overset{u}{\longrightarrow} F \overset{\varphi}{\longrightarrow} J \longrightarrow 0,$$

where $\varphi = (\varphi_1, \ldots, \varphi_r)$. In particular, $\operatorname{Coker} u = J$.

iii) $\operatorname{Coker} u$ is the saturated ideal of an ACM curve or is equal to R.

iv) $\operatorname{Coker} u$ is a torsion free R-module (i.e., if $ax = 0$ with $a \in R$ and $x \in \operatorname{Coker} u$, then a or x is zero).

Proof.

1) After passage to the fraction field $k(X,Y,Z,T)$ we reduce to the case of vector spaces, which is well known.

2) Let us prove that $\varphi u = 0$. We have to show that, for all j, $\sum_{i=1}^{r} a_{ij}\varphi_i = 0$. We introduce a matrix B obtained by adding to A on the left a column of entries a_{ij}, $i = 1, \ldots, r$. We then have $\det B = \sum_{i=1}^{r} a_{ij}\varphi_i$, but as this matrix has two identical columns, its determinant is 0.

i) $\Rightarrow$ ii): we have to show that $\operatorname{Ker}\varphi \subset \operatorname{Im} u$. Consider $a = (a_1, \ldots, a_r) \in \operatorname{Ker}\varphi$. We then have $\sum_{i=1}^{r} a_i\varphi_i = 0$. Let B be the square matrix $r \times r$ obtained by adding to A the column of elements a_i. The above relation proves that $\det B = 0$ and hence B is not injective. Let $(\lambda_0, \lambda_1, \ldots, \lambda_{r-1})$ be a non-zero vector in $\mathrm{Ker}\,B$ and let λ be the column vector ${}^t(\lambda_1, \ldots, \lambda_{r-1})$. We have a matrix equality $A\lambda = -\lambda_0 a$. If we can show that λ_0 divides the other components λ_j, then we are done since the vector $(a_1, \ldots, a_r)$ is then in the image of u. Multiplying the above equality on the left by the matrix Δ_i and applying Formula (1) of 2.6 we get the following equation

$$\Delta_i A \lambda = \varphi_i \lambda = -\lambda_0 \Delta_i a.$$

We see that λ_0 divides all the products $\varphi_i \lambda_j$, and hence divides their gcd. As the polynomials φ_i have no common factor their gcd is 1, and hence the gcd of the products $\varphi_i \lambda_j$ is λ_j, and we are done.

(Remember that the coefficients are contained in the polynomial ring R which is factorial. In particular, the gcd is well defined.)

ii $\Rightarrow$ i): let g be the gcd of the polynomials φ_i: g is a homogeneous polynomial and we denote its degree by d. Our aim is to show that d is zero. Set $\varphi_i = g\varphi_i'$: let J' be the ideal generated by the polynomials φ_i' and let φ' be the degree zero homogeneous homomorphism $\varphi' : F(d) = \bigoplus_{i=1}^r R(-n_i + d) \to J'$ given by the formula $\varphi'(a_1, \ldots, a_r) = \sum_{i=1}^r a_i \varphi_i'$. As R is an integral domain, it is clear that the kernel of φ' is the same as that of φ, so we have the following exact sequence:

$$0 \longrightarrow E(d) \overset{u}{\longrightarrow} F(d) \longrightarrow J' \longrightarrow 0.$$

Let C be the scheme defined by J'. As J' is saturated (cf. 1.9) $J' = I(C)$. As the polynomials f_i' have no common factor and $r \geqslant 2$, $\dim C \leqslant 1$. By 1.9.3 the scheme C is either an ACM curve with saturated ideal J' or empty. If C is a curve, we apply 2.4.1 to the degrees of $E(d)$ and $F(d)$ and we get $\sum_{j=1}^{r-1}(m_j + d) = \sum_{i=1}^r (n_i + d)$, which, since $\sum_{j=1}^{r-1} m_j = \sum_{i=1}^r n_i$, implies $d = 0$. If C is empty, then $I_C = J' = R$, and an Euler characteristic calculation similar to the one given in the proof of 2.4.1 shows that d vanishes.

We note that the above proof also shows that ii) $\Rightarrow$ iii). The fact that iii) $\Rightarrow$ iv) is obvious, and it remains to show that iv) $\Rightarrow$ ii). The fact that $\varphi u = 0$ implies there is a surjective map $\pi : \mathrm{Coker}\,u \to J$ such that the following diagram commutes:

$$
\begin{array}{ccc}
F & \overset{p}{\longrightarrow} \mathrm{Coker}\,u \longrightarrow 0 \\
\| & \quad \downarrow \pi \\
F & \overset{\varphi}{\longrightarrow} \; J \; \longrightarrow 0
\end{array}
$$

It remains to show that π is injective. Consider $\bar{a} \in \mathrm{Ker}\,\pi$ such that $\bar{a} = p(a)$, $a \in F$. We therefore have $\sum_{i=1}^r a_i \varphi_i = 0$, and, as in the proof of i) $\Rightarrow$ ii), it follows that there exists a $\lambda_0 \in R$ and $\lambda \in E$ which are not both zero such that $\lambda_0 a + u(\lambda) = 0$, so $\lambda_0 a \in \mathrm{Im}\,u$ and $\lambda_0 \bar{a} = 0$ in $\mathrm{Coker}\,u$. As λ_0 does not vanish (because u is injective) and $\mathrm{Coker}\,u$ is torsion free $\bar{a} = 0$.

Remark 2.8. In the above lemma, the case $J = R$ (corresponding to the empty set) arises if and only if one of the minors φ_i is a non-zero constant. Of course, this can only happen if u is not minimal (otherwise the entries a_{ij} and hence also the polynomials φ_i are of degree > 0). This also implies that the corresponding integer n_i is zero, and on setting

$$F' = \bigoplus_{k \neq i} R(-n_k),$$

u induces an isomorphism from E to F'. In particular, the integers m_j and the integers n_k for $k \neq i$ are the same.

Our second lemma will play a key role in calculating resolutions by liaison. We start by recalling some results on the duality of graded modules.

We recall (*cf.* Chapter II, 7.5) that if $E = \bigoplus_{n \in \mathbf{Z}} E_n$ and $F = \bigoplus_{n \in \mathbf{Z}} F_n$ are two graded R-modules, then a homomorphism $u : E \to F$ is said to be homogeneous of degree d if $u(E_n) \subset F_{n+d}$. We say that u is graded if u is a finite sum of homogeneous homomorphisms. When E is an R-module of finite type (which we will assume henceforth is the case), every homomorphism is automatically graded. This provides a natural grading on the R-module $\mathrm{Hom}_R(E, F)$.

If E is a graded R-module, then we denote by $E^\vee$ the dual R-module to E, *i.e.*, the graded R-module $\mathrm{Hom}_R(E, R)$ of graded homomorphisms from E to R. Let E be a graded free R-module with basis $e_1, \ldots, e_s$, with e_j of degree m_j. We then have $E = \bigoplus_{j=1}^s R(-m_j)$. The module $E^\vee$ is also free, and more precisely we obtain a basis of $E^\vee$ by taking the dual basis $e_1^*, \ldots, e_s^*$ defined by $e_j^*(e_i) = \delta_{ij}$, where δ_{ij} is the Kronecker delta: e_j^* is then of degree $-m_j$, so that $E^\vee = \bigoplus_{j=1}^s R(m_j)$.

The duality operation is functorial and contravariant: given a homomorphism $u : E \to F$ which is homogeneous of degree 0 we deduce the existence of a homomorphism $^t u : E^\vee \to F^\vee$ called the transposition of u and defined by the formula $^t u(f) = fu$. This homomorphism is homogeneous of degree 0 and we have $^t(uv) = {}^t v \, {}^t u$. If E and F are free with bases (e_j) and (ε_i) and u is given with respect to this basis by a matrix A, then $^t u$ is given by the matrix $^t A$ with respect to the dual bases.

Lemma 2.9. *We use the notation of 2.6. We assume that the minors φ_i are not all zero and have no common factors. Let J be the ideal generated by the minors φ_i. Let $^t u$ (resp. $^t\varphi$) be the transposition of u (resp. φ). (Here, we consider φ as a homomorphism from F to R.) There is an exact sequence*

$$0 \longrightarrow R \xrightarrow{\,^t\varphi\,} F^\vee \xrightarrow{\,^t u\,} E^\vee \longrightarrow \mathrm{Coker}\,{}^t u \longrightarrow 0,$$

and the annihilator of the R-module $\mathrm{Coker}\,{}^t u$ is equal to J.

Proof. We prove first that the sequence is exact. It is clear that ${}^t\varphi$ is injective and ${}^tu{}^t\varphi = 0$. (This follows by functoriality of duality from the equation $\varphi u = 0$ seen in 2.7.)

It remains to show that the sequence is exact at $F^\vee$. Set $x = \sum_{i=1}^r x_i \varepsilon_i^* \in \mathrm{Ker}\,{}^tu$, which we represent by a column matrix $x = {}^t(x_1, \ldots, x_r)$. We have a matrix equality ${}^tAx = 0$ and, multiplying on the left by the matrix Δ (*cf.* 2.6), we see that for every pair (i_1, i_2)

$$(*) \qquad\qquad x_{i_2}\varphi_{i_1} - x_{i_1}\varphi_{i_2} = 0.$$

We fix an index $i_1 \in [1, r]$. If φ_{i_1} is zero, then so is x_{i_1}. Otherwise, as we are in the (factorial) ring R we can decompose φ_{i_1} as a product of irreducible elements, $\varphi_{i_1} = \prod p_m^{\alpha_m}$, where the elements p_m are prime and distinct. As the elements φ_i have no common factor, for any given m there is an φ_{i_2} such that p_m does not divide φ_{i_2}. But then $(*)$ shows that $p_m^{\alpha_m}$ divides x_{i_1}, and applying this for every m it follows that φ_{i_1} divides x_{i_1}, and hence $x \in \mathrm{Im}\,{}^t\varphi$, QED.[5]

Let us now calculate the annihilator $\mathrm{Ann}(\mathrm{Coker}\,{}^tu)$. Transposing (1) from 2.6 (that is, $\Delta_i A = \varphi_i I_{r-1}$), we get ${}^tA\,{}^t\Delta_i = \varphi_i I_{r-1}$. We can interpret ${}^t\Delta_i$ as being a degree 0 homomorphism from $E^\vee(-n_i)$ to $F^\vee$ and the above formula shows that the vectors $\varphi_i e_j^*$ in $E^\vee$ are in the image of ${}^tu$, and hence vanish in $\mathrm{Coker}\,{}^tu$, so φ_i annihilates $\mathrm{Coker}\,{}^tu$ and J is contained in the annihilator of $\mathrm{Coker}\,{}^tu$.

For the converse, we consider the following diagram:

$$
\begin{array}{ccccccccc}
0 & \longrightarrow & R & \xrightarrow{\;{}^t\varphi\;} & F^\vee & \xrightarrow{\;{}^tu\;} & E^\vee & \xrightarrow{\;p\;} & \mathrm{Coker}\,{}^tu \longrightarrow 0 \\
& & \downarrow{\scriptstyle \nu_a} & & \downarrow{\scriptstyle aI_{F^\vee}} & & \downarrow{\scriptstyle aI_{E^\vee}} & & \downarrow{\scriptstyle \mu_a} \\
0 & \longrightarrow & R & \xrightarrow{\;{}^t\varphi\;} & F^\vee & \xrightarrow{\;{}^tu\;} & E^\vee & \xrightarrow{\;p\;} & \mathrm{Coker}\,{}^tu \longrightarrow 0
\end{array}
$$

in which the vertical maps are induced by multiplication by the element $a \in R$. Consider $a \in \mathrm{Ann}(\mathrm{Coker}\,{}^tu)$. We then have $\mu_a = 0$, and hence $\mu_a p = p a I_{E^\vee} = 0$, so the image of $a I_{E^\vee}$ is in $\mathrm{Ker}\,p = \mathrm{Im}\,{}^tu$. As $E^\vee$ is free, this implies there is a homogeneous homomorphism $f : E^\vee \to F^\vee$ such that $a I_{E^\vee} = {}^tu f$.

Multiplying this equation on the right by ${}^tu$ we get $a I_{E^\vee}{}^tu = {}^tu a I_{F^\vee} = {}^tu f {}^tu$, or, alternatively, ${}^tu(a I_{F^\vee} - f {}^tu) = 0$, so the image of $a I_{F^\vee} - f {}^tu$ is in $\mathrm{Ker}\,{}^tu = \mathrm{Im}\,{}^t\varphi$. As $F^\vee$ is free, there is a $g : F^\vee \to R$ such that $a I_{F^\vee} - f {}^tu = {}^t\varphi g$. We multiply this equation on the right by ${}^t\varphi$ and we see that $a I_{F^\vee}{}^t\varphi = f {}^tu {}^t\varphi + {}^t\varphi g {}^t\varphi = {}^t\varphi \nu_a$. As ${}^tu {}^t\varphi = 0$ and ${}^t\varphi$ is injective, it follows that $\nu_a = g {}^t\varphi$, which means exactly that a is in $J = \mathrm{Im}\,\varphi$.

---

[5] We could also have passed to the fraction field of R and reduced the problem to ordinary linear algebra.

Remarks 2.10. The following remarks may shed some light on the above proof for those readers who are familiar with commutative algebra.

1) The sophisticated reader will have recognised $\operatorname{Coker}{}^t u$ as being the module $\operatorname{Ext}_R^1(J, R) \simeq \operatorname{Ext}_R^2(R/J, R)$. The exactness of the above sequence follows from the fact that R/J has projective dimension 2 and is hence of depth 2 by the Auslander-Buchsbaum formula (*cf.* [H] Chapter III, 6.12A), so

$$\operatorname{Ext}_R^1(R/J, R) = 0, \ i.e., \ \operatorname{Ker}{}^t u = \operatorname{Im}{}^t \varphi.$$

2) It is clear that the annihilator of $\operatorname{Ext}_R^2(R/J, R)$ contains J by the functoriality of Ext. The converse follows from the formula $\operatorname{Ext}_R^2(\operatorname{Ext}_R^1(J, R), R) = R/J$, which follows from the exact sequence 2.9. Indeed, if a annihilates $\operatorname{Ext}_R^1(J, R)$, then it annihilates the Ext^2 of the module in R, hence it annihilates R/J, hence[6] it is in J.

3 Liaison of space curves

In this section we will study the liaison operation. We will essentially restrict ourselves to ACM curves. For the general case, which requires levels of commutative algebra which we prefer to avoid, we refer the reader to [PS], which also contains the case of liaison in arbitrary dimension.

a. Introduction and definition

Definition 3.1. *Let C and Γ be two space curves, with saturated ideals I_C and I_Γ. The ideal $I_C \cap I_\Gamma$ is the saturated ideal of a curve X which is called the* scheme-theoretic union *of C and Γ.*

Remark 3.2. Set-theoretically, the scheme-theoretic union is simply the union of C and Γ. This set-theoretic union can also be defined by the product ideal $I_C I_\Gamma$, but from a scheme-theoretic point of view, the intersection ideal is a kind of sup of the two structures, whereas the product ideal is a sum (consider the extreme example where $I_C = I_\Gamma$).

Definition 3.3. *Let C and Γ be two space curves with saturated ideals I_C and I_Γ. We assume the following:*

1) C and Γ have no common irreducible component.

2) The scheme-theoretic union X of C and Γ is a schematic complete intersection curve, i.e., (cf. 2.3) the ideal $I_C \cap I_\Gamma$ is generated by two elements f, g.

We then say that C and Γ are geometrically linked *by X (or by the surfaces with equations f, g).*

---

[6] This proof, and its elementary version, was pointed out to me by Mireille Martin-Deschamps.

Proposition 3.4. *Let C and Γ be two* reduced *space curves without isolated points, with saturated ideals $I_C = I(C)$ and $I_\Gamma = I(\Gamma)$, geometrically linked by X. We have*

$$I_\Gamma = \{a \in R \mid aI_C \subset I_X\},$$
$$I_C = \{a \in R \mid aI_\Gamma \subset I_X\}.$$

We say that I_Γ, for example, is the transporter *of I_C in I_X and we write $I_\Gamma = (I_X : I_C)$.*

Proof. It will of course be enough to prove the formula for I_C. As $I_C I_\Gamma$ is contained in $I_C \cap I_\Gamma = I_X$, it is clear that I_C is contained in the transporter. To prove the converse we decompose C into irreducible components, $C = C_1 \cup \cdots \cup C_r$: $I_C = I_{C_1} \cap \cdots \cap I_{C_r}$, where the ideals I_{C_k} are prime (which would no longer be the case if C were not reduced). Consider $a \notin I_C$: our aim is to show that a is not in the transporter of I_Γ in I_X. As $a \notin I_C$, there is an index k such that $a \notin I_{C_k}$. As C and Γ have no common components $I_\Gamma \not\subset I_{C_k}$, so there is an $f \in I_\Gamma$ such that $f \notin I_{C_k}$. As this last ideal is prime, $af \notin I_{C_k}$, so $af \notin I_C$ and a is not contained in the transporter.

We will use Proposition 3.4 to give a more general definition of liaison.

Proposition-Definition 3.5. *Let C be a space curve with saturated ideal I_C and let X be a scheme-theoretic complete intersection containing C (i.e., such that $I_X \subset I_C$). Let J be the transporter of I_C in I_X. Then the ideal J is saturated and the subscheme Γ in $\mathbf{P}^3$ defined by J is a space curve (or empty) and we say that Γ is* algebraically linked *to C by X (or by the surfaces defining X).*

Remarks 3.6.

0) Proposition 3.4 shows that two reduced space curves (*i.e.*, varieties) which are geometrically linked are also algebraically linked. In fact, this property holds for arbitrary curves in the sense of 3.15 below, *i.e.*, with no point components, embedded or otherwise. In particular, it holds for ACM curves. The proof is similar, using a primary decomposition of the ideals I_C and I_Γ (*cf.* [Pes] § 3 or [Bbki] AC Chapter IV).

1) The geometrical linking relation is obviously symmetric but with no extra hypotheses on C, the algebraic linking relation may not be. This difficulty comes from our unwise choice of definition of space curves. To get around this problem we have to assume the curves satisfy the conditions of Definition 3.15.

2) We will see below (*cf.* 3.9) that if C is ACM, this problem vanishes.

3) It is immediate that Γ is empty if and only if $C = X$ (this follows from the fact that I_X is saturated).

Examples 3.7 (Explicit calculation of a liaison).

1) If we link the smooth conic C of ideal $(Y, X^2 - ZT)$ using the surfaces with equations YZ and $X^2 - ZT$, the curve we get is the double line with

ideal (X^2, Z). We want to find all the $a \in R$ such that $aY \in (YZ, X^2 - ZT)$ or, alternatively, $aY = bYZ + c(X^2 - ZT)$. It is immediate that Y divides c: $c = Yc'$ and it follows that $a = bZ + c'(X^2 - ZT) \in (X^2, Z)$. This example shows that even if we start with a smooth curve, we can get a non-reduced curve by liaison.

2) We start with the curve C which is a disjoint union of two conics $Z = P = 0$ and $T = Q = 0$, where P and Q are two homogeneous polynomials of degree 2, and we assume that Z (resp. T) does not divide P (resp. Q). That these conics are disjoint means that on writing $P = P_0(X, Y) + ZP_1 + TP_2$ and $Q = Q_0(X, Y) + ZQ_1 + TQ_2$ the polynomials P_0 and Q_0 have no common zeros. The ideal I_C is (ZT, ZQ, TP, PQ).

We link C to Γ via the surfaces ZT and $ZQ + TP$. We leave it as an exercise for the reader to show that $I_\Gamma = (ZQ + TP, Z^2, ZT, T^2)$. Once again, even though the initial curve is (in general) smooth, the linked curve is not reduced. This time, the curves in question are not ACM.

b. Changing resolutions by liaison: the ACM case

Theorem 3.8. *Let C be an ACM curve and, using the notations of 2.6, let*

$$(*) \qquad\qquad 0 \longrightarrow E \xrightarrow{\;u\;} F \xrightarrow{\;p\;} I_C \longrightarrow 0$$

be a (not necessarily minimal) resolution of its saturated ideal. Let A be the matrix of u in the canonical bases. By 2.7 $I_C = (f_1, \ldots, f_r)$, where the polynomials f_i are the $(r-1)$-minors of A. Let X be a complete intersection curve of two surfaces f, g of respective degrees s and t which contains C and is not equal to C. We set $f = \sum_{i=1}^{r} \lambda_i f_i$ and $g = \sum_{i=1}^{r} \mu_i f_i$. Let Γ be the scheme which is algebraically linked to C by X, defined by the ideal J which is the transporter of I_C in I_X. Then the following hold.

1) The ideal J has projective dimension 1, with resolution

$$0 \longrightarrow F^\vee(-s-t) \xrightarrow{\;v\;} E^\vee(-s-t) \oplus R(-t) \oplus R(-s) \longrightarrow J \longrightarrow 0,$$

where the matrix B of v in the dual bases of the canonical bases is the $(r+1) \times r$ matrix obtained by adding the rows formed by the coefficients λ_i and the coefficients μ_i to the bottom of A.

2) The ideal I_Γ is equal to J.

3) The scheme Γ is an ACM curve.

Proof. It is enough to prove 1) since J will then be saturated by 1.9 and Γ, which is non-empty because X is not equal to C, will be an ACM curve by 2.7. Let us therefore calculate the transporter ideal J of I_C in I_X. Consider $a \in J$. This means that, for all $i = 1, \ldots, r$, $af_i \in (f, g)$, or in other words $uf_i = \alpha_i f + \beta_i g$. Set $\lambda = \sum_{k=1}^{r} \lambda_k \varepsilon_k$ and $\mu = \sum_{k=1}^{r} \mu_k \varepsilon_k$. The above equation can be written as $p(a\varepsilon_i - \alpha_i \lambda - \beta_i \mu) = 0$. The exact sequence $(*)$ proves that there is an $x^i \in E$ such that $a\varepsilon_i = u(x^i) + \alpha_i \lambda + \beta_i \mu$.

Let $u' : E(s+t) \oplus R(t) \oplus R(s) \to F(s+t)$ be the homomorphism defined on blocks by $u' = (u, \lambda, \mu)$ (we leave it to the reader to check that the degrees are compatible with the given shiftings). The above calculation shows that $a \in J \Leftrightarrow \forall i \ a\varepsilon_i \in \mathrm{Im}\, u'$, which means exactly that a annihilates $\mathrm{Coker}\, u'$. Let us now consider the transposition homomorphism $v = {}^t u' : F^\vee(-s-t) \to E^\vee(-s-t) \oplus R(-t) \oplus R(-s)$.

We note that the degrees of the modules are the same, so we can apply 2.7 and 2.9. The polynomials f and g are already contained (up to sign) in the set of r-minors of the matrix B of V, obtained by suppressing one of the last two rows. As f and g have no common factor, the same is true of the r-minors of B, and it follows from Lemma 2.9 that J is the ideal generated by the r-minors of B and by 2.7 we have a resolution

$$0 \longrightarrow F^\vee(-s-t) \xrightarrow{\ {}^t u'\ } E^\vee(-s-t) \oplus R(-t) \oplus R(-s) \xrightarrow{\ (\varphi, g, f)\ } J \longrightarrow 0,$$

where $\varphi = (\varphi_1, \ldots, \varphi_{r-1})$, the polynomials φ_i being the r-minors of B obtained by removing a line of ${}^t\!A$.

Corollary 3.9. *With the above notations, the curve C is algebraically linked to Γ by X.*

Proof. We calculate the resolution of J, the transporter of I_Γ in I_X, by the method given in the theorem. The matrix A'' we are interested in is an $(r+2) \times (r+1)$-matrix, and as f and g are in the set of generators of I_Γ, it can be written in the form

$$A'' = \begin{pmatrix} A & \lambda & \mu \\ 0 & 1 & 0 \\ 0 & 0 & 1 \end{pmatrix}.$$

The generators of J' are the $r+1$-minors of this matrix, which can be easily calculated; the minors obtained by removing one of the last two lines are f and g, and those obtained on removing one of the first r lines are simply the r-minors of A, *i.e.*, the polynomials f_i. Hence the ideal J' is equal to I_C. QED.

Remark 3.10. Even starting with the minimal resolution of C the resolution of Γ given in 3.8 is not necessarily minimal. This is the case if one of the coefficients λ_i or μ_i is a non-zero constant (we then say that f or g is a minimal generator of I_C). For example, if we start with a space cubic and link with two quadrics, then we obtain a line, but the above resolution is not minimal (it has four generators rather than two).

Proposition 3.11 (Calculating invariants). *With the notations of 3.8, if d, g (resp. d', g') are the degrees and genus of C (resp. Γ), then*
  1) $d + d' = st$,
  2) $g' - g = (\frac{1}{2}(s+t) - 2)(d' - d)$.

Proof. This follows from the calculation of the resolution of the linked curve and the formulas in 2.4.

c. The Peskine-Szpiro theorem

We can now answer the question asked in the introduction.

Theorem 3.12 (Apéry-Gaéta-Peskine-Szpiro). *Let C be a curve in $\mathbf{P}^3$, I_C its saturated ideal and $\mathcal{J}_C$ the ideal sheaf defining C. The following are equivalent.*

1) There is an integer $n \geqslant 0$ and a sequence of curves $C_0, C_1, \ldots, C_n = C$ such that C_i and C_{i+1} are algebraically linked[7] and C_0 is a complete intersection.

2) C is an ACM curve (i.e., $H^1 \mathcal{J}_C(n) = 0$ for all n or dp $I_C = 1$, cf. 2.2).

Proof. The fact that 1) implies 2) follows from the description of the resolution of a complete intersection (*cf.* Chapter III, 10.1) and the calculation of the resolution of the linked curves (3.8).

The converse is proved by induction on r, the minimal number of generators of I_C (that is to say, the rank of F in the minimal resolution (*) of I_C). If $r = 2$, C is a complete intersection and the result follows. Otherwise, let $(f_1, \ldots, f_r)$ be the generators of I_C, *i.e.*, the images of the basis vectors of the free module F. We have the following lemma.

Lemma 3.13. *With the above notation, assume that $(f_1, \ldots, f_r)$ are of degrees $n_1 \leqslant \cdots \leqslant n_r$ respectively. There are then homogeneous polynomials a_i for $i = 2, \ldots, r$, of degree $n_r - n_i$ such that $a_r \in k^*$ and $f = f_1$ and $g = \sum_{i=2}^{r} a_i f_i$ have no common factor.*

Proof. Let h be a prime factor of f_1. One of the polynomials f_i, $i = 1, \cdots, r$, is not a multiple of h since otherwise C would contain the surface $V(h)$. We consider the subspace W_h of $H^0 \mathcal{J}_C(n_r)$ formed of multiples of h. It is not equal to the whole of $H^0 \mathcal{J}_C(n_r)$ (if f_i is not a multiple of h and l is a linear form $l \neq h$, then $f_i l^{n_r - n_i} \notin W_h$). It follows that the union of the sets W_h is a proper closed subset of $I_{C, \leqslant n_r} = \bigoplus_{n=0}^{n_r} H^0 \mathcal{J}_C(n)$, so its complement meets the open set of all polynomials g of the form $g = \sum_{i=1}^{r} a_i f_i$ with $a_r \in k^*$ and the result follows.

We now complete the proof of 3.12. Lemma 3.13 proves that up to change of the basis vector ε_r in F to $\sum_{i=2}^{r} a_i \varepsilon_i$ we can assume that f_1 and f_r have no common factor. We link C to Γ using f_1 and f_r. Calculating I_Γ as in 3.8 we can show (*cf.* above) that I_Γ has $r - 1$ generators and we are done by the induction hypothesis. To see that I_Γ is generated by $r - 1$ elements we note that it is generated by the r-minors of the $r + 1 \times r$ matrix B obtained by adding the rows $(1, 0, \ldots, 0)$ and $(0, \ldots, 0, 1)$ to ${}^t\!A$. On writing ${}^t\!A = (u \ A' \ v)$,

---

[7] As all the curves are ACM, the algebraic liaison relation is symmetric (*cf.* 3.9).

where A' is a $r - 1 \times r - 2$ matrix and u and v are column vectors, we note on expanding the minors of B that the generators we seek are in the ideal generated by the $r - 2$ minors of A'. There are $r - 1$ such generators. Using the non-zero constant coefficients in the matrix B, we can also simplify two terms in the resolution of I_Γ as in the proof of 1.8.

Remarks 3.14.

1) NB: it really is sometimes necessary to change the set of minimal generators. This is the case if we start with the curve C of ideal $I_C = (XY, YZ, XZ)$ and three generators XY, YZ, ZX. An arbitrary pair of these generators have a common factor. To carry out a liaison we have to replace one of them by a linear combination: XY and $Z(X + Y)$, for example.

2) Moreover, Lemma 3.13 is not generally true if g is of degree $< n - r$, cf. $I_C = (XZ, YZ, XT^2, YT^2)$.

d. Discussion of the non-ACM case

In this section we will give a sketch (without proof) of what happens in the non-ACM case. The interested reader will find more details in the references.

We begin by giving a more suitable definition of space curves for our purposes.

Definition 3.15. *A space curve is a subscheme of $\mathbf{P}^3$ which is of dimension 1 and which has no (embedded or not) point components (cf. the Appendix below).*

Proposition 3.16. *Let C be a space curve (in the above sense), X a complete intersection containing C and Γ the subscheme algebraically linked to C by X, which is the complete intersection of two surfaces of degrees s and t. Then:*

 1) Γ is a space curve (in the above sense).
 2) C is linked to Γ by X,
 3) The ideal I_Γ is the transporter of I_C in I_X,
 4) Formulae 3.11 for the degrees and genus are still valid,
 5) For all $n \in \mathbf{Z}$,

$$h^1 \mathcal{J}_C(n) = h^1 \mathcal{J}_\Gamma(s + t - n - 4),$$
$$h^0 \mathcal{J}_C(n) - h^0 \mathcal{J}_X(n) = h^1 \mathcal{O}_\Gamma(s + t - n - 4),$$
$$h^0 \mathcal{J}_\Gamma(n) - h^0 \mathcal{J}_X(n) = h^1 \mathcal{O}_C(s + t - n - 4).$$

Proof. See [PS] or [Rao].

Remark 3.17. We have a formula for resolutions which is similar to, but more complicated than, the formula given in 3.8. In particular, we have to consider separately two different types of resolutions (resolutions of E-type and resolutions of N-type cf. [MDP]).

The liaison relation is symmetric and generates an equivalence relation ("step-by-step" liaison). The main problem is describing the liaison classes. The class of ACM curves is one such equivalence class (this is essentially the Peskine-Szpiro theorem). For non-ACM curves a fundamental invariant was introduced by R. Hartshorne and studied by A.P. Rao; it is the module

$$M_C = \bigoplus_{n \in \mathbf{Z}} H^1 \mathcal{J}_C(n),$$

which we call the Rao module of C. This is a graded R-module with only a finite number of non-zero terms (this is clear for $n \gg 0$ by Serre's Theorem VII, 4.6, and for $n \ll 0$ this can be proved by Serre duality, cf. [H] Chapter III, 7.6) and every component is finite dimensional. This module transforms in a very simple way under liaison; if Γ is linked to C by two surfaces of degrees s and t, then $M_\Gamma = M_C^*(4 - s - t)$. (The star indicates the graded dual module of M_C.) A theorem of Rao's shows that any such module is, up to shift, the Rao module of a curve which can be assumed to be smooth, and that two curves are in the same liaison class if and only if they have the same Rao module, up to duality and shift.

A possibly even more interesting relation is the biliaison (or even liaison) relation: in this case the relevant invariant is simply the Rao module up to shift only. Hence in a biliaison class all the curves have shifted Rao module M, i.e., $M(n)$ with $n \in \mathbf{Z}$. It can be shown that there is a minimal such n, and the curves corresponding to this n are called the minimal curves of the class. They are essentially unique and allow us to determine all the other curves in the class. For more information see cf. [MDP] or [BBM]. For information on the use of the Rao module in classifying space curves problems, cf. [MDP].

e. Appendix: embedded points on curves

We will here explain via an example the phrase "embedded component," which appears in Definition 3.15.

Let C be a curve in $\mathbf{P}^3$ (in the weak sense of the word: a one-dimensional subscheme). This curve can have irreducible point components, which may or may not be reduced. It can also have special points called embedded points (or components) which, as we will see, play a role very similar to that of point components.

Definition 3.18. *Let C be a curve in $\mathbf{P}^3$ and let $x \in C$ be a (non-isolated) point. We denote by $\mathcal{O}_{C,x}$ the local ring of C at x and by $m_{C,x}$ its maximal ideal. We say that x is an embedded point of C if there is an $a \in \mathcal{O}_{C,x}$, $a \neq 0$, such that the annihilator of a is the ideal $m_{C,x}$. (We say that $m_{C,x}$ is a prime ideal associated to $\mathcal{O}_{C,x}$, cf. [Bbki] AC Chapter IV or [Pes] § 3. In a reduced ring only the minimal prime ideals, corresponding to irreducible components, are associated.)*

Example 3.19. Consider $C = \operatorname{Proj} k[X, Y, T]/(XY, X^2)$. This plane curve is the intersection of two curves, V which is the union of two lines, $X = 0$ and $Y = 0$, and W, which is the double line $X^2 = 0$. Set-theoretically, C is simply the line $X = 0$, but it has a singular point at $(0, 0, 1)$. Indeed, it is immediate that the maximal ideal (x, y) is the annihilator of y. We note that C is not reduced $(x^2 = 0)$ but $C - \{(0, 0, 1)\}$ is, so the nilpotents of C are concentrated at the embedded point. Another way of understanding this is to note that the ideal (XY, X^2) is the intersection of the ideal (X) and (X^2, Y), so C is the scheme-theoretic union of a line $X = 0$ and the double point whose equations are (X^2, Y) and which, as a scheme, is not contained in the line and plays the part of an extra point component.

This embedded point can also arise as the "limit" of a real point component. Indeed, consider the scheme C_λ defined by the ideal $(XY, X(X - \lambda T))$, with $\lambda \in k$. For $\lambda \neq 0$ this scheme is the disjoint union of a line $X = 0$ and an isolated point $(\lambda, 0, 1)$. For $\lambda = 0$ we obtain C, which is therefore the limit of the schemes C_λ, the embedded point being the limit of the isolated points.

And finally, if we carry out an (algebraic) liaison on C (resp. C_λ) (considered as a curve in $\mathbf{P}^3$ on adding the equation $Z = 0$) with respect to the surfaces XY, Z, we get in either case Γ, whose equations are $I_\Gamma = (Y, Z)$. If we now carry out liaison on Γ using the same surfaces, we do not obtain C or C_λ but the curve $X = Z = 0$: in other words, double liaison forgets the embedded or isolated point. We see on this example that without the hypothesis in 3.15 (that the curves should not have point components, even embedded ones) the liaison operation is not generally symmetric.

Exercises

1 Resolutions

1) Let R be a graded k-algebra, $R = \bigoplus_{n \in \mathbf{N}} R_n$, with $R_0 = k$. Let $m = R^+$ be the maximal ideal of elements of degree > 0, so R/m is isomorphic to k. We equip k with the quotient graded R-module structure. Determine the minimal resolution of the R-module k in the following cases: $R = k[\varepsilon]$, with $\varepsilon^2 = 0$ (the ring of dual numbers), $R = k[T]$, $R = k[X, T]$, $R = k[X, Y, T]$, $R = k[X, Y, Z, T]$.

2) Determine the minimal resolution of the (non-saturated) ideal $I = (X^2, XY, XT)$ in $k[X, Y, T]$. Compare this resolution with that of the associated saturated ideal.[8]

2 Constructing curves by liaison

1) Prove that there exist in $\mathbf{P}^3$ curves C which are ACM of degree d and genus g for the following choices of d and g: $d = 7, g = 5$; $d = 8, g = 7$; $d = 24, g = 64$; $d = s^2 - 1, g = s^3 - 2s^2 \quad 2s + 4$, for any integer $s \geqslant 2$.

---

[8] There is a program, called Macaulay, which enables us to calculate without any effort more or less all the resolutions we might possibly want.

Give the resolution of the ideals of the given curves. (Use Riemann-Roch to determine the degrees of the surfaces on which such curves could lie, then use liaison to reduce d and g via the formulas in 3.11).

2) Let n be an integer ≥ 1. Prove by induction on n using liaison that there is an ACM curve whose ideal has resolution

$$0 \longrightarrow R(-n-1)^n \longrightarrow R(-n)^{n+1} \longrightarrow I \longrightarrow 0.$$

3) Let D be a line.
   a) Prove that there exist two surfaces S_1 and S_2 of degrees 2 and 5 containing D which have no common components.
   b) Let C be the curve linked to D by S_1 and S_2. Prove that C is an ACM curve whose degree is 9 and whose genus is 12 and give its minimal resolution.
   c) Prove there are surfaces S_1' and S_2' of degrees 3 and 6 containing C which have no common components.
   d) Let C' be a curve linked to C by S_1' and S_2'. Prove that C' is an ACM curve of degree 9 and genus 12 whose minimal resolution is different from that of C.

3 The rational quartic

Let Γ be the union in $\mathbf{P}^3$ of the lines of equation (X,Y) and (Z,T). We link Γ using the surfaces $XT - YZ$ and $XZ^2 - Y^2T$ to a curve C. Prove that the curve C has degree 4 and genus 0. Is it ACM? Calculate its ideal I_C and prove that C is smooth and connected. (The ideal I_C contains, besides the equations of the above surfaces, the equations $YT^2 - Z^3$ and $ZX^2 - Y^3$.)

Appendices

A

Summary of useful results from algebra

The contents of this summary are used throughout this book, even in the very first chapters. They must therefore be rapidly mastered. We will mostly need only the definitions and the statements of the theorems, but it may be useful to try to prove them for exercise's sake. Those theorems whose proof is not obvious have been marked with a ¶. In this case the reader may refer to the bibliography.

1 Rings

We assume the reader is familiar with rings (which will always be assumed commutative with a unit), polynomial rings, ideals, quotient rings, fields, and modules. We denote by (x) or xA the ideal generated by x in A, *i.e.*, the set of all elements of the form xa with $a \in A$.

1.1 Rings

a. The isomorphism theorem. Let $f : A \to B$ be a ring homomorphism and set $I = \operatorname{Ker} f$. Let J be an ideal of J contained in I and let $p : A \to A/J$ be the canonical projection. Then:

1) There is a unique homomorphism $\overline{f} : A/J \to B$ such that $f = \overline{f}p$ (we say that f factors through A/J),
2) $\overline{f}$ is injective if and only if $J = I$,
3) $\overline{f}$ is surjective if and only if f is.
In particular, $\operatorname{Im} f \simeq A/\operatorname{Ker} f$.

b. Universal property of rings of polynomials. Let A and B be two rings. Giving a homomorphism $f : A[X_1, \ldots, X_n] \to B$ is equivalent to giving its restriction to A (*i.e.*, a homomorphism from A to B) and the images of the variables X_i (*i.e.*, n elements in B) (*cf.* [L] Chapter V).

c. Euclidean division. Let A be a ring and consider a polynomial $P \in A[X]$, $P \neq 0$, whose leading coefficient is a *unit*. For all $F \in A[X]$ there are $Q, R \in A[X]$ such that $F = PQ + R$ and $\deg R < \deg P$ or $R = 0$ (*cf.* [P] Chapter II, [L] Chapter V).

For example, in $k[X, Y]$ we can divide by $Y^2 - X^3$ with respect to Y, but not by $XY - 1$.

As an application of a,b and c, prove that the ring of Gaussian integers $\mathbf{Z}[i]$ is isomorphic to the quotient ring $\mathbf{Z}[X]/(X^2 + 1)$.

d. Products of rings. The direct product of two rings A and B is the product set $A \times B$ (*i.e.*, pairs (a, b) with product laws: $(a, b) + (a', b') = (a + a', b + b')$, $(a, b)(a', b') = (aa', bb')$).

Example: if p and q are two coprime integers, prove that $\mathbf{Z}/(pq) \simeq \mathbf{Z}/(p) \times \mathbf{Z}/(q)$ (Chinese remainder theorem) (*cf.* [L] Chapter II).

1.2 Ideals

a. Operations on ideals. An arbitrary intersection of ideals is an ideal.

The *sum* of a family I_k of ideals of A is the set of *finite* sums $\sum x_k$ with $x_k \in I_k$. This is an ideal which is the upper bound of the ideals I_k for inclusion. We denote it by $\sum I_k$. In particular, if $I_k = (f_k)$, then we obtain the *ideal generated* by the elements f_k. In $\mathbf{Z}$ the sum of the ideals generated by (x) and (y) is the ideal generated by the gcd of x and y.

The *product* of two ideals I and J is the ideal denoted by IJ *generated* by products xy with $x \in I$ and $y \in J$. We have $IJ \subset I \cap J$, but the converse is false: in $\mathbf{Z}$ the product ideal (resp. intersection) of (x) and (y) is the ideal generated by xy (resp. by the lcm of x and y).

b. Prime ideals. An *integral domain* is a ring $A \neq \{0\}$ such that $\forall a, b \in A$, $ab = 0 \Rightarrow a = 0$ or $b = 0$.

For example, a field is an integral domain, a subring of an integral domain is an integral domain, and a ring of polynomials over an integral domain is an integral domain.

An ideal $\mathfrak{p}$ of A is said to be *prime* if $A/\mathfrak{p}$ is an integral domain. We note that the inverse image of a prime ideal under a homomorphism is a prime ideal.

An ideal $\mathfrak{m}$ in A is said to be *maximal* if it is maximal for inclusion amongst the ideals of A different from A. Equivalently, $A/\mathfrak{m}$ is a field, called the *residue field* of $\mathfrak{m}$. It follows that any maximal ideal is prime, but the converse is not generally true. (The maximal ideals of $\mathbf{Z}$ are the ideals generated by a prime number p, whereas the prime ideals are all these ideals plus the ideal (0).)

It can be proved using Zorn's lemma that any ideal is contained in a maximal ideal.

c. Ideals of a quotient. Let A be a ring, I an ideal and p the canonical projection from A to A/I. The ideals of A/I are in (increasing) bijection with the ideals of A containing I via the maps p and p^{-1}. Moreover, the prime ideals (resp. maximal ideals) correspond under this bijection (*cf.* [P] Chapter II).

Application: describe the ideals of $\mathbf{Z}/n\mathbf{Z}$.

d. Nilpotent elements. An element $a \in A$ is said to be *nilpotent* if there is an integer $n > 0$ such that $a^n = 0$. The set of nilpotent elements form an ideal called the *nilradical* of A. This ideal is the intersection of all the prime ideals of A (¶ *cf.* Exercise 4.2).

A ring without non-zero nilpotent element is said to be *reduced*. Example: which ones of the rings $\mathbf{Z}/n\mathbf{Z}$ are reduced, are integral domains, are fields?

1.3 Noetherian rings

An ideal I in a ring A is said to be *finitely generated* if it is generated by a finite number of elements, *i.e.*, if we can find $x_1, x_2, \ldots, x_n \in I$ such that any $x \in I$ can be written in the form $x = \sum a_i x_i$ with $a_i \in A$.

A ring A is said to be *Noetherian* if it satisfies one of the following three equivalent properties (*cf.* [P] Chapter II, [L] Chapter VI):

1) Any ideal in A is finitely generated.

2) Any increasing sequence of ideals in A is eventually stable.

3) Any non-empty set of ideals in A has a maximal element for the inclusion relation.

Examples. A field, $\mathbf{Z}$ and more generally any principal ring (*i.e.*, a ring all of whose ideals are principal), and any quotient of a Noetherian ring are all Noetherian. If A is Noetherian, then $A[X]$ is also Noetherian (¶) (and hence so is $A[X_1, \ldots, X_n]$ and all its quotients).

In a Noetherian ring the set of minimal prime ideals is finite (¶ *cf.* Exercise 4.3).

1.4 Factorial rings

In this paragraph all rings are assumed to be integral domains. We denote by A^* the set of invertible elements of A.

An element $p \in A$ is said to be *irreducible* if $p \notin A^*$ and $p = ab \Rightarrow a \in A^*$ or $b \in A^*$.

For example, the irreducible elements in $\mathbf{Z}$ are the prime numbers.

An integral domain is said to be *factorial* if every element in A can be written uniquely as a product of irreducible elements.

A principal ring is factorial, so $\mathbf{Z}$ and $k[X]$ (where k is a field) are factorial rings.

If A is factorial, then so is $A[X]$ (¶) (and $A[X_1, \ldots, X_n]$). For example, if k is a field, then $k[X, Y]$ is factorial but not principal (*cf.* [P] Chapter II).

1.5 Finite type algebras

Let A be a ring. An A-*algebra* is a ring B equipped with a homomorphism $f : A \to B$ (which is often but not always injective). It is said to be *of finite type* if it is generated as an algebra by a finite number of elements $x_1 \dots, x_n$ of B, *i.e.*, if every element of B is a polynomial function of the elements x_i with coefficients in A. This is equivalent to asking that (*cf.* 1.b) B should be isomorphic to a quotient of a polynomial ring $A[X_1, \dots, X_n]$.

NB: "of finite type" as an A-module and as an A-algebra are not the same thing. In the first case, every element can be written as a linear combination of the elements x_i, in the second as a polynomial function of the elements x_i. We can also talk about being "of finite type as a field," in which case we allow ourselves rational fractions in the elements x_i. In each case, the structure generated by the elements x_i is the smallest substructure containing them.

1.6 Localisation

a. Local rings. A ring A is said to be *local* if it has a unique maximal ideal $\mathfrak{m}$, called its Jacobson radical. Every element of $A - \mathfrak{m}$ is then invertible.

Example: the ring of formal series $k[[X]]$. The maximal ideal consists of series with vanishing constant terms.

b. Localisation: definition. A subset S in A is said to be *multiplicative* if $1 \in S$ and $\forall\, a, b \in S, ab \in S$.

Let A be an integral domain and S a multiplicative subset of $A - \{0\}$. We define the ring A_S (or $S^{-1}A$), called the localisation of A along S, to be the quotient of the set $A \times S$ by the equivalence relation $\mathcal{R}$ defined by

$$(a, s)\mathcal{R}(a', s') \iff as' = a's.$$

The class of (a, s) is written a/s and we define the laws of composition on the quotient to be like those on the field $\mathbf{Q}$ of rational numbers. Moreover, when $S = A - \{0\}$, A_S is the field of fractions $\mathrm{Fr}(A)$, just as $\mathbf{Q} = \mathrm{Fr}(\mathbf{Z})$. Otherwise, A_S is a subring of this field.

There is a homomorphism $i : A \to A_S$ into the localisation given by $i(a) = a/1$. The image under i of an element of S is invertible and A_S is universal for this property (intuitively, A_S is the smallest ring containing A and inverses for all the elements of S).

When A is not a domain, we can still use this construction, but the equivalence relation must be modified as follows: $(a, s)\mathcal{R}(a', s') \iff \exists t \in S, \ t(as' - a's) = 0$. The homomorphism i is no longer injective in general: its kernel consists of all those elements which are annihilated by an element of S. There is no fraction field in this case (*cf.* [L] II.3).

c. Examples.

1) Consider $f \in A$ and $S = \{f^n \mid n \in \mathbf{N}\}$. We write $A_S = A_f$ and we have $A_f \simeq A[T]/(fT - 1)$. Example: what is $\mathbf{Z}_{10}$?

2) Let $\mathfrak{p}$ be a prime ideal of A and set $S = A - \mathfrak{p}$. Then S is a multiplicative subset of A and we set $A_S = A_{\mathfrak{p}}$. This is a local ring whose maximal ideal is $\mathfrak{p}A_{\mathfrak{p}} = \{a/s \mid a \in \mathfrak{p}\}$.

The prime ideals of A_S are in bijective correspondence via i^{-1} with the prime ideals of A, which do not meet S. In the case of $A_{\mathfrak{p}}$, these are exactly the prime ideals of A contained in $\mathfrak{p}$.

1.7 Integral elements

Let $f : A \to B$ be an A-algebra and consider $x \in B$. We say that x is *integral* over A if it satisfies a unitary equation

$$x^n + f(a_{n-1})x^{n-1} + \cdots + f(a_0) = 0,$$

with $a_i \in A$. (If f is the inclusion of A in B, we omit f.)

Example: i and $\sqrt{2}$ are integral over $\mathbf{Z}$, but $1/2$, $1/\sqrt{2}$ and π are not.

The algebra B is said to be integral over A if all its elements are integral over A. It is enough to check this for a system of generators of B (as an algebra) (¶). In all cases, the set of elements of B which are integral over A is a subring of B (¶) called the *integral closure* of B over A (*cf.* [L] Chapter X, 2.3.4).

Example: the integral closure of $\mathbf{Z}$ in $\mathbf{Q}(i\sqrt{3})$ is $\mathbf{Z}[j]$ (¶, *cf.* [S] 2.5).

If B is an A-algebra of finite type, then B is integral over A if and only if B is an A-module of finite type. We then say that B is a *finite* A-algebra.

An integral domain is said to be *integrally closed* (or normal) if the integral closure of A in its field of fractions is simply A. Otherwise, the integral closure in question A' is called the *integral closure* of A (or the *normalisation* of A) and is an integrally closed ring. If A is an algebra of finite type over a field, the integral closure A' is an A-module of finite type (*cf.* [S]).

Example: what is the integral closure of $\mathbf{Z}[i\sqrt{3}]$?

Any factorial ring is integrally closed. The converse is false; see, for example $\mathbf{Z}[i\sqrt{5}]$.

1.8 Some topology, for a change

Let X be a topological space and let $\mathcal{U}$ be a family of open sets of X. We say that $\mathcal{U}$ is a *basis* of open sets of X if every open set of X is a union of open sets in the family $\mathcal{U}$.

Let X be a topological space. We say that X is *quasi-compact* if any open cover of X has a finite subcover. (This is simply compactness without the Hausdorff condition.)

2 Tensor products

2.1 Definition: modules

Let A be a ring and M, N two A-modules. The tensor product of M and N over A is an A-module which we denote by $M \otimes_A N$ (or $M \otimes N$ when there is no ambiguity) generated by the symbols $x \otimes y$ with $x \in M$ and $y \in N$ (that is to say, any element of $M \otimes_A N$ is a finite linear combination $\sum_i a_i(x_i \otimes y_i)$ with coefficients in A), with the following rules of calculation: $(x + x') \otimes y = x \otimes y + x' \otimes y$, the same relation holds on exchanging the roles of x and y and $(ax) \otimes y = x \otimes (ay) = a(x \otimes y)$, for $a \in A$, $x \in M$ and $y \in N$.

We do not need to know how to construct this object. On the other hand, it is important to know that it has the following universal property.

Given an A-bilinear map $f : M \times N \to P$ between A-modules, there is a unique A-linear map $\overline{f} : M \otimes_A N \to P$ such that $\overline{f}(x \otimes y) = f(x, y)$.

2.2 Properties

We recall that an A-module M is said to be *free* if it possesses a basis (just like a vector space: the difference is that modules are not always free).

Proposition. *If M and N are free A-modules with bases (e_i) with $i \in I$ and (f_j) with $j \in J$, then $M \otimes N$ is free with basis $e_i \times f_j$, where $(i, j) \in I \times J$. In particular, this is the case if A is a field.*

The tensor product is *functorial*. This means, that given a homomorphism $f : M \to M'$, there is an induced homomorphism $f \otimes \mathrm{Id} : M \otimes N \to M' \otimes N$ given by the formula $(f \otimes \mathrm{Id})(x \otimes y) = f(x) \otimes y$. (This follows from the universal property.)

The tensor product is *right exact*: given an exact sequence $0 \to M' \to M \to M'' \to 0$, there is an induced exact sequence

$$M' \otimes N \longrightarrow M \otimes N \longrightarrow M'' \otimes N \longrightarrow 0.$$

NB: in general, the first map is not injective. It is, however, injective if N is free.

2.3 Extension of scalars

Let A be a ring and $f : A \to B$ an A-algebra. We note first that if N is a B-module, then we can also consider it as an A-module on setting, $ay = f(a)y$ for all $a \in A$ and $y \in N$. We will sometimes denote the A-module thus obtained by *restriction of scalars* from B to A, by $N_{[A]}$.

Conversely, if M is an A-module, then the tensor product $M \otimes_A B$ has a canonical B-module structure given by $b(x \otimes c) = x \otimes bc$. We say that $M \otimes_A B$ is obtained from M by *extension* of scalars from A to B.

The module $M \otimes_A B$ has the following universal property: if N is a B-module, then any A-linear homomorphism $f : M \to N_{[A]}$ induces a unique B-linear homomorphism $f \otimes_A B : M \otimes_A B \to N$ defined by $(f \otimes_A B)(x \otimes b) = bf(x)$. In other words, we have an isomorphism

$$\mathrm{Hom}_A(M, N_{[A]}) \simeq \mathrm{Hom}_B(M \otimes_A B, N).$$

This universal property often enables us to calculate modules $M \otimes_A B$. For example, there are isomorphisms

$$M \otimes_A A/I \simeq M/IM$$

(IM is the submodule of M generated by elements of the form ax, where $a \in I$ and $x \in M$),

$$S^{-1}A \otimes_A M \simeq S^{-1}M$$

(where $S^{-1}M$ is defined in a similar way to $S^{-1}A$).

2.4 Tensor products of algebras

Let A be a ring and B, C two A-algebras. In particular, B and C are A-modules, and we can construct their tensor product $B \otimes_A C$, which is equipped with an algebra structure on setting $(b \otimes c)(b' \otimes c') = bb' \otimes cc'$. This tensor product has the following universal property: given two homomorphisms of A-algebras $u : B \to D$ and $v : C \to D$, there is a unique homomorphism $u \otimes v : B \otimes C \to D$ given by $(u \otimes v)(b \otimes c) = u(b)v(c)$.

As above, we calculate most tensor products of algebras using this universal property. For example, there are isomorphisms

$$A[X_1, \ldots, X_n] \otimes_A A[Y_1, \ldots, Y_m] \simeq A[X_1, \ldots, X_n, Y_1, \ldots, Y_m],$$
$$B \otimes_A A/I \simeq B/IB.$$

(Examples: calculate $\mathbf{C} \otimes_{\mathbf{R}} \mathbf{C}$ and $\mathbf{Z}/p\mathbf{Z} \otimes_{\mathbf{Z}} \mathbf{Z}/q\mathbf{Z}$. Show that

$$S^{-1}A \otimes_A B \simeq S^{-1}B.)$$

2.5 Exercise: Nakayama's lemma

a) Let A be a ring, M an A-module and $x_1, \ldots, x_n$ elements of M. We assume that the elements x_i are solutions of a linear system

$$\forall j = 1, \ldots, n, \qquad \sum_{i-1}^{n} u_{ij}x_i = 0,$$

where the coefficients u_{ij} are in A. If U is the matrix with entries u_{ij}, then $\det U x_i = 0$. (NB: even if $\det(U)$ does not vanish, it does not follow that the elements x_i are zero: a module can have torsion, *i.e.*, we can have $ax = 0$ for a, x such that $a \neq 0$ in A and $x \neq 0$ in M. Consider the **Z**-module $\mathbf{Z}/n\mathbf{Z}$.)

b) Let A be a local ring with maximal ideal m and set $k = A/m$. Let M be an A-module of finite type. We assume $M \otimes_A k \; (= M/mM) = 0$. Prove that $M = 0$ (Nakayama's lemma). (Take generators $x_1, \ldots, x_n$ in M and write that they are in mM: now use a).)

c) Let A be a ring, m a maximal ideal in A and M an A-module of finite type. Assume $M/mM = 0$. Prove that $M_m = M \otimes_A A_m = 0$ and there is an $f \in A, .f \notin m$ such that $M_f = M \otimes_A A_f = 0$.

Application: if $x_1 \ldots, x_n$ are elements of M whose images generate M/mM over $k = A/m$, then there is an $f, f \notin m$, such that the elements x_i generate M_f as an A_f-module.

3 Transcendence bases

3.1 Definitions

The notions below are analogous to the notions of linear independence and basis for vector spaces.

Let $K \subset L$ be a field extension. A subset B in L is said to be *algebraically free* over K (we also say that its elements are algebraically independent) if for any finite subset $\{x_1, \ldots, x_n\} \subset B$ and any polynomial $P \in K[X_1, \ldots, X_n]$, the equality $P(x_1, \ldots, x_n) = 0$ implies $P = 0$. (Otherwise, we say that the elements of B are algebraically dependent.) For example, if $B = \{x\}$, B is free if and only if x is transcendental over K.

Let $K \subset L$ be a field extension. A subset $B \subset L$ is said to be an *algebraic generating set* over K if L is *algebraic* over the subfield $K(B)$ generated by B. (We note the difference with vector spaces: as far as transcendence bases are concerned, algebraic elements don't count.)

Let $K \subset L$ be a field extension. A subset B in L is a *transcendence basis* for L over K if it is both algebraically free and an algebraic generating set.

a. Existence. Let $K \subset L$ be a field extension. There are transcendence bases B for L over K. These bases all have the same cardinality, called the *transcendence degree* of L over K. We denote it by $\partial_K(L)$.

The proof of existence is identical to that for vector spaces: we use Zorn's lemma to construct a maximal free subset.

b. Examples.

1) If L is algebraic over K, then $B = \varnothing$ and $\partial_K(L) = 0$.

2) If $L = K(X_1, \ldots, X_n)$, a field of rational fractions in variables $X_1, \ldots, X_n$, then we can take $B = \{X_1, \ldots, X_n\}$ and $\partial_K(L) = n$. When n is equal to 1, any element of $L - K$ is a transcendence basis.

3) If $A = K[X, Y]/(F)$, where F is non-constant in Y and $L = \mathrm{Fr}(A)$, then it is easy to see that if x is the image of X in A, $\{x\}$ is a transcendence basis for L over K.

c. Other results. There are a certain number of other results which resemble basic results in linear algebra.

Extending base theorem. *If $x_1 \ldots, x_n$ is a free subset of L over K, then it can be extended to a transcendence basis, and hence $n \leqslant \partial_K(L)$.*

Conversely, if $x_1, \ldots x_n$ is a generating set (in the above sense: L is algebraic over $K(x_1, \ldots, x_n)$), then we can extract from it a transcendence basis. It follows that $n \geqslant \partial_K(L)$.

Of course, in both cases equality holds if and only if the subsets are both free and generating.

Given three fields $K \subset L \subset M$, $\partial_K(M) = \partial_K(L) + \partial_L(M)$ on taking the union of two transcendence bases.

4 Some algebra exercises

4.1 Prime ideals: exercises

a) Let A be a ring and $\mathfrak{p}$ a prime ideal of A. Assume that $\mathfrak{p}$ contains a product $I_1 \cdots I_n$ of n ideals. Show that $\mathfrak{p}$ contains one of the ideals I_k.

b) (The avoiding prime ideals lemma.)

Let A be a ring and let I be an ideal. Assume that I is contained in the union $\mathfrak{p}_1 \cup \cdots \cup \mathfrak{p}_n$ of n prime ideals of A. Prove that I is contained in one of the ideals $\mathfrak{p}_k$. (Argue by contradiction: assume that n is minimal and consider a suitable element of the form $a_1 + a_2 \cdots a_n$.)

4.2 Nilradical and prime ideals

Let A be a ring and N its nilradical (*cf.* 2.d).

a) Prove that N is an ideal contained in the intersection I of all the prime ideals of A.

b) We will now show that conversely $I \subset N$. We consider an element $s \notin N$ and the set $S = \{1, s, s^2, \ldots, s^n, \ldots\}$. We note that $0 \notin S$.

Show using Zorn's lemma (or the ring $S^{-1}A$) that there is an ideal $\mathfrak{p}$ which is maximal amongst the ideals of A not meeting S.

Prove that $\mathfrak{p}$ is prime (if $ab \in \mathfrak{p}$ consider the ideals $\mathfrak{p} + (a)$ and $\mathfrak{p} + (b)$). Complete the proof of the theorem.

4.3 Minimal prime ideals of a Noetherian ring

Let A be a Noetherian ring.

a) Let I be an ideal of A. Prove that there are a finite number of prime ideals $\mathfrak{p}_1, \ldots, \mathfrak{p}_n$ such that the product $\mathfrak{p}_1 \ldots \mathfrak{p}_n$ is contained in I. (Argue by contradiction; assume there exists an I which does not have this property and take a maximal such I.)

b) Show that A has only a finite number of minimal prime ideals. (Use a) in the case where $I = (0)$ plus Exercise 1.a.)

4.4 A non-Noetherian ring

Let $A = \mathcal{H}(\mathbf{C})$ be the ring of holomorphic functions on the complex plane.

a) Prove that A is an integral domain. What is its fraction field? Determine A^*.

b) Prove that A is not Noetherian (consider the ideals

$$I_k = \{f \in A \mid \forall z \in \mathbf{N} - \{0, 1, \cdots k\} \ f(z) = 0\}$$

for all $k \in \mathbf{N}$ and the function $\sin 2\pi z$).

c) Prove that $f \in A$ is irreducible if and only if f has a unique zero z in $\mathbf{C}$ and this zero is a simple zero (*i.e.*, $f'(z) \neq 0$).

d) Deduce that A is not factorial. Prove that A is, however, integrally closed.

(It is possible to show that A satisfies Bézout's theorem: every ideal of A of finite type is principal, *cf.* [R] Chapter 15.)

B

Schemes

The aim of this chapter is to present a partial (in both senses of the word) introduction to scheme theory. NB: the theory of schemes developed here is not exactly the theory developed by Grothendieck, for which we refer the interested reader to [H], Chapter II. However, in the case we are interested in (schemes of finite type over an algebraically closed base field k) the two theories are essentially equivalent (*cf.* [EGA] IV, 10.9). The power of the more general definition is important in arithmetic, for example, but this is another story.

Throughout the following discussion we will be working over an algebraically closed base field k.

0 Introduction

When discussing Bézout's theorem we used finite (*i.e.*, zero-dimensional) schemes. We will now define schemes in any dimension. The essential difference between varieties and schemes is the presence of nilpotent elements in the rings associated to schemes (as we already saw for finite schemes). Schemes are useful because they enable us to take multiplicity into account. For example, the line in the plane of equation $X = 0$ (resp. X^2) is a simple (resp. double) line and in scheme theory this difference will be visible in their rings, which are $k[X, Y]/(X)$ and $k[X, Y]/(X^2)$, respectively. In the second ring, the image of X is not zero, but its square is. One way of understanding this nilpotent element x is to think of it as being infinitely small, as in the definition of the tangent space.

1 Affine schemes

a. Definition

Let A be a k-algebra of finite type. We can write $A = k[X_1, \ldots, X_n]/I$, where I is an ideal. We consider the algebraic set $X = V(I) \subset k^n$ with its Zariski topology. There is a standard basis of open sets $D(f)$ with $f \in A$. We define a sheaf of rings $\mathcal{O}_X$ on X by setting $\Gamma(D(f), \mathcal{O}_X) = A_f$. In particular, $\Gamma(X, \mathcal{O}_X) = A$.

NB: this sheaf $\mathcal{O}_X$ is not exactly the sheaf used to define the usual variety structure on X. The difference is that we have used the ideal I and not $\mathrm{rac}(I)$. If I is not equal to its radical, then the ring A is not reduced—in other words, it contains nilpotents.

The affine scheme associated to the ring A is then defined to be the ringed space $(X, \mathcal{O}_X)$. We denote it by $\mathrm{Spm}(A)$ (or $\mathrm{Spec}(A)$ if we are sure Grothendieck is not listening).

b. Examples

Using this method we can construct a scheme $\mathrm{Spm}(k[X, Y]/(F))$, for example, which describes the plane curve of equation $F = 0$ without neglecting the multiplicities of the factors of F. Alternatively, if F and G have no common factors we can construct $\mathrm{Spm}(k[X, Y]/(F, G)$ which is the intersection scheme of F and G with multiplicities (*cf.* Bézout's theorem).

2 Schemes

a. Definition

A scheme is a ringed space $(X, \mathcal{O}_X)$ which is locally isomorphic to an affine scheme. We suppose moreover that X is quasi-compact.

A scheme is said to be reduced if for any open set U in X the ring $\Gamma(U, \mathcal{O}_X)$ is reduced (*i.e.*, does not contain any non-zero nilpotent element). A *reduced scheme is simply a variety*. If X is a scheme, then we associate to X in a canonical way a reduced scheme denoted by X_{red} which has the same underlying topological space as X and whose rings $\mathcal{O}_{X_{\mathrm{red}}}(U)$ are the reductions of the rings $\mathcal{O}_X(U)$.

b. Examples

1) Projective schemes. Let I be a homogeneous ideal of $R = k[X_0, \ldots, X_n]$ and let $S = R/I$ be its quotient ring. Let $X = V(I)$ be the projective algebraic set defined by I with its Zariski topology. In particular, we have as usual a basis of open sets $D^+(f)$, where $f \in S$ is a homogeneous element. We define a sheaf of rings on X by setting $\Gamma(D^+(f), \mathcal{O}_X) = S_{(f)}$. We check as for varieties that $(D^+(f), \mathcal{O}_X|_{D^+(f)}) \simeq \operatorname{Spec} S_{(f)}$, so $(X, \mathcal{O}_X)$ is a scheme which (by abuse of notation, $cf.$ Grothendieck) we denote by $\operatorname{Proj}(S)$ and call the projective scheme associated to S. If $I = 0$, then $\operatorname{Proj}(S)$ is simply $\mathbf{P}^n$; for other choices of I we get the closed subschemes of $\mathbf{P}^n$. Once again, the difference between projective varieties and projective schemes is that the rings of projective schemes may contain nilpotent elements.

2) Finite schemes. With the above notation we assume in addition that X is finite. There is then a hyperplane H which does not meet X and we can assume $H = V(X_0)$. We then have $X = \operatorname{Proj}(S) \simeq \operatorname{Spec} S_{(X_0)}$, so the finite scheme X is both projective and affine. Conversely, it is possible to prove that the finite schemes are the only schemes which are both affine and projective. The ring $R = S_{(X_0)}$ is a finite-dimensional k-vector space which is a direct product of its local rings $\mathcal{O}_{X,P}$ with $P \in X$. These local rings are also finite dimensional and their maximal ideals are nilpotent (*cf.* the rings of type $k[\eta]$ with $\eta^n = 0$, for example). If a local ring is reduced, then it is simply k and we say that the associated point is a simple point.

3 What changes when we work with schemes

First of all, nothing changes at all when we work with reduced schemes: they are simply varieties. Furthermore, most of the notions (such as irreducibility, components, dimension, products, separatedness, morphisms, subschemes and cohomology...) which we have defined for varieties can also be defined for schemes. This is obvious for topological properties such as the number of components or the dimension, since the topological properties of a scheme X are the same as those of the variety X_{red}.

The tangent space of a scheme can be calculated just like the tangent space of a variety: we consider (for an affine scheme, say) the (infinitesimal) deformations of the algebra $\Gamma(X, \mathcal{O}_X)$ at the point in question. Of course, a priori this ring might not be reduced. In particular, we can describe the tangent space in the following way.

Let I be an ideal of $k[X_1, \ldots, X_n]$ (which is not necessarily radical), set $A = k[X_1, \ldots, X_n]/I$, set $X = \operatorname{Spm}(A)$ and let x be a point of X. We suppose that $I = (F_1, \ldots, F_r)$. Let $d_x(F_1, \ldots, F_r) : k^n \to k^r$ be the Jacobian matrix of the polynomials F_i at the point x. Then $T_x(X) = \operatorname{Ker} d_x(F_1, \ldots, F_r)$.

We have the following smoothness criterion which is analogous to the criterion for smooth varieties.

Theorem 1. *Let I be an ideal of the ring $k[X_1, \ldots, X_n]$ and set $X = \operatorname{Spm}(k[X_1, \ldots, X_n]/I)$. Assume that X is irreducible and of dimension n Let x be a point of X and let $T_x(X)$ be the tangent space to X at x. Assume $\dim T_x(X) = n$. Then X is a variety, X is smooth at x and $I(X) = I$.*

It is easy to give a refined version of this criterion for non-irreducible X. Moreover, the same result holds for $X = \mathrm{Proj}(k[X_0, \ldots, X_n]/I)$ when I is a homogeneous ideal. Finally, we note that if X is a finite scheme, then it is smooth if it is reduced, *i.e.*, if it is a variety. Indeed, this implies that the tangent space at every point vanishes, so $m_x = 0$ for every $x \in X$ by Nakayama's lemma.

4 Why working with schemes is useful

There are several types of problems in which schemes are indispensable.

a. Intersection problems

We saw a lot of scheme-theoretic intersections when talking about Bézout's theorem. More generally, given two subschemes X and Y in $\mathbf{P}^n$ (for example) defined by homogeneous ideals I and J, their intersection scheme is the subscheme defined by the ideal $I + J$. Even if X and Y are varieties, the scheme structure on the intersection is fundamentally important: it explains intersection multiplicities and all contact phenomena.

b. Fibres

Let $\varphi : X \to Y$ be a morphism of schemes (or varieties). We have proved several results on the fibre $\varphi^{-1}(y)$ at a point of Y, considered as a variety. The dimension theorems are the same whether we work with schemes or varieties, but certain more subtle results can only be proved using the scheme structure on the fibre. This can be done in the following way. Assume Y is affine with ring A (which we are allowed to do) and that X is affine with ring B. There is therefore a map $\varphi^* : A \to B$ and y corresponds to a maximal ideal m in A. We have seen that set-theoretically $\varphi^{-1}(y) = V(mB)$. We define the scheme $\varphi^{-1}(y)$ to be $\mathrm{Spm}(B/mB)$. (The variety $\varphi^{-1}(y)$ corresponds to the radical of the ideal mB.)

If X is not affine, then we glue together the fibres obtained by this method in each of the open affine sets of X (*cf.* Exercise VII, 2.1).

Using this definition of a fibre it follows that, for $x \in X$ and $y \in \varphi(x)$, $T_x(\varphi^{-1}(y)) = \mathrm{Ker}\, T_x(\varphi)$. (Compare this result with Problem VI, 1.2.)

Let's look at an example to understand why this construction is useful. Consider the morphism $\varphi : V \to W$, where $V = V(Y^3 + XY + X) \subset k^2$, $W = k$ and $\varphi(x, y) = x$. Looking at rings, we get a map $\varphi^* : k[X] \to k[X, Y]/(Y^3 + XY + X)$ given by $\varphi^*(X) = X$. Consider $x \in k$. The point x corresponds to the maximal ideal $(X - x)$ in $k[X]$ and hence (as a scheme) the fibre over x has associated ring $k[Y]/(Y^3 + xY + x)$.

Set-theoretically, this fibre has three distinct points if $x \neq 0, -27/4$ (and in this case its ring is isomorphic to $k \times k \times k$), two points if $x = -27/4$ (but with ring $k \times k[Y]/(Y^2)$, $i.e.$, with one double and one simple point) and one point if $x = 0$ (whose ring is $k[Y]/(Y^3)$, so this point is a triple point). In every case, the dimension of the ring of the fibre as a k-vector space is always equal to 3 (which cannot be seen looking only at the variety structure): we say that φ is a degree 3 covering of Y ramified at the points 0 and $-27/4$ (*cf.* [H] Chapter IV, 2).

c. Differential calculus

We have already seen how to use the scheme $\mathrm{Spec}\, k[\varepsilon]$ with $\varepsilon^2 = 0$ to calculate tangent spaces. We can generalise this to higher-order differential calculus by copying the techniques of differential geometry (jets and so forth) using nilpotent elements of order > 2.

5 A scheme-theoretic Bertini theorem

We will use the notations of Problem VI. Using the above remarks on tangent spaces it is easy to prove the following theorem.

Theorem 2. *Let* $\varphi : X \to Y$ *be a dominant morphism between irreducible varieties. Assume that* X *is* smooth. *There is a non-empty subset* $V \in Y$ *such that* $\varphi|_{\varphi^{-1}(V)} : \varphi^{-1}(V) \to V$ *is smooth. In particular, the fibres* $\varphi^{-1}(y)$ *(which a priori have a* scheme *structure) are smooth varieties for all* $y \in V$.

When the dimensions of X and Y are the same, the general fibres are finite and formed of simple points, as in Example 4.b.

We also have the following version of Bertini's theorem, which is proved using the methods of Problem VI.

Theorem 3. *Let* $X \subset \mathbf{P}^n$ *be an irreducible smooth projective variety. If* H *is a general hyperplane of* $\mathbf{P}^n$, *the scheme* $X \cap H$ *is a smooth (projective) variety.*

When X is a curve, this shows that in general $X \cap H$ is finite and consists of d simple distinct points (where d is the degree of the curve X). In the plane, for example, this means that we can always find lines which are not tangent to a given curve. The notion of a scheme is indispensable for understanding these phenomena as $X \cap H$, which is finite, is always smooth as a variety, so the version of Bertini's theorem given in Problem VI does not help us.

C

Problems

Problem I

The aim of this problem is to study products of algebraic varieties. We work over an algebraically closed base field k.

1 Products of affine algebraic sets

a) Let $V \subset k^n$, $W \subset k^m$ be two affine algebraic sets. Prove that $V \times W$ is an affine algebraic set in k^{n+m}. Prove that the projections p and q from $V \times W$ to V and W are regular maps and give the associated ring morphisms.
   We denote by x (resp. y) a point in k^n (resp. k^m) and by $k[X]$ (resp. $k[Y]$) the rings of polynomials $k[X_1, \ldots, X_n]$ (resp. $k[Y_1, \ldots, Y_m]$).

b) Prove that the formula $\varphi(\sum_i f_i \otimes g_i)(x, y) = \sum_i f_i(x) g_i(y)$ defines a k-algebra homomorphism $\varphi : \Gamma(V) \otimes_k \Gamma(W) \to \Gamma(V \times W)$ (cf. Summary 2). Prove that φ is an isomorphism. (To prove the injectivity of φ, consider bases for the k-vector spaces $\Gamma(V)$ and $\Gamma(W)$.)
   Deduce that the ideal $I(V \times W)$ is generated by polynomials $f(X)$ and $g(Y)$ such that $f \in I(V)$ and $g \in I(W)$. (Reduce the problem to calculating the kernel of the natural map from $k[X] \otimes_k k[Y]$ to $\Gamma(V) \otimes_k \Gamma(W)$.)

c) Prove that the projections p and q are open with respect to the Zariski topologies (i.e., the image of an open set is open). (Consider an open set $D(f)$ in the product and write $f = \varphi(\sum_i f_i \otimes g_i)$, where the elements g_i are part of a basis of $\Gamma(W)$ over k. NB: the Zariski topology on the product is not the product of the Zariski topologies.)

d) Prove that if V and W are irreducible, then so is $V \times W$ (use c)).

e) Deduce from d) the following purely algebraic result: if A and B are two integral domain k-algebras, then the algebra $A \otimes_k B$ is an integral domain. Prove that this is not the case if k is not algebraically closed (cf. Summary 2.3).

2 Products of varieties

Let $\mathcal{C}$ be a category (*i.e.*, in simple terms, a collection of objects and maps). Let X and Y be two objects in $\mathcal{C}$. A product of X and Y in $\mathcal{C}$ is an object Z in $\mathcal{C}$ with two maps in $\mathcal{C}$, $p : Z \to X$ and $q : Z \to Y$ (called projections) with the following (universal) property: for any object T in $\mathcal{C}$ and maps $p' : T \to X$ and $q' : T \to Y$ there is a unique map f from T to Z such that $p' = pf$ and $q' = qf$.

a) Prove that if X and Y have a product in $\mathcal{C}$, then this product is unique up to canonical isomorphism. We denote this product by $X \times Y$.

b) Prove that if X and Y are affine algebraic sets (from which it follows that they are affine algebraic varieties), then the affine algebraic variety $X \times Y$ is a product in the category of algebraic varieties.

c) Prove that products exist in the category of algebraic varieties: if X and Y are algebraic varieties, then their product is defined as follows:

   i) The underlying set is the product set $X \times Y$.
   ii) A basis of open sets of $X \times Y$ is given by considering all open affine sets U in X and V in Y, and then taking all the affine open sets in the product $U \times V$ (NB: this topology is finer than the product topology).
   iii) The sheaf of rings is defined on this basis of open sets in the only reasonable way.

d) Generalise c) and d) of 1 to products of arbitrary varieties.

3 Products of projective varieties

a) Prove that we can define a morphism (called the Segre morphism) $\varphi : \mathbf{P}^r \times \mathbf{P}^s \to \mathbf{P}^{rs+r+s}$ by setting

$$\varphi((x_0, \ldots, x_r), (y_0, \ldots, y_s)) = (x_0 y_0, \ldots, x_i y_j, \ldots, x_r y_s).$$

(Here i and j vary from 0 to r and 0 to s respectively.)

b) Prove that the image of φ is the closed subvariety $V(I)$ in $\mathbf{P}^{rs+r+s}$, where I is the kernel of the homomorphism from $k[Z_{i,j}]$ ($i = 0, \ldots, r$, $j = 0, \ldots, s$) to $k[X_i, Y_j]$ (with the same indices) associating $X_i Y_j$ to $Z_{i,j}$.

c) Prove that φ is an isomorphism from $\mathbf{P}^r \times \mathbf{P}^s$ to $V(I)$. (Prove first that φ is injective, then restrict to the open affine sets $Z_{i,j} \neq 0$, $X_i \neq 0$, $Y_j \neq 0$.)

d) Deduce that the product of two projective varieties is a projective variety.

e) Give the equations of the Segre embedding of $\mathbf{P}^1 \times \mathbf{P}^1$ in $\mathbf{P}^3$.

4 Separated varieties

Let X be an algebraic variety.

We say that X is separated if the diagonal $\Delta = \{(x, y) \in X \times X \mid x = y\}$ is closed in $X \times X$.

a) Prove that any affine algebraic variety is separated.

b) Let X be an algebraic variety. We suppose that for any $x, y \in X$ there is an open affine set U containing x and y. Prove that X is separated.

c) Prove that any projective variety is separated.

d) Let X be a separated algebraic variety. Prove that the intersection of two open affine sets of X is an open affine set.

e) Let $f : X \to Y$ be a morphism and let Y be separated. Prove that the graph of f, $G(f) = \{(x, y) \in X \times Y \mid y = f(x)\}$ is closed in $X \times Y$.

Problem II

The aim of this problem is to study complete (or proper) algebraic varieties and in particular to show that projective varieties are complete. We work over an algebraically closed base field k.

1 Generalities on complete varieties

An algebraic variety X is said to be *complete* if for every variety Y the second projection $p : X \times Y \to Y$ is closed, *i.e.*, sends closed sets to closed sets. (See Problem I for the definition of the product.)

a) Let $f : X \to Y$ be a morphism. We assume that X is complete and Y is separated. Prove that the image $f(X)$ is closed and that it is a complete variety (*cf.* Problem I, 4.e).

b) Assume that X and Y are complete. Prove that $X \times Y$ is complete.

c) Assume that X is complete. Let Y be a closed subvariety of X. Prove that Y is complete.

d) Let n be greater than or equal to 1. Prove that the affine space $\mathbf{A}^n(k)$ is not a complete variety (use a)). More generally, it is possible to prove that any complete affine variety is finite.

2 Completeness of $\mathbf{P}^n$

In this section we will prove that $\mathbf{P}^n$ is a complete variety. Let Y be a variety, let p be the projection $p : \mathbf{P}^n \times Y \to Y$ and let Z be a closed set in $\mathbf{P}^n \times Y$. We have to prove that $p(Z)$ is a closed set in Y.

a) Prove that we can reduce the problem to the case where Y is an affine variety with associated ring R.

b) Set $U_i = D^+(X_i) \times Y$. Prove that the sets U_i form an open affine cover of $\mathbf{P}^n \times Y$ and that $\Gamma(U_i) = R[X_0/X_i, \ldots, X_n/X_i]$. (We will denote this ring by A_i.)

c) Let J be the homogeneous ideal generated in $S = R[X_0, \ldots, X_n]$ by the homogeneous polynomials F such that for all i

$$F(X_0/X_i, \ldots, X_n/X_i) \in I(Z \cap U_i).$$

Let S_r (resp. J_r) be the degree r homogeneous part of S (resp. J). Prove that for all i and all $f \in I(Z \cap U_i)$ there are integers k, r such that $F = X_i^k f \in J_r$.

d) Consider $y \in Y - p(Z)$ corresponding to the maximal ideal m in R. Determine the closed set $V(mA_i)$ in U_i. Prove the equality $A_i = mA_i + I(Z \cap U_i)$. (Use the Nullstellensatz.)

e) Prove there is an integer t such that, for all i, $X_i^t \in J_t + mS_t$. Prove there is an integer r such that $S_r = J_r + mS_r$.

f) Deduce that there is an $f \in R$, $f \notin m$ such that $fS_r \subset J_r$. (Use Nakayama's lemma, cf. Summary 2.4.c, applied to the ring R and the module S_r/J_r).

g) Prove that f is contained in $I(Z \cap U_i)$ for every i. Complete the proof of the theorem.

3 Applications

a) Prove that every projective variety is complete.

b) Let V be an irreducible projective variety. Prove that $\Gamma(V, \mathcal{O}_V) = k$ or, alternatively, every morphism $f : V \to k$ is constant (use 3.a, 1.a and 1.d).

Problem III

1 Notation

If A is an integral domain, we denote its fraction field by $\mathrm{Fr}(A)$.

If K is a field extension of a field k, we denote its transcendence degree over k by $\partial_k(K)$. We refer to Summary 3 for more information on algebraic independence, transcendence degree and so forth.

If A is a ring we denote the set of its prime ideals by $\mathrm{Spec}(A)$ and its Krull dimension by $\dim_K(A)$.

If $x_1, \ldots, x_n$ are elements of a k-algebra A, we denote the subalgebra of A generated by the elements x_i by $k[x_1, \ldots, x_n]$, but we reserve the notation $k[X_1, \ldots, X_n]$ with capital letters for the ring of polynomials in the variables X_i.

2 Noether's normalisation lemma

The aim of this section is to prove the following result.

Theorem 1. *Let k be a field and let A be a k-algebra of finite type which is an integral domain. Set $K = \mathrm{Fr}(A)$ and $n = \partial_k(K)$. There exist elements $x_1, \ldots, x_n \in A$, algebraically independent over k, such that A is integral over $k[x_1, \ldots, x_n]$.*

a) Write A as a quotient $k[Y_1, \ldots, Y_m]/I$. Prove that $m \geqslant n$ and prove the theorem when $m = n$.

b) Assume $m > n$. Let $y_1, \ldots, y_m$ be the images of the variables Y_i in A. Prove that they satisfy an algebraic equation $F(y_1, \ldots, y_m) = 0$, where F is a non-zero polynomial with coefficients in k.

c) Choose positive integers $r_2, \ldots, r_m$ and set

$$z_2 = y_2 - y_1^{r_2}, \ldots, z_m = y_m - y_1^{r_m}.$$

Prove that $y_1, z_2, \ldots, z_m$ also satisfy a non-trivial algebraic equation with coefficients in k.

¶ Prove that, for large enough r_i with large enough growth (*i.e.*, $0 \ll r_2 \ll \cdots \ll r_m$), y_1 is integral over the subring of A generated by the elements z_i.

d) Complete the proof of the theorem by induction on m.

3 The Cohen-Seidenberg going-up theorem

We aim to prove the following theorem.

Theorem 2. *Let A and B be two rings such that $A \subset B$ and B is integral over A. The following properties hold.*

1) *The map $\mathfrak{q} \mapsto \mathfrak{q} \cap A$ from $\operatorname{Spec} B$ to $\operatorname{Spec} A$ is surjective. Moreover, the following hold.*
   For all $\mathfrak{p}, \mathfrak{p}' \in \operatorname{Spec} A$ such that $\mathfrak{p} \subset \mathfrak{p}'$ and all $\mathfrak{q} \in \operatorname{Spec} B$ such that $\mathfrak{q} \cap A = \mathfrak{p}$ there is a $\mathfrak{q}' \in \operatorname{Spec} B$ such that $\mathfrak{q}' \cap A = \mathfrak{p}'$ and $\mathfrak{q} \subset \mathfrak{q}'$.
2) *The map $\mathfrak{q} \mapsto \mathfrak{q} \cap A$ is "almost" injective: given $\mathfrak{q}, \mathfrak{q}' \in \operatorname{Spec} B$ such that $\mathfrak{q} \subset \mathfrak{q}'$ and $\mathfrak{q} \cap A = \mathfrak{q}' \cap A$, $\mathfrak{q} = \mathfrak{q}'$.*
3) $\dim_K(A) = \dim_K(B)$.

a) We use the notations and hypotheses of Theorem 2; moreover, we assume that B is integral over A. Prove that the following are equivalent: A is a field and B is a field.

b) Let J be an ideal of B and set $I = J \cap A$. Prove that A/I is a subring of B/J and that B/J is integral over A/I.

c) We now further assume that A is local and its maximal ideal is $\mathfrak{m}$. Prove that the prime ideals of B over $\mathfrak{m}$ (*i.e.*, the ideals $\mathfrak{q}$ such that $\mathfrak{q} \cap A = \mathfrak{m}$) are exactly the maximal ideals of B (use b)).

d) Let $\mathfrak{p}$ be a prime ideal of A and set $S = A - \mathfrak{p}$. We denote by $A_\mathfrak{p}$ and $B_\mathfrak{p}$ the localisations $S^{-1}A$ and $S^{-1}B$. Prove that $A_\mathfrak{p}$ is local and contained in $B_\mathfrak{p}$ and that $B_\mathfrak{p}$ is integral over $A_\mathfrak{p}$.

e) Prove 1) and 2) of the theorem using d) and c).

f) Complete the proof of the theorem.

4 Dimensions of k-algebras of finite type

Our aim is to prove the following fundamental theorem on dimensions.

Theorem 3. *Let k be a field and let A be a k-algebra of finite type which is an integral domain. Set $K = \operatorname{Fr}(A)$. Then $\dim_K(A) = \partial_k(K)$.*

a) Prove that

$$\partial_k \operatorname{Fr}(k[X_1, \ldots, X_n]) = n \text{ and } \dim_K(k[X_1, \ldots, X_n]) \geqslant n.$$

b) Let $\mathfrak{p}$ be a non-zero prime ideal of $k[X_1, \ldots, X_n]$. Prove that

$$\partial_k \operatorname{Fr}(k[X_1, \ldots, X_n]/\mathfrak{p}) \leqslant n - 1.$$

c) Prove Theorem 3 when $\partial_k(K) = 0$.

d) Prove Theorem 3 by induction on $\partial_k(K)$. (Use the normalisation lemma and going-up to reduce to the case where $A = k[X_1, \ldots, X_n]$. Then take a chain of prime ideals $(0) \subset \mathfrak{p}_1 \subset \cdots \subset \mathfrak{p}_r$ of $k[X_1, \ldots, X_n]$ and use b), plus the induction hypothesis.)

5 Applications to the Nullstellensatz

Let K be a k-algebra of finite type. Assume K is a field. Prove that K is algebraic over k (use theorems 1 and 2.a). Deduce a proof of the weak Nullstellensatz (*cf.* Chapter I, 4.1) in the general case.

Problem IV

1 Discrete valuation rings

1) Let A be an integral domain and K its fraction field. Assume $A \neq K$. Prove that the following are equivalent.

   i) A is local and principal.

   ii) A is local and Noetherian and its unique maximal ideal $\mathfrak{m}$ is principal.

   iii) There exists an irreducible $\pi \in A$, $\pi \neq 0$ such that $\forall x \in A$, $x \neq 0$, $x = u\pi^n$, where $n \in \mathbf{N}$ and $u \in A^*$.

   iv) There is a map $v : K \to \mathbf{Z} \cup \{\infty\}$ such that:
- $v(0) = \infty$,
- $\forall x, y \in K$, $v(xy) = v(x) + v(y)$,
- $\forall x, y \in K$, $v(x + y) \geqslant \inf(v(x); v(y))$,
- $v(K) \neq \{0, \infty\}$,

and such that $A = \{x \in K \mid v(x) \geqslant 0\}$. (Arithmetic in $\mathbf{Z} \cup \{\infty\}$ is as one would expect: $n + \infty = \infty$, etc.)

A ring satisfying the above properties is said to be a *discrete valuation ring*; v is its valuation and π is said to be a uniformising parameter.

2) Prove that any regular (local Noetherian integral domain) ring of dimension 1 is a discrete valuation ring. Prove that the ring of formal series $k[[T]]$ is a discrete valuation ring.

3) Assume that A is both a k-algebra and a discrete valuation ring and the natural map from k to $A/\mathfrak{m}$ is an isomorphism. Consider $a \in A$. Prove that $\dim_k A/(a) = v(a)$.

2 The link with Dedekind rings

Our aim is to prove the following theorem.

Theorem. *Let A be a Noetherian domain. The following are equivalent.*

1) A is one-dimensional and is integrally closed.
2) For any maximal ideal $\mathfrak{m}$ in A, $A_\mathfrak{m}$ is a discrete valuation ring.

We then say that A is a *Dedekind* domain.

A) We prove that 2) $\Rightarrow$ 1). Assume that all the rings $A_\mathfrak{m}$ are discrete valuation rings.

A1) Prove that A is one-dimensional.
A2) Prove that any discrete valuation ring is integrally closed.
A3) Prove the formula

$$A = \bigcap_{\mathfrak{m} \in \mathrm{Max}\, A} A_\mathfrak{m}.$$

(To prove $A \supset \bigcap_{\mathfrak{m} \in \mathrm{Max}\, A} A_\mathfrak{m}$, consider the conductor of an element x in K,

$$I_x = \{a \in A \mid ax \in A\}.)$$

Deduce that A is integrally closed.

B) We now prove 1) $\Rightarrow$ 2). Assume A satisfies 1). Let $\mathfrak{m}$ be a maximal ideal in A.

B1) Prove that $A_\mathfrak{m}$ is a local domain of dimension 1 and that it is integrally closed.
B2) Let R be a non-zero Noetherian ring. Prove there is an $a \in R$ such that $\mathrm{Ann}\, a = \{x \in R \mid ax = 0\}$ is a prime ideal of R (called an associated prime ideal of R). (Take a maximal element amongst the annihilating ideals of R.)
B3) Let M be an $n \times n$ matrix with coefficients in an integral domain R and let $P(X) = \det(XI - M)$ be its characteristic polynomial. Prove that $P(X)$ is a unitary polynomial with coefficients in R and $P(M) = 0$ (the Cayley-Hamilton theorem). (Pass to fraction fields to reduce to the usual Cayley-Hamilton theorem.)
B4) We now prove that if A is a local Noetherian domain with maximal ideal $\mathfrak{m}$ which is one-dimensional and integrally closed, then it is a discrete valuation ring.

We consider an element $f \in \mathfrak{m}$, $f \neq 0$.

a) Prove there is a $g \in A$, $g \notin fA$ such that $(g/f)\mathfrak{m} \subset A$. (Apply B2 to the ring A/fA.)
b) Prove $(g/f)\mathfrak{m} \subset \mathfrak{m}$ or $(g/f)\mathfrak{m} = A$. In the latter case, prove that $\mathfrak{m}$ is principal and complete the proof of the theorem in this case.
c) Prove that $(g/f)\mathfrak{m} \subset \mathfrak{m}$ is impossible. (Otherwise, g/f would define an endomorphism of the finite-type A-module $\mathfrak{m}$: writing out the Cayley-Hamilton equation for this endomorphism we can show that g/f is then integral over A and produce a contradiction.)

3 An example

Set $A = \mathbf{C}[X,Y]/(Y^2 - X^3 + X)$.

1) Prove that the plane curve of equation $Y^2 - X^3 + X$ is irreducible and non-singular. Deduce that A is integrally closed.
2) Prove that A is not factorial. (Denoting the images of X and Y in A by x and y, prove that y is irreducible but the ideal (y) is not prime because $y^2 = x^3 - x = x(x-1)(x+1)$.)

Problem V

The aim of this problem is to prove the following theorem.

Theorem. *Let X be an irreducible variety.*
1) For every $x \in X$, $\dim T_x(X) \geqslant \dim X$.
2) There is a non-empty open set in X on which equality holds (i.e., X has a smooth open set).
(If X is not irreducible, then $\dim T_x(X) \geqslant \dim_x X$.)

1) Prove that the map from X to $\mathbf{N}$ which associates $\dim T_x(X)$ to x is upper semi-continuous (i.e., if $\dim T_a(X) = n$, then $\dim T_x(X) \leqslant n$ for x close to a) (cf. Exercise IV, 3).
2) Prove that the theorem holds for an irreducible hypersurface $V(F)$ in k^n. (NB: the proof is somewhat trickier in positive characteristic.)
3) Let X and Y be two irreducible varieties. We say that X and Y are birationally equivalent if there is a non-empty open set of X, U, and an open set of Y, V, which are isomorphic. Prove that if X and Y are affine, then they are birationally equivalent if and only if their rational fraction fields $K(X)$ and $K(Y)$ (i.e., the fraction fields of $\Gamma(X)$ and $\Gamma(Y)$) are k-isomorphic.
4) We recall the primitive element theorem: if $K \subset L$ is a *separable* (which is always the case if K is of characteristic zero, for example) finite algebraic extension, then there is an $x \in L$ (called the primitive element) such that $L = K(x)$.
   Prove that X is birationally equivalent to a hypersurface V in k^n. (Reduce to the case where X is affine and consider the field of functions $K(X)$. Take a transcendence basis $x_1, \ldots, x_n$ in $K(X)$ over k and then take a primitive element for the extension $k(x_1, \ldots, x_n) \subset K(X)$. The reader may prefer to consider only the characteristic zero case and refer to Shafarevitch p. 29 for the general case.)
5) Complete the proof of the theorem.
6) *Application to algebraic groups.* Let G be an algebraic group (i.e., an algebraic variety equipped with a group structure such that the multiplication $\mu : G \times G \to G$ and the inverse $\sigma : G \to G$ given by $\mu(x,y) = xy$ and $\sigma(x) = x^{-1}$ respectively are morphisms). For example, the usual classical groups—the linear group, the orthogonal groups and so forth—are algebraic groups.
   a) Prove that the translations $x \mapsto ax$ are variety isomorphisms from G to itself.
   b) Prove that if G is connected, then it is irreducible (argue by contradiction assuming that G is not irreducible and using a)).
   c) Prove that G is a smooth variety. (Reduce to the case where G is connected and use both a) and the above theorem.)

Problem VI

The aim of this problem is to study smooth morphisms between irreducible smooth varieties and in particular to prove the generic smoothness theorem and one of the many versions of Bertini's theorem. An integral morphism will always be assumed to be dominant and hence surjective.

1 Smooth morphisms

Let X and Y be *smooth* irreducible varieties and let $\varphi : X \to Y$ be a dominant map. We say that φ is smooth if for any $x \in X$ the linear tangent map $T_x(\varphi) : T_x(X) \to T_y(Y)$, where $y = \varphi(x)$ is surjective.

NB: if X and Y are not smooth this definition is not the right one (*cf.* [H] Chapter III, 10). In what follows we will only consider smooth morphisms between smooth varieties.

1) Prove that the composition of two smooth morphisms is again a smooth morphism. Prove that if Z and Y are smooth, then the projection from $Y \times Z$ to Y is smooth. If U is an open subset of a smooth variety X, prove that the inclusion of U in X is smooth.

2) Let $\varphi : X \to Y$ be a morphism. We denote by $\varphi^{-1}(y)$ the fibre of $y \in Y$ with its variety structure. Prove that for all $x \in \varphi^{-1}(y)$, $T_x(\varphi^{-1}(y)) \subset \operatorname{Ker} T_x(\varphi)$.

3) Deduce that if φ is smooth, the non-empty fibres of φ are smooth of dimension $\dim(X) - \dim Y$ and that $T_x(\varphi^{-1}(y)) = \operatorname{Ker} T_x(\varphi)$ for all $x \in X$. (Use Problem V and the dimension of fibres theorem.)

2 An example

Let $\varphi : X \to Y$ be a dominant morphism of irreducible affine varieties. We set $A = \Gamma(Y)$ and $B = \Gamma(X)$: there is then an injective map $\varphi^* : A \to B$. We suppose that $B = A[\xi] \simeq A[T]/(P)$, where $P(T) = T^n + a_{n-1}T^{n-1} + \cdots + a_0$ with $a_i \in A$. The morphism φ is then integral. Furthermore, we assume that $f = P'(\xi) = n\xi^{n-1} + (n-1)a_{n-1}\xi^{n-2} + \cdots + a_1$ is $\neq 0$ in B.

1) Prove that if $x \in D(f)$, then the linear tangent map $T_x(\varphi) : T_x(X) \to T_y(Y)$, where $y = \varphi(x)$, is surjective. (Lift a deformation $\chi : A \to k[\varepsilon]$ of the form $\chi(a) = a(y) + v(a)\varepsilon$ to a deformation ψ of B of the form $\psi(\xi) = \xi(x) + w(\xi)\varepsilon$; the trick is to find $w(\xi)$.)

2) Deduce that if $D(f)$ and Y are smooth, then the morphism $\varphi|_{D(f)} : D(f) \to Y$ is smooth.

3 The generic smoothness theorem, first version

Henceforth we assume the field k is of *characteristic zero*. We will prove the following theorem.

Theorem 1. *Let* $\varphi : X \to Y$ *be a dominant map between irreducible varieties (which are not assumed to be smooth). There are non-empty (smooth) open sets U in X and V in Y such that $\varphi(U) \subset V$ and $\varphi|_U : U \to V$ is smooth.*

1) Prove that we can assume that X and Y are affine and smooth (*cf.* Problem V).

2) Prove that after replacing X by one of its open sets we can write $\varphi = pi\psi$: $X \xrightarrow{\psi} \Omega \xrightarrow{i} Y \times k^n \xrightarrow{p} Y$, where ψ is integral, i is the inclusion of the open set Ω in $Y \times k^n$, and p is projection. (Use the arguments seen in the proof of the dimension of fibres theorem given in Section 3 of Chapter IV.) Deduce that it will be enough to prove the theorem for integral φ.

3) Assume that φ is integral. Using the primitive element theorem (*cf.* Problem V: this is where we need the hypothesis that k is of characteristic zero) prove that, after possibly modifying X, we can reduce to the case where φ is of the form given in Example II. Complete the proof of the theorem.

4 The generic smoothness theorem, second version

Theorem 2. *Let* $\varphi : X \to Y$ *be a dominant morphism between irreducible varieties such that X is smooth. There is then a non-empty open set V in Y such that $\varphi|_{\varphi^{-1}(V)} : \varphi^{-1}(V) \to V$ is smooth. In particular, the fibres $\varphi^{-1}(y)$ are smooth for all $y \in V$.*

1) Prove that it is enough to deal with the case where Y is smooth.

2) Set $X_r = \{x \in X \mid \operatorname{rank}(T_x(\varphi)) \leqslant r\}$. Prove that $\dim \overline{\varphi(X_r)} \leqslant r$. (Consider the restriction of φ to suitable irreducible components of $\overline{X_r}$ and $\varphi(X_r)$ and use Theorem 1 and the dimension of fibres theorem.)

3) Complete the proof of the theorem.

5 Bertini's theorem

Let E be a vector space of dimension $n+1$. A hyperplane H in $\mathbf{P}(E) \simeq \mathbf{P}^n$ is defined by a non-zero linear form $f \in E^*$ and two such forms define the same hypersurface if and only if they are proportional. The set G of hyperplanes in $\mathbf{P}(E)$ is therefore in canonical correspondence with the projective space $\mathbf{P}(E^*) \simeq \mathbf{P}^n$. Henceforth we identify these two spaces.

Let $X \subset \mathbf{P}^n$ be an irreducible smooth projective variety. Our aim is to prove Bertini's theorem.

Theorem 3. *If H is a general hyperplane in $\mathbf{P}^n$, $X \cap H$ is a smooth (projective) variety.*

(Here, the word general means that there is a dense open set U in G such that if $H \in U$, then $X \cap H$ is smooth.)

We consider the incidence variety $V \subset X \times G$:

$$V = \{(x, H) \mid x \in H\}.$$

1) Prove that V is a closed subset of $X \times G$ or $\mathbf{P}^n \times G$ and give its equations in terms of the equations for X.

2) Prove that V is irreducible (use Exercise 7 of Chapter IV) and smooth (calculate its dimension and the dimension of its tangent space).

3) Complete the proof of the theorem by applying Theorem 2 to the projection from V to G.

We refer to the appendix on schemes for more details of these two theorems.

Problem VII

The aim of this problem is to establish the existence and uniqueness of the intersection multiplicity of two plane curves in a point. This multiplicity will be constructed as a function verifying certain natural properties. The following is largely inspired by Fulton [F], Chapter 3, Section 3.

1 Statement of the theorem

There is a unique map $\mu : k^2 \times (k[X,Y] - \{0\})^2 \to \mathbf{N} \cup \{\infty\}$ associating to the point $P \in k^2$ and the non-zero polynomial $F, G \in k[X, Y]$ a number $\mu_P(F, G)$ called the *intersection multiplicity of F and G at the point P* satisfying the following seven axioms.

1) $\mu_P(F, G) = \infty$ if and only if F and G have a common factor H passing through P (*i.e.*, such that $H(P) = 0$).

2) $\mu_P(F, G) = 0$ if and only if $P \notin V(F) \cap V(G)$.

3) $\mu_P(F, G) = \mu_P(G, F)$ for all P, F, G.

4) If $u : k^2 \to k^2$ is a bijective affine map, then for all P, F, G: $\mu_P(F, G) = \mu_{u^{-1}(P)}(F^u, G^u)$. (Recall that F^u is the polynomial such that for all $x, y \in k$, $F^u(x, y) = F(u(x, y))$.)

5) We have $\mu_P(F, G) \geqslant \mu_P(F)\mu_P(G)$ with equality if and only if F and G have no common tangents at P.

5') Variant: if $P = (0, 0)$, then $\mu_P(X, Y) = 1$.

6) If $F = \prod_{i=1}^m F_i^{r_i}$ and $G = \prod_{j=1}^n G_j^{s_j}$, then $\mu_P(F, G) = \sum_{i,j} r_i s_j \mu_P(F_i, G_j)$.

7) We have $\mu_P(F, G) = \mu_P(F, G + AF)$ for any polynomial $A \in k[X, Y]$ (in other words $\mu_P(F, G)$ essentially depends only on the ideal (F, G)).

Moreover, when F and G have no common factor passing through P this number is given by the formula

$$\mu_P(F, G) = \dim_k \mathcal{O}_{k^2, P}/(F, G).$$

2 Uniqueness

Let μ and μ' be two maps satisfying Axioms 1) to 7) (with variant 5'). Our aim is to show that $\mu = \mu'$.

a) Prove that if $\mu_P(F, G) = \infty$, then $\mu'_P(F, G) = \infty$.

b) Prove that if $\mu_P(F, G) = 0$, then $\mu'_P(F, G) = 0$.

c) Set $P = (0, 0)$ and let m, n be positive integers. Prove that $\mu_P(X^m, Y^n) = \mu'_P(X^m, Y^n) = mn$.

We now proceed by induction on $\mu_P(F, G)$. Consider $n \in \mathbf{N}$ such that $n > 0$. We assume that uniqueness has been proved for $\mu_P(F, G) < n$. Consider P, F, G such that $\mu_P(F, G) = n$. We will show that $\mu'_P(F, G) = n$.

d) Prove that F and G have no common factors passing through P and $P \in V(F) \cap V(G)$. Prove we can assume $P = (0, 0)$.

We consider the polynomials $F(X, 0)$ and $G(X, 0)$; let the integers (r, s) be their respective degrees. We proceed by induction on $\inf(r, s)$.

e) Assume $\inf(r, s) = 0$. Prove that $\mu'_P(F, G) = n$. (Factorise one of the polynomials by Y and use Axioms 6) and 7), c) above and the induction hypothesis on n.)

f) Assume $\inf(r, s) > 0$. Prove that we can assume $r \leqslant s$. Prove there is a polynomial A such that if $H = G + AF$ the degree of $H(X, 0)$ is $< s$. Prove further that there is an A such that the degree of $H(X, 0)$ is $< r$ and complete the proof by applying the induction hypothesis on $\inf(r, s)$ to F and H.

We note that this uniqueness proof gives us an algorithm for calculating $\mu_P(F, G)$ in a finite number of operations. Of course, the calculation is much easier using 5) than 5').

g) Apply the above to calculate $\mu_P(F, G)$ in the following cases:

1) $P = (0, 0)$ and F, G are arbitrary polynomials in the following list

$$Y^2 - X^3, \; X^3 + Y^3 + XY, \; (X^2 + Y^2)^2 + 3X^2Y - Y^3,$$
$$(X^2 + Y^2)^3 - 4X^2Y^2, \; 2X^4 + Y^4 - Y(3X^2 + 2Y^2) + Y^2,$$
$$X^2Y^3 + X^2 + Y^2, \; Y^2 + X^2 + X^2Y^2 - 2XY(X + Y + 1).$$

2) $P = (1, i, 0)$ and F, G are homogeneous polynomials

$$(X^2 + Y^2)^2 + T(3X^2Y - Y^3), \; (X^2 + Y^2)^3 - 4X^2Y^2T^2.$$

(Use the affine open set $X \neq 0$.)

3 Existence with variant 5'

When F and G have no common factors passing through P, we set $\mu_P(F, G) = \dim_k \mathcal{O}_{k^2, P}/(F, G)$ (otherwise we set $\mu_P(F, G) = \infty$). Our aim is to prove that μ satisfies the seven axioms (with variant 5'). We will occasionally write $\mathcal{O}_{k^2, P} = \mathcal{O}$ for short.

a) Assume $F = F_1 F_2$ and $F_1(P) \neq 0$. Prove there is an isomorphism $\mathcal{O}_{k^2,P}/(F,G) \simeq \mathcal{O}_{k^2,P}/(F_2,G)$. Deduce that $\mu_P(F,G)$ is unchanged if we multiply F or G by a polynomial which does not vanish at P.

b) Prove that μ satisfies Property 1). When F and G have no common factor passing through P, use the isomorphism

$$ k[X,Y]/(F,G) \simeq \prod_{P \in V(F) \cap V(G)} \mathcal{O}_{k^2,P}/(F,G); $$

otherwise, if H is an irreducible factor of F and G such that $H(P) = 0$, prove that $\mathcal{O}_{k^2,P}/(F,G)$ is of dimension larger than $\Gamma(V(H))$ and this latter ring is of infinite dimension.

c) Prove Properties 2), 3), 4), 5'), 7). (For 4), note that if u is a morphism from k^2 to itself it induces an isomorphism of the local rings $\mathcal{O}_{k^2,P}$ and $\mathcal{O}_{k^2,u^{-1}(P)}$.)

d) To prove 6) reduce the problem first to establishing the formula $\mu_P(F,GH) = \mu_P(F,G) + \mu_P(F,H)$ when F and GH have no common factor and then prove there is an exact sequence

$$ 0 \longrightarrow \mathcal{O}/(F,H) \xrightarrow{\psi} \mathcal{O}/(F,GH) \xrightarrow{\pi} \mathcal{O}/(F,G) \longrightarrow 0, $$

where π is the canonical projection and ψ is induced by multiplication by G.

4 Existence: Property 5)

We define μ as in 2. We will prove 5), which is somewhat trickier. We set $m = \mu_P(F)$ and $n = \mu_P(G)$.

a) Prove we can reduce to the case where $P = (0,0)$.

b) Let I be the ideal (X,Y) in $k[X,Y]$. Calculate $\dim_k(k[X,Y]/I^r)$ for all $r \in \mathbf{N}$.

c) Prove there is an exact sequence

$$ k[X,Y]/I^n \times k[X,Y]/I^m \xrightarrow{\psi} k[X,Y]/I^{m+n} \xrightarrow{\varphi} k[X,Y]/(I^{m+n}, F, G) \longrightarrow 0, $$

where φ is the canonical projection and ψ is given by $\psi(\overline{A}, \overline{B}) = \overline{AF + BG}$. Deduce that $\dim_k k[X,Y]/(I^{m+n}, F, G) \geqslant mn$ with equality if and only if ψ is injective.

d) Prove there is an isomorphism $\alpha : k[X,Y]/(I^{m+n}, F, G) \to \mathcal{O}/(I^{m+n}, F, G)$. (Start by proving that the first ring is local.) Deduce that $\mu_P(F,G) \geqslant mn$. Prove that equality holds if and only if we have the following two conditions:

   1) ψ is injective;
   2) the projection $\pi : \mathcal{O}/(F,G) \to \mathcal{O}/(I^{m+n}, F, G)$ is an isomorphism, or, alternatively, $I^{m+n} \subset (F,G)\mathcal{O}$ (the ideal generated by F and G in $\mathcal{O}$).

We now assume that F and G have no common tangent at P.

e) Prove that ψ is injective. (If $\psi(\overline{A}, \overline{B}) = \overline{AF + BG} = \overline{0}$, then consider the lowest degree terms of A, B, F, G.)

f) Prove that, for large enough r, $I^r \subset (F, G)\mathcal{O}$ (use a polynomial H which vanishes at all the points of $V(F) \cap V(G)$ except P and apply the Nullstellensatz).

g) Let $L_1, \ldots, L_m$ (resp. $M_1, \ldots, M_n$) be the tangents to $V(F)$ (resp. $V(G)$) at P (some of the L_i (resp. M_j) can be equal, but $L_i \neq M_j$ for all i, j). For all $i, j \geqslant 0$, we set $A_{i,j} = L_1 \cdots L_i M_1 \cdots M_j$ with the convention that if $i > m$ (resp. $j > n$), $L_i = L_m$ (resp. $M_j = M_n$).
Prove that for $t \geqslant 0$ the set of polynomials $A_{i,j}$ such that $i + j = t$ form a basis for the vector space of homogeneous polynomials of degree t.

h) Prove that if $i + j \geqslant m + n - 1$, then $A_{i,j} \in (F, G)\mathcal{O}$ (use f) and g)).

i) Complete the proof of the theorem.

Problem VIII

The aim of this problem is to establish the following theorem.

Theorem. *Let C be an irreducible projective plane curve. The curve C is then birationally equivalent to an irreducible projective plane curve X which has only ordinary singular points (i.e., singular points with distinct tangent lines).*

As we know that every irreducible curve is birationally equivalent to a projective plane curve (*cf.* Problem V, for example), we see that every irreducible curve is birationally equivalent to a projective plane curve with ordinary singularities.

What follows is taken from Fulton [F], Chapter 7, paragraph 4.

We work in the projective plane $\mathbf{P}^2 = \mathbf{P}^2(k)$ over an algebraically closed base field k of characteristic zero. (For the characteristic p case, see Fulton [F], appendix.) The homogeneous coordinates of the plane are denoted x, y, z. We consider the three points (called the *fundamental* points)

$$P = (0, 0, 1), \quad P' = (0, 1, 0) \quad \text{and} \quad P'' = (1, 0, 0)$$

and the three lines (called the *exceptional* lines)

$$L = V(Z), \quad L' = V(Y) \quad \text{and} \quad L'' = V(X).$$

These three lines form a triangle whose vertices are the fundamental points. We denote the open set $\mathbf{P}^2 - V(XYZ)$ in $\mathbf{P}^2$ by U.

If F is a homogeneous polynomial on $\mathbf{P}^2$ and S is a point in $\mathbf{P}^2$, we denote by $\mu_S(F)$ the multiplicity of F at S. If F and G are two homogeneous polynomials and S is a point of $\mathbf{P}^2$ we denote the intersection multiplicity of the curves whose equations at S are F and G by $\mu_S(F, G)$. Finally, if d is the degree of the curve $C = V(F)$, then we call the integer

$$g^*(C) = (d-1)(d-2)/2 - \sum_{P \in \mathbf{P}^2} \mu_P(\mu_P - 1)/2 \quad \text{where} \quad \mu_P = \mu_P(F)$$

the *expected genus* of C. Recall that if C is irreducible, then $g^*(C) \geqslant 0$ (*cf.* Exercises VI).

We define the map $Q : \mathbf{P}^2 - \{P, P', P''\} \to \mathbf{P}^2$ by the formula $Q(x, y, z) = (yz, zx, xy)$. This map is called the *standard quadratic transformation* on $\mathbf{P}^2$.

1) Prove that Q is a morphism of varieties (and is hence a rational map from $\mathbf{P}^2$ to itself). Why do we have to restrict the domain of this map?

2) Prove that Q is an involution of the open set U (*i.e.*, $Q^2 = \mathrm{Id}_U$). Deduce that Q is a birational map of $\mathbf{P}^2$. Determine the image of the exceptional lines and the image of Q.

We consider an irreducible projective curve $C = V(F) \subset \mathbf{P}^2$ of degree d. We assume that C is not one of the exceptional lines. We denote by C' the closure in $\mathbf{P}^2$ of $Q^{-1}(C \cap U)$.

3) Prove that C' is an irreducible projective curve which is birationally equivalent to C and $(C')' = C$.

4) We set $F^Q(X, Y, Z) = F(YZ, ZX, XY)$. We assume $\mu_P(C) = r$ (resp. $\mu_{P'}(C) = r'$, resp. $\mu_{P''}(C) = r''$). Prove that Z^r (resp. $Y^{r'}$, resp. $X^{r''}$) is the highest power of Z (resp. Y, resp. X) dividing F^Q.

We set $F^Q = X^{r''} Y^{r'} Z^r F'$.

5) Prove that F' is a homogeneous polynomial of degree $2d - r - r' - r''$. Prove that $\mu_P(C') = d - r' - r''$ and establish similar formulae for P' and P''. Prove that $(F')' = F$, F' is irreducible and $C' = V(F')$.

We say that C is *in good position* if no exceptional line is tangent to C at a fundamental point.

6) Assume that C is in good position. Prove that C' is also in good position. (Consider $\mu_{P'}(F', Z)$.)

7) Prove that if C is in good position and $P_1, \ldots, P_s$ are the non-fundamental points in $C' \cap L$, then

$$\mu_{P_i}(C') \leqslant \mu_{P_i}(C', L) \qquad \text{and} \qquad \sum_{i=1}^{s} \mu_{P_i}(C', L) = r.$$

We say that C is in *excellent position* if C is in good position, L meets C (transversally) in d distinct non-fundamental points and L' and L'' both meet C (transversally) in $d - r$ non-fundamental points.
In questions 8), 9) and 10) we assume that C is in excellent position. We denote the non-fundamental points of $C' \cap L$ by $P_1, \ldots, P_s$.

8) Prove that the singular points of C' are the following:
   a) The points in $C' \cap U$ whose image under Q is a singular point of $C \cap U$. Show that these are of the same kind (*i.e.*, ordinary or otherwise) and have the same multiplicity in C and C' (*cf.* Fulton, Problem 3.24).
   b) The points P, P', P'', which are *ordinary* singular points of C' with multiplicities d, $d - r$ and $d - r$ respectively.
   c) Possibly some of the points P_i.

9) Prove that $C' \cap L'$ and $C' \cap L''$ contain no non-fundamental points.

10) Prove that we have the following formula for expected genuses:

$$g^*(C') = g^*(C) - \sum_{i=1}^{s} r_i(r_i - 1)/2, \qquad \text{where} \quad r_i = \mu_{P_i}(C').$$

11) Let $C = V(F)$ be an arbitrary irreducible curve in $\mathbf{P}^2$ and let A be a point of C. Prove that there is a homography h of $\mathbf{P}^2$ such that $h(C)$ is in excellent position and $h(A) = P$. (Prove that if P is a point of C of multiplicity r, there are an infinite number of lines passing through P meeting C in $d - r$ distinct points (*cf.* Fulton, Problem 5.26): at this point we need to use the fact we are in characteristic zero.)

We call the composition Qh of a standard quadratic transformation and a homography h a *quadratic transformation*. This is a birational transformation of $\mathbf{P}^2$.

12) Prove Theorem 1. (Use quadratic transformations and proceed by induction on $N + g^*(C)$, where N is the number of non-ordinary singular points of C.)

13) ¶ Prove that the curve of equation $(X^2 - YZ)^2 + Y^3(Y - Z)$ is rational.

Problem IX

The aim of this problem is to prove certain results quoted in Chapter X.

1 The snake lemma

This is an algebraic lemma which is extremely useful in many diagram chases.

We assume given a commutative diagram

$$
\begin{array}{ccccccccc}
0 & \to & A' & \xrightarrow{i} & A & \xrightarrow{p} & A'' & \to & 0 \\
& & \downarrow{u'} & & \downarrow{u} & & \downarrow{u''} & & \\
0 & \to & B' & \xrightarrow{j} & B & \xrightarrow{q} & B'' & \to & 0
\end{array}
$$

where the objects are commutative groups, the maps are group homomorphisms and the two horizontal rows are exact.

Prove there is an exact sequence

$$ 0 \longrightarrow \operatorname{Ker} u' \longrightarrow \operatorname{Ker} u \longrightarrow \operatorname{Ker} u'' \longrightarrow \operatorname{Coker} u' \longrightarrow \operatorname{Coker} u \longrightarrow \operatorname{Coker} u'' \longrightarrow 0. $$

(Start by defining the maps: the only problematic one is the map linking $\operatorname{Ker} u''$ and $\operatorname{Coker} u'$.)

Give a version of this lemma for exact sequences with more than three terms.

(Of course, this statement also holds for exact sequence of A-modules, $\mathcal{O}_X$-modules, etc.)

2 Projective dimension of modules over polynomial rings

In this section our aim is to prove by induction on $n + 1$, the number of variables, Proposition X, 1.6—or, more precisely, to prove the following proposition.

Proposition. *Set $R = k[X_0, X_1, \ldots, X_n]$ and let M be a graded R-module of finite type. We assume given an exact sequence of graded R-modules (with degree zero homomorphisms):*

$$0 \longrightarrow E_{n+1} \longrightarrow L_n \longrightarrow \cdots \longrightarrow L_i \xrightarrow{u_i} L_{i-1} \longrightarrow \cdots$$

$$\longrightarrow L_1 \xrightarrow{u_1} L_0 \xrightarrow{u_0} M \longrightarrow 0.$$

If the modules L_i are free graded R-modules, then so is E_{n+1}.

1) We start by proving a graded analogue of Nakayama's lemma. Let R be a graded ring, M a graded R-module of finite type, N a graded submodule of M and $f \in R$ a homogeneous element of degree $d > 0$. Assume $M = N + fM$. Prove that $M = N$.

2) Prove that Proposition 1 is true if $n + 1 = 0$.

3) Assume the proposition holds for n variables: we now prove it for $n+1$ variables. We set $E_{i+1} = \operatorname{Ker} u_i$.

   a) We recall that an R-module F is said to be torsion free if the equation $ax = 0$ for $a \in R$ and $x \in F$ can only be satisfied if a or x is zero. Prove that for $i = 0, \ldots, n$ the R-modules L_i and E_{i+1} are torsion free.

   b) Let F be a torsion free R-module. Prove that multiplication by X_n induces an exact sequence of graded modules

   $$0 \longrightarrow F(-1) \xrightarrow{\cdot X_n} F \longrightarrow F/X_n F \longrightarrow 0.$$

   c) Set $\overline{R} = R/(X_n) = k[X_0, \ldots, X_{n-1}]$. If F is a graded R-module, then set $\overline{F} = F/X_n F$. Prove that $\overline{F}$ is a graded $\overline{R}$-module and that the $\overline{R}$-modules $\overline{L}_i$ are free.

   d) Using the snake lemma, prove there is an exact sequence of graded $\overline{R}$-modules

   $$0 \longrightarrow \overline{E}_{n+1} \longrightarrow \overline{L}_n \longrightarrow \cdots \longrightarrow \overline{L}_i \xrightarrow{\overline{u}_i} \overline{L}_{i-1} \longrightarrow \cdots$$

   $$\longrightarrow \overline{L}_1 \xrightarrow{\overline{u}_1} \overline{E}_1 \longrightarrow 0.$$

   e) Deduce that $\overline{E}_{n+1}$ is a free graded $\overline{R}$-module.
   We consider a basis $\overline{e}_1, \ldots, \overline{e}_r$ of $\overline{E}_{n+1}$ over $\overline{R}$, where $\overline{e}_i$ is the image of $e_i \in E_{n+1}$ which is homogeneous of degree d_i.

   f) Prove that the elements e_i generate E_{n+1} over R (use Nakayama's lemma).

   g) Prove that the elements e_i are independent over R. (Argue by contradiction using an equation of minimal degree linking the elements e_i.)

3 Minimal resolutions

We now return to Proposition 1.8 in Chapter X.

Let R be a graded Noetherian ring, $R = \bigoplus_{i \in \mathbb{N}} R_i$. We assume $R_0 = k$ is a field and we set $m = R^+ = \bigoplus_{i > 0} R_i$. This is a maximal ideal of R whose quotient is isomorphic to k. (The classical example of such a ring is $k[X_0, \ldots, X_n]$. The reader may restrict him or herself to this case if he or she wishes.)

Let M be a graded R-module of finite type. A *minimal cover* of M is a homogeneous degree zero map $\varphi : L_0 \to M$ such that L_0 is free and graded, φ is surjective and the induced homomorphism

$$\overline{\varphi} = \varphi \otimes_R k : L_0 \otimes_R k = L_0/mL_0 \longrightarrow M \otimes_R k = M/mM$$

is an isomorphism.

0) Let $\varphi : L_0 \to M$ be a minimal cover. Prove that the rank of L_0 is the minimal number of generators of M.

1) Given an R-module, M, prove that there exists a minimal cover $\varphi : L_0 \to M$. (Lift a k-basis of $M \otimes_R k$.)

2) Let $\varphi : L_0 \to M$ be a minimal cover and let $\psi : L \to M$ be a surjective homomorphism such that L is free. Prove that φ is a "direct summand" of ψ, i.e., that there is a direct sum decomposition $L = L_0' \oplus L_0''$ with L_0' and L_0'' free and an isomorphism $\theta : L_0 \simeq L_0'$ such that $\varphi = (\psi|_{L_0'})\theta$. Deduce that if φ and ψ are two minimal covers of M, then there is an isomorphism $\theta : L_0 \to L$ such that $\varphi = \psi\theta$.

3) Let $L_1 \xrightarrow{u_1} L_0 \xrightarrow{u_0} M \to 0$ be an exact sequence of graded R-modules such that the modules L_i are free. Prove that u_0 is a minimal cover if and only if u_1 is minimal in the sense of definition 1.7 of Chapter X. (We note that this is equivalent to saying that $u_1 \otimes_R k$ vanishes.)

4) We consider a free resolution $L^{\bullet}$ of a graded module M:

$$0 \longrightarrow L_n \longrightarrow \cdots \longrightarrow L_i \xrightarrow{u_i} L_{i-1} \longrightarrow \cdots \longrightarrow L_1 \xrightarrow{u_1} L_0 \xrightarrow{u_0} M \longrightarrow 0$$

and we set $E_{i+1} = \operatorname{Ker} u_i$. Prove that this resolution is minimal if and only if for all $i = 1, \ldots, n$ the natural map $L_i \to E_i$ induced by u_i is a minimal cover of E_i. Prove that this establishes the existence of minimal resolutions.

5) Let $L^{\bullet}$ and $L'^{\bullet}$ be two minimal resolutions of M. Prove that these resolutions are isomorphic or, more precisely, that there exist isomorphisms $\theta_i : L_i \to L_i'$ such that $\theta_{i-1} u_i = u_i' \theta_i$.

6) Assume that M is of finite projective dimension. Prove that $\operatorname{dp}(M)$ is the length of any minimal resolution of M.

Midterm, December 1991

The two parts of the exam paper are independent.

Problem 1

Let k be an algebraically closed field and let p, q, r be integers $\geqslant 1$. We denote by $\mathbf{M}_{p,q}(k)$ (or $\mathbf{M}_{p,q}$ for short) the space of $p \times q$ matrices (*i.e.*, matrices with p lines and q columns) with coefficients in k. This is an affine space of dimension pq which we equip with its affine algebraic variety structure, particularly its Zariski topology. We recall that the rank function is then a lower semi-continuous function of $\mathbf{M}_{p,q}$:

in other words, the set of matrices of rank $\geqslant n$ is open. In particular, the set $\mathrm{GL}_p(k)$ of $p \times p$ invertible matrices is an open subset of $\mathbf{M}_{p,p}(k)$. We recall also that the closed set K_n of matrices of rank $\leqslant n$ is irreducible.

A $p \times q$ matrix is said to be of maximal rank if it is of rank $\inf(p, q)$.

It will occasionally be convenient to identify a linear map from k^q to k^p and its $(p \times q)$ matrix with respect to the canonical bases.

We consider the set

$$C_{p,q,r} = \{(A, B) \in \mathbf{M}_{p,q}(k) \times \mathbf{M}_{q,r}(k) \mid AB = 0\}.$$

This set is an affine algebraic variety, and the aim of this problem is to study its properties: irreducible components, dimension and singular points.

We denote by π_1 (resp. π_2) the projection from $C_{p,q,r}$ onto $\mathbf{M}_{p,q}$ (resp. $\mathbf{M}_{q,r}$).

1) Prove that if $(A, B) \in C_{p,q,r}$, then $\mathrm{rank}(A) + \mathrm{rank}(B) \leqslant q$.

2) Prove that any irreducible component of $C_{p,q,r}$ is of dimension $\geqslant pq + qr - pr$. Determine the fibre of π_1 (resp. π_2) at the point A (resp. B) and calculate its dimension as a function of the rank of A (resp. B).

3) Assume $q < p + r$. Prove that $C_{p,q,r}$ is reducible. (Consider the inverse images under π_1 and π_2 of the open sets of $\mathbf{M}_{p,q}$ and $\mathbf{M}_{q,r}$ consisting of matrices of maximal rank.)

4) Assume $q \geqslant p + r$. Let Ω be the open set of $\mathbf{M}_{p,q}$ consisting of matrices of rank p and let U be the open set (contained in Ω) of matrices written in block form $A = (A_1 A_2)$, where $A_1 \in \mathrm{GL}_p(k)$ and $A_2 \in \mathbf{M}_{p,q-p}$.

a) Determine the inverse image $\pi_1^{-1}(U)$ by writing the matrices B in the form

$$B = \begin{pmatrix} B_1 \\ B_2 \end{pmatrix},$$

where $B_1 \in \mathbf{M}_{p,r}$ and $B_2 \in \mathbf{M}_{q-p,r}$. Prove that $\pi_1^{-1}(U)$ is isomorphic to an open subset of $\mathbf{M}_{p,q} \times \mathbf{M}_{q-p,r}$. Deduce that $\pi_1^{-1}(U)$ is irreducible and then that $\pi_1^{-1}(\Omega)$ is irreducible. (Start by proving that a finite union of irreducible open sets whose intersection is non-empty is irreducible.)

b) Prove that $\pi_1^{-1}(\Omega)$ is dense in $C_{p,q,r}$. (Prove that for any $(A, B) \in C_{p,q,r}$ there is an $A_0 \in \Omega$ such that $(A_0, B) \in C_{p,q,r}$ and work in the fibre of π_2 over B.)

c) Prove that $C_{p,q,r}$ is irreducible and calculate its dimension.

d) Consider $(A, B) \in C_{p,q,r}$. Assume A is of rank p and B is of rank r. Prove that (A, B) is a smooth point of $C_{p,q,r}$. (Determine the tangent space at this point.)

5) We now again assume $q < p + r$. Let (m, n) be integers such that $0 \leqslant m \leqslant p$, $0 \leqslant n \leqslant r$, $m + n = q$. We set:

$$F_{m,n} = \{(A, B) \in C_{p,q,r} \mid \mathrm{rank}(A) = m, \ \mathrm{rank}(B) = n\},$$
$$G_{m,n} = \{(A, B) \subset C_{p,q,r} \mid \mathrm{rank}(A) \leqslant m, \ \mathrm{rank}(B) \leqslant n\}.$$

a) Prove that $F_{m,n}$ is a non-empty open set of $C_{p,q,r}$.

b) ¶ Using an argument similar to that given in 4), prove that $F_{m,n}$ is irreducible. (Consider first those matrices A whose top left-hand $m \times m$ minor, A_1, is invertible and calculate the fibre of π_1 over A using the block form expression.)

c) Prove that the irreducible components of $C_{p,q,r}$ are the varieties $G_{m,n}$ and calculate their dimensions.

d) Prove that every point of $F_{m,n}$ is a smooth point of $C_{p,q,r}$. (Give the tangent space of $C_{p,q,r}$ at this point.)

Problem II

We recall that an integral domain A with fraction field K is said to be integrally closed if $\forall\, x \in K$, x integral over $A \Rightarrow x \in A$.

We recall further that a local Noetherian domain of dimension 1 is regular if and only if it is integrally closed.

0) Let A be an integrally closed ring and let S be a multiplicative subset of A. Prove that $S^{-1}A$ is integrally closed.

Let X and Y be two irreducible affine algebraic varieties defined over an algebraically closed base field k and let $\varphi : X \to Y$ be a dominant map. We set $A = \Gamma(Y)$ and $B = \Gamma(X)$, and we therefore have an injective map $\varphi^* : A \to B$. We identify A and its image under φ^*. We assume that B is integral over A and that A and B have the same fraction field (*i.e.*, φ is an integral birational map). We set $I = \{a \in A \mid aB \subset A\}$.

1) Prove that I is a non-zero ideal of A.

2) Set $V = Y - V(I)$. This is a non-empty open set in Y and we set $U = \varphi^{-1}(V)$. Prove that the restriction of φ to U is an isomorphism from U to V.

3) Let y be a point of Y, m_y the corresponding maximal ideal of A and $A_{m_y} = \mathcal{O}_{Y,y}$ the associated localised ring. Assume that A_{m_y} is integrally closed. Prove that $y \in V$.

4) Assume that B is integrally closed. Prove that, conversely, if y is contained in V, then the ring A_{m_y} is integrally closed.

5) Assume Y is a curve (*i.e.*, $\dim Y = 1$) and B is integrally closed. Prove that X is a smooth curve and φ is an isomorphism away from the singular points of Y.

Exam, January 1992

The aim of this problem is to study some properties of certain curves in $\mathbf{P}^3$ which are linked to graded resolutions of their ideals. The hardest questions are marked with one or more symbols ¶.

Notation

In what follows k is an algebraically closed field. We denote the ring of polynomials $k[X, Y, Z, T]$ by R and the vector space of degree n homogeneous polynomials by R_n. We denote by $R(d)$ the graded R-module which is equal to R with a shifted grading, $R(d)_n = R_{d+n}$. This is a free rank-1 R-module having the constant polynomial 1 as a basis: this polynomial has degree $-d$. If $\mathcal{F}$ is a coherent sheaf over $\mathbf{P}^3$, then we denote by $h^i(\mathcal{F})$ the dimension of the k-vector space $H^i(\mathbf{P}^3, \mathcal{F})$. We denote by $(F_1, \ldots, F_n)$ the ideal of R generated by the polynomials $F_1, \ldots, F_n$.

Let C be a curve in $\mathbf{P}^3$ (*i.e.*, a closed subvariety all of whose components are of dimension 1). We denote by $\mathcal{O}_C$ the structure sheaf of C and by I_C the homogeneous ideal of C, ($I_C = \{F \in R \mid \forall P \in C \quad F(P) = 0\}$). We denote by $\mathcal{J}_C$ the sheaf $\widetilde{I_C}$ associated to I_C, d the degree of C and g its arithmetic genus. We recall there is an exact sequence

$$0 \longrightarrow \mathcal{J}_C \longrightarrow \mathcal{O}_{\mathbf{P}^3} \longrightarrow \mathcal{O}_C \longrightarrow 0.$$

1 Cohomological study of ACM curves

We now assume that C is a curve whose ideal has a resolution of the following form:

$$(*) \qquad\qquad 0 \longrightarrow E \overset{\varphi}{\longrightarrow} F \overset{p}{\longrightarrow} I_C \longrightarrow 0,$$

where E and F are free graded R-modules, $E = \bigoplus_{j=1}^s R(-m_j)$, $F = \bigoplus_{i=1}^r R(-n_i)$. Here, the integers m_j and n_i are such that $0 < n_1 \leqslant \cdots \leqslant n_r$ and $0 < m_1 \leqslant \cdots \leqslant m_s$, and φ and p are R-linear graded maps of degree zero (*i.e.*, sending an element of degree n to an element of degree n). We denote the elements of the canonical bases of E and F by e_j ($j = 1, \ldots, s$) and ε_i ($i = 1, \ldots, r$) respectively.

Such a curve will be called an ACM (arithmetically Cohen-Macaulay) throughout the following discussion.

We also have an associated resolution of sheaves

$$(**) \qquad\qquad 0 \longrightarrow \mathcal{E} \overset{\widetilde{\varphi}}{\longrightarrow} \mathcal{F} \overset{\widetilde{p}}{\longrightarrow} \mathcal{J}_C \longrightarrow 0,$$

where $\mathcal{E} = \bigoplus_{j=1}^s \mathcal{O}_{\mathbf{P}^3}(-m_j)$, and $\mathcal{F} = \bigoplus_{i=1}^r \mathcal{O}_{\mathbf{P}^3}(-n_i)$.

0) Prove that the Euler-Poincaré characteristic of $\mathcal{O}_{\mathbf{P}^3}(n)$ is a polynomial function of n to be determined.

1) Prove that $h^1 \mathcal{J}_C(n) = 0$ for all $n \in \mathbf{Z}$. Deduce that the curve C is connected.

2) Calculate the Euler-Poincaré characteristic $\chi(\mathcal{J}_C(n))$ as a function of d and g, then as a function of the integers n_i and m_j. Prove that $s = r-1$ and $\sum_{j=1}^s m_j = \sum_{i=1}^r n_i$; calculate d and g as a function of the integers n_i and m_j.

3) Prove that φ is given by an $r \times (r-1)$ matrix whose φ_{ij} term is a homogeneous polynomial whose degree is to be determined. What is φ_{ij} when $m_j < n_i$? We say that the resolution $(*)$ is *minimal* if φ_{ij} is zero whenever $n_i = m_j$.

4) ¶ Prove that if the resolution $(*)$ is not minimal, then there is a resolution

$$0 \longrightarrow E' \longrightarrow F' \longrightarrow I_C \longrightarrow 0$$

of the same form with $r' < r$. (Hint: there is a term $n_i = m_j = n$ such that $\varphi_{ij} = \lambda \neq 0$. Perform a change of basis on E to obtain $E = R(-n) \oplus E'$, $F = R(-n) \oplus F'$ in such a way that φ is in block form

$$\varphi = \begin{pmatrix} \lambda & 0 \\ u & \varphi' \end{pmatrix}.)$$

5) Assume that the resolution $(*)$ is minimal. Prove that $n_1 < m_1$. Let s_0 be the smallest degree of any surface containing C.

$$s_0 = \inf\{n \in \mathbf{N} \mid h^0 \mathcal{J}_C(n) > 0\}.$$

Calculate s_0 as a function of the integers n_i and the m_j.

6) Assume the resolution $(*)$ is minimal.
   a) ¶ Prove that $n_r < m_{r-1}$.
   b) Let e by the speciality index of C, i.e.,

   $$e = \sup\{n \in \mathbf{N} \mid h^1 \mathcal{O}_C(n) > 0\}.$$

   Calculate e as a function of the integers n_i and the m_j. Prove there is no *smooth* ACM curve such that the integers n_i are $(2, 2, 4)$ and the integers m_j are $(3, 5)$.

7) Calculate d, g, s_0, e when $r = 4$ and the integers n_i (resp. the integers m_j) are all equal to 3 (resp. 4).

2 Curves linked to complete intersections

Let C and Γ be two curves without common components. Assume $W = \Gamma \cup C$ is a scheme-theoretic complete intersection, which means that $I_W = (F, G)$ (and hence in particular, $W = V(F, G)$) for two homogeneous polynomials F and G of degrees s and t respectively). We then say that C and Γ are *linked* by the surfaces whose equations are F and G.

1) Prove that F and G are non-zero, non-constant and have no common non-constant common factors. Prove that $I_W = I_\Gamma \cap I_C$.

2) Prove that $I_C = \{U \in R \mid \forall K \in I_\Gamma, UK \in I_W\}$ and that the analogous equation in which the roles of C and Γ are interchanged also holds.

   We now assume that Γ is also a scheme-theoretic complete intersection: in other words, $I_\Gamma = (A, B)$, where A and B are two homogeneous polynomials of degrees a and b respectively. We note that A and B are non-zero, non-constant and have no common non-constant factors. We set $F = F'A + F''B$ and $G = G'A + G''B$ and $H = F'G'' - F''G'$. Give the degrees of the homogeneous polynomials F', F'', G', G'', H.

3) Our aim is to prove that $I_C = (F, G, H)$.
   a) Prove that $(F, G, H) \subset I_C$.

b) ¶ Let U be a polynomial such that $UA = \alpha F + \beta G$ and $UB = \gamma F + \delta G$. Prove there are polynomials φ and ψ such that

$$\beta F' + \delta F'' = \varphi F \qquad \text{and} \qquad U - \alpha F' - \gamma F'' = \varphi G,$$
$$\alpha G' + \gamma G'' = \psi G \qquad \text{and} \qquad U - \beta G' - \delta G'' = \psi F.$$

Prove that $U \in (F, G, H)$. (Start with the case where F' and F'' have no common factors.) Complete the proof of the proposition.

c) Prove that $C = V(F, G, H)$.

4) Our aim is to prove there is a resolution of I_C of the following form:

$$0 \longrightarrow R(-s - t + a) \oplus R(-s - t + b) \xrightarrow{\varphi}$$
$$\xrightarrow{\varphi} R(-s - t + a + b) \oplus R(-t) \oplus R(-s) \xrightarrow{p} I_C \longrightarrow 0.$$

a) Determine the map p and prove that it is surjective.

b) Give a reasonable guess for φ (our aim is to find all the relations between H, G and F; they have already appeared above). Check that φ is injective.

c) ¶ Let a, β, γ be polynomials such that $\alpha F + \beta G + \gamma H = 0$. Prove that $\gamma \in (A, B)$, $\gamma = \gamma_1 A + \gamma_2 B$. Expressing the fact that γH is contained in the ideal (F, G) in two different ways, prove that α is contained in (G', G'') and β is contained in (F', F'') and complete the proof of the proposition.

5) ¶¶ Assume k is of characteristic zero. Let Γ be the line whose equations are $A = X$ and $B = Y$ and let $s = t$ be an integer > 1. Using the notations above, prove that for a suitable choice of polynomials F', F'', G', G'' of degree $s - 1$, the curve $C = V(F, G, H)$ is irreducible and smooth. (Use the generic smoothness theorem.)

3 An example

1) Let C_0 be the plane curve of equation $Y^4 + X^3 T + XY T^2$. Prove that C_0 is irreducible. Determine its singular points and its geometric genus. (Assume that the characteristic of k is not 5. ¶ What happens when the characteristic of k is 5?)

2) Set $C = V(F, G, H)$, the subvariety of $\mathbf{P}^3$ defined by the equations $F = XT - YZ$, $G = X^2 Z + Y^3 + YZT$ and $H = XZ^2 + Y^2 T + ZT^2$. Prove that C is an irreducible curve (use the projection onto the (x, z, t) plane, for example). Using the projection π onto the (x, y, t) plane, prove that C and C_0 are birationally equivalent. (Give the morphisms and their domain of definition carefully.)

3) Assume the characteristic of k is not 5. Prove that C is a smooth curve. (You may use π to shorten the calculations.) Prove that $V(F, G) = C \cup \Gamma$, where Γ is the line $V(X, Y)$. Prove (¶¶) that C and Γ are linked by the surfaces of equations F and G. Deduce the ideal I_C and prove that C is ACM. Calculate the invariants d, g, s_0 and e. (You may either use the results from Parts I and II or prove this result directly.)

Exam, June 1992

The two parts of the exam paper are independent.

Problem I

We work over an algebraically closed base field k.

Recall that a variety X is said to be separated if the diagonal

$$\Delta = \{(x,y) \in X \times X \mid x = y\}$$

is closed in the product $X \times X$. We also recall that every projective variety is separated and complete: for any variety Y the second projection $p : X \times Y \to Y$ is closed, *i.e.*, sends closed sets to closed sets.

1) Let X, Y, Z be three varieties. We assume that X and Y are irreducible, that X is projective and that Z is separated. We denote by π, p and q the projections $\pi : X \times Y \times Z \to X$, $p : X \times Y \times Z \to Y \times Z$ and $q : Y \times Z \to Y$.

Let $\varphi : X \times Y \to Z$ be a morphism. We assume there is a point y_0 in Y such that $\varphi(X \times \{y_0\})$ is a point. Our aim is to prove that for any $y \in Y$, $\varphi(X \times \{y\})$ is a point.

Let Γ be the graph of φ,

$$\Gamma = \{(x,y,z) \in X \times Y \times Z \mid z = \varphi(x,y)\}.$$

    a) Prove that Γ is closed in $X \times Y \times Z$, $\Gamma' = p(\Gamma)$ is closed in $Y \times Z$ and Γ' is irreducible.

    b) Prove that $\dim \Gamma' = \dim Y$ (use the projection q).

    c) Complete the proof of this result. (Consider, for any $x_0 \in X$, the subvariety $\Gamma_{x_0} = \pi^{-1}(\{x_0\}) \cap \Gamma$ in Γ and its image under p.) Is this result still true if X is not assumed to be projective?

2) Let G be an algebraic group (*i.e.*, an algebraic variety with a group structure such that multiplication $\mu : G \times G \to G$ and inverse $s : G \to G$ given respectively by $\mu(x,y) = xy$ and $\sigma(x) = x^{-1}$ are morphisms).

    a) Assume G is an irreducible projective variety. Prove that G is a commutative group. (Use the map $\varphi : G \times G \to G$ given by $\varphi(g,h) = g^{-1}hg$.) Is this result still valid if G is not assumed to be projective? If G is not assumed to be irreducible?

    b) Let G and H be two irreducible projective algebraic groups and let $\varphi : G \to H$ be a morphism of varieties. Prove there is an element $a \in H$ and a morphism $\psi : G \to H$ which is both a morphism of varieties and a morphism of groups such that for all $g \in G$, $\varphi(g) = a\psi(g)$.

Problem II

We work over an algebraically closed base field k. If X is a variety, we denote its structure sheaf by $\mathcal{O}_X$.

1) Let F be a finite subset of $\mathbf{P}^2$. We equip F with its natural algebraic subvariety of $\mathbf{P}^2$ structure and we denote by $\mathcal{J}_F$ the sheaf of ideals of $\mathcal{O}_{\mathbf{P}^2}$ defining F. Prove there is a line Δ which does not meet F. We denote the equation of this line by δ. Prove that multiplication by δ gives us an exact sequence of sheaves

$$0 \longrightarrow \mathcal{J}_F(-1) \xrightarrow{\ \cdot\delta\ } \mathcal{J}_F \longrightarrow \mathcal{O}_\Delta \longrightarrow 0.$$

(Work in standard open affine sets.)
Deduce that if $H^1(\mathbf{P}^2, \mathcal{J}_F(n)) = 0$ for some $n \geqslant -2$, then

$$H^1(\mathbf{P}^2, \mathcal{J}_F(k)) = 0$$

for all $k \geqslant n$.

Throughout the following problem, we work in projective space $\mathbf{P}^3$.

2) Let C be an irreducible smooth curve of degree d in $\mathbf{P}^3$ and let H be a plane of equation h. We assume that $C \cap H$ is finite and consists of d distinct points. We denote the sheaf of ideals defining C in $\mathbf{P}^3$ by $\mathcal{J}_C$ and that defining $C \cap H$ in H by $\mathcal{J}_{C\cap H, H}$. Prove that multiplication by h gives us an exact sequence of sheaves

$$0 \longrightarrow \mathcal{J}_C(-1) \xrightarrow{\ \cdot h\ } \mathcal{J}_C \longrightarrow \mathcal{J}_{C\cap H, H} \longrightarrow 0.$$

Assume that for some integer $n \geqslant -2$

$$H^1(\mathbf{P}^3, \mathcal{J}_C(n)) = H^1(C, \mathcal{O}_C(n-1)) = 0.$$

Prove that, for all $k \geqslant n$, $H^1(\mathbf{P}^3, \mathcal{J}_C(k)) = 0$.
Let $\varphi : \mathbf{P}^1 \to \mathbf{P}^3$ be the map given by

$$\varphi(u,v) = (u^4, u^3 v, uv^3, v^4).$$

3) a) Prove that φ is an isomorphism from $\mathbf{P}^1$ to an irreducible smooth projective curve C in $\mathbf{P}^3$. Determine the degree and genus of C.

   We denote the ideal sheaf defining C in $\mathcal{O}_{\mathbf{P}^3}$ by $\mathcal{J}_C$.
   b) Prove that $H^1(C, \mathcal{O}_C(n)) = 0$ whenever $n \geqslant 0$. Calculate the dimensions of the spaces $H^1(C, \mathcal{O}_C(n))$ for all $n \in \mathbf{Z}$.
   c) Determine explicitly the spaces

$$H^0(\mathbf{P}^3, \mathcal{J}_C(1)) \text{ and } H^0(\mathbf{P}^3, \mathcal{J}_C(2)).$$

   Calculate the dimensions of $H^1(\mathbf{P}^3, \mathcal{J}_C(n))$ for $n \in \mathbf{Z}$. (Start with $n = 1$ and 2 and use 2).)
   d) Calculate the dimensions of the spaces $H^i(\mathbf{P}^3, \mathcal{J}_C(n))$ for $i = 0, 1, 2, 3$ and $n \in \mathbf{Z}$.
   ¶ Determine explicitly the space $H^0(\mathbf{P}^3, \mathcal{J}_C(3))$.

Exam, January 1993

The aim of this problem is to prove a theorem of Castelnuovo's which gives an upper bound for the genus of a curve in $\mathbf{P}^3$ in terms of its degree.

Notations

Throughout the following problem, k is an algebraically closed field and $\mathbf{P}^N$ is projective space of dimension N over k. We denote by R the space of polynomials $k[X, Y, T]$. If $\mathcal{F}$ is a coherent sheaf on a closed subvariety Z of $\mathbf{P}^N$ and $j : Z \to \mathbf{P}^N$ is the canonical injection, then we identify $\mathcal{F}$ and its direct image $j_*(\mathcal{F})$. We know that this identification does not alter the cohomology of the sheaf in question. If $\mathcal{F}$ is a coherent sheaf over $\mathbf{P}^N$, we denote by $h^i(\mathcal{F})$ the dimension of the k-vector space $H^i(\mathbf{P}^N, \mathcal{F})$.

If X is a closed subvariety of $\mathbf{P}^N$, we denote by $\mathcal{O}_X$ the structure sheaf of X, I_X the homogeneous ideal of X and $\mathcal{J}_X$ the ideal sheaf in $\mathcal{O}_{\mathbf{P}^N}$ which defines X. On identifying $\mathcal{O}_X$ and its direct image we have an exact sequence

$$0 \longrightarrow \mathcal{J}_X \longrightarrow \mathcal{O}_{\mathbf{P}^N} \longrightarrow \mathcal{O}_X \longrightarrow 0.$$

For all $n \in \mathbf{Z}$ there are also analogous exact sequences

$$0 \longrightarrow \mathcal{J}_X(n) \longrightarrow \mathcal{O}_{\mathbf{P}^N}(n) \longrightarrow \mathcal{O}_X(n) \longrightarrow 0,$$

obtained by shifting.

The cardinal of a finite set A is denoted by $|A|$. The integral part of a real number $x > 0$ is denoted by $[x]$.

1 Cohomological study of finite sets in $\mathbf{P}^2$

Let Z be a finite set of points in $\mathbf{P}^2$ of cardinality $d > 0$. We equip Z with its natural algebraic variety structure. The structure sheaf $\mathcal{O}_Z$ of Z is then simply the sheaf of functions from Z to k. We recall that $d = h^0\mathcal{O}_Z$ and that $\mathcal{O}_Z$ is isomorphic to $\mathcal{O}_Z(n)$ for all n.

Let D be a line in $\mathbf{P}^2$ whose equation is δ. We set $Z' = Z \cap D$ and $Z'' = Z - Z'$ and we equip Z' and Z'' with their natural algebraic variety structures. We denote by $I_{Z'/D}$ the ideal of Z' in $R/(\delta)$ and by $\mathcal{J}_{Z'/D}$ the sheaf of ideals defining Z' in D.

0) Calculate $h^1 \mathcal{J}_Z(n)$ for $n \leqslant 0$.

1) Consider $n \in \mathbf{Z}$. Prove that multiplication by δ induces an exact sequence

$$0 \longrightarrow \mathcal{J}_{Z''}(n-1) \longrightarrow \mathcal{J}_Z(n) \longrightarrow \mathcal{J}_{Z'/D}(n) \longrightarrow 0.$$

(Start by showing there is an exact sequence of graded R-modules

$$0 \longrightarrow I_{Z''}(n-1) \xrightarrow{\;\cdot\,\delta\;} I_Z(n) \longrightarrow I_{Z'/D}(n).)$$

2) Assume $\mid Z' \mid = l \geqslant 0$. Prove that $h^1 \mathcal{J}_{Z'/D}(n)$ vanishes for $n \geqslant l - 1$. (Prove that $\mathcal{J}_{Z'/D}(n)$ is isomorphic to the sheaf $\mathcal{O}_{\mathbf{P}^1}(n-l)$, for example.)

3) Prove that the function $n \mapsto h^1 \mathcal{J}_Z(n)$ is decreasing in $\mathbf{N}$. (Apply 1) to a line which does not meet Z.)

4) Prove that, for $n \geqslant 0$, $h^1 \mathcal{J}_Z(n) \leqslant \sup(0, d - n - 1)$. (Argue by induction on d, applying 1) to a line D containing only one point of Z.)

5) Assume Z has no trisecant (*i.e.*, no three distinct points of Z are ever collinear). Prove that for $n \geqslant 0$, $h^1 \mathcal{J}_Z(n) \leqslant \sup(0, d - 2n - 1)$.

2 Existence of good plane sections

Let C be an irreducible smooth curve in $\mathbf{P}^3$ of degree d. We assume that C is not a plane curve. The aim of this section is to prove the following result.

There is a plane H such that $Z = C \cap H$ is a finite set of cardinal d without a trisecant (*cf.* Chapter I, 5).

We denote the vector space k^4 by E. We have $\mathbf{P}(E) = \mathbf{P}^3$. We consider the projective space $\mathbf{P}(E^*)$ associated to the dual space. The points of $\mathbf{P}(E^*)$ correspond to non-zero linear forms on E up to multiplication by a scalar or, alternatively, to planes in $\mathbf{P}^3$.

1) Assume that $Z = C \cap H$ contains d distinct points. Prove that Z is not contained in a line.

   You may quote the result that there is a non-empty open set Ω in $\mathbf{P}(E^*)$ such that any plane $H \in \Omega$ has the following two properties.
   a) $Z = C \cap H$ consists of d distinct points (*cf.* Exercise VIII, 1).
   b) There are two distinct points P, Q of $Z = C \cap H$ such that the line $\langle PQ \rangle$ is not a trisecant of Z.

2) We set $F = \{(P, Q, R) \in C \times C \times C \mid P, Q, R \text{ collinear}\}$. Prove that F is a closed set of $C \times C \times C$.

3) Let V be the open subvariety of $C \times C$ consisting of those points P, Q such that $P \neq Q$. Let V' be the subset of V corresponding to the trisecants of C:

$$ V' = \{(P, Q) \in V \mid |C \cap \langle PQ \rangle| \geqslant 3\}. $$

   Prove that if V' is not contained in a proper closed set of V, then it contains a non-empty open set of V. (Write V' as the projection of an open set in F.)

4) Prove that V' is contained in a proper closed subset of V. (If V' contains a non-empty open subset of V, then we get a contradiction by considering the closed subvariety of $V \times \Omega$

$$ W = \{(P, Q, H) \in V \times \Omega \mid P, Q \in H\} $$

   and its projections p and π to V and Ω respectively.)

5) We denote by M the set of planes of Ω whose intersection with C contains three collinear points. Prove using W that $\dim M \leqslant 2$. Complete the proof of the theorem.

3 The upper bound on the genus of a curve in $\mathbf{P}^3$

Let C be an irreducible smooth curve in $\mathbf{P}^3$ of degree d and genus g. We set

$$ e(C) = \sup\{n \in \mathbf{Z} \mid h^1 \mathcal{O}_C(n) \neq 0\}. $$

Let H be a plane not containing C and let h be its equation. We assume that $Z = C \cap H$ contains d distinct points. We identify H and $\mathbf{P}^2$ and we use the notations of the first section.

1) Prove that, for all n, $h^1 \mathcal{O}_C(n) = h^2 \mathcal{J}_C(n)$.

2) Prove that multiplication by h induces the following commutative diagram of exact sequences

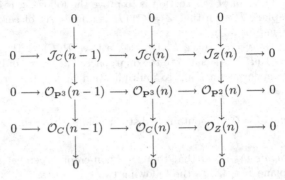

3) Prove that $h^1 \mathcal{O}_C(n-1) - h^1 \mathcal{O}_C(n) \leqslant h^1 \mathcal{J}_Z(n)$. Deduce that for all $n \in \mathbf{Z}$

$$h^1 \mathcal{O}_C(n) \leqslant \sum_{k \geqslant n+1} h^1 \mathcal{J}_Z(k).$$

4) Prove that $g \leqslant (d-1)(d-2)/2$. Prove that equality holds if and only if C is a plane curve.

5) Assume that C is not a plane curve. Prove that

$$g \leqslant \begin{cases} (m-1)^2, & \text{if } d = 2m; \\ m(m-1), & \text{if } d = 2m+1. \end{cases}$$

(This is Castelnuovo's theorem.)

6) Prove that $e(C) \leqslant d - 3$ (resp. $e(C) \leqslant [d/2] - 2$ if C is not a plane curve). For which curves is $e(C) = d - 3$?

4 An example

Let a, b be integers > 0 such that $a + 2 \leqslant b$. Let C be an irreducible smooth curve in $\mathbf{P}^3$ whose ideal sheaf has a resolution

$$0 \longrightarrow \mathcal{O}_{\mathbf{P}^3}(-b-2)^{b-a-1} \longrightarrow \mathcal{O}_{\mathbf{P}^3}(-b-1)^{2(b-a)}$$
$$\longrightarrow \mathcal{O}_{\mathbf{P}^3}(-2) \oplus \mathcal{O}_{\mathbf{P}^3}(-b)^{b-a+1} \longrightarrow \mathcal{J}_C \longrightarrow 0.$$

Calculate the degree and genus of C. Compare this calculation with the bounds given in 3, 5). Propose resolutions for curves for which these bounds may be sharp.

Exam, June 1993

The problem and the two exercises are independent.

Problem

We work over an algebraically closed base field k and denote by $\mathbf{P}^N$ the projective space of dimension N over k. We denote by R the ring of polynomials $k[X, Y, Z, T]$.

We recall the following result (*cf.* Exercise IV, 7): let $\varphi : X \to Y$ be a dominant morphism of projective varieties. If Y is irreducible and the fibres of φ are irreducible and of constant dimension, then X is irreducible.

1 Constructing the variety of lines in $\mathbf{P}^3$

Let $P = (x, y, z, t)$ and $P' = (x', y', z', t')$ be two distinct points of $\mathbf{P}^3$. We denote by l, m, n, l', m', n' the six 2×2 minors of the matrix

$$\mu(P, P') = \begin{pmatrix} x & y & z & t \\ x' & y' & z' & t' \end{pmatrix};$$

more precisely, we set

$$l = xy' - x'y, \; l' = zt' - z't, \; m = xz' - x'z,$$
$$m' = yt' - y't, \; n = xt' - x't, \; n' = yz' - y'z.$$

0) Prove that $ll' - mm' + nn' = 0$.

We consider projective space $\mathbf{P}^5$ with homogeneous coordinates l, m, n, l', m', n'. Let G be the subset of $\mathbf{P}^5$ defined by the equation

$$ll' - mm' + nn' = 0.$$

1) Prove that G is a smooth irreducible projective variety of $\mathbf{P}^5$ and give its dimension.

2) Let φ be the map associating to two distinct points P, P' of $\mathbf{P}^3$ the six minors l, m, n, l', m', n' of $\mu(P, P')$ defined as above. Let Δ be the diagonal in $\mathbf{P}^3 \times \mathbf{P}^3$ (*i.e.*, the set of pairs (P, P) for $P \in \mathbf{P}^3$). Prove that φ induces a morphism from $\mathbf{P}^3 \times \mathbf{P}^3 - \Delta$ to $\mathbf{P}^5$ whose image is contained in G. Prove that $\varphi(P, P')$ only depends on the line $D = \langle P, P' \rangle$ (if Q, Q' are two different distinct points of D, then compare $\mu(Q, Q')$ and $\mu(P, P')$). Prove that φ induces a bijection from the set of projective lines in $\mathbf{P}^3$ to G. (Consider the intersection points of lines and coordinate planes.)

Henceforth we will identify the set of projective lines in $\mathbf{P}^3$ and the variety G (called the *Grassmannian*). The coordinates l, m, n, l', m', n' are called the *Plücker coordinates* of the line $\langle P, P' \rangle$.

3) Study the fibres of the morphism φ and, in particular, determine their dimension.

2 The incidence variety and applications

Let d be a positive integer and let $R_d = H^0(\mathbf{P}^3, \mathcal{O}_{\mathbf{P}^3}(d))$ be the vector space of homogeneous polynomials of degree d in X, Y, Z, T. We identify the space of degree d surfaces in $\mathbf{P}^3$ with the projective space $\mathbf{P}(R_d)$.

0) What is the dimension of $\mathbf{P}(R_d)$?

We set $V_d = \{(D, F) \in G \times \mathbf{P}(R_d) \mid D \subset F\}$ and denote by π (resp. p) the projection of V_d onto $\mathbf{P}(R_d)$ (resp. onto G).

1) Let H be a plane of equation $\alpha X + \beta Y + \gamma Z + \delta T = 0$ and let D be a line which is not contained in H. Determine both the homogeneous coordinates of the point of intersection of D and H as a function of $\alpha, \beta, \gamma, \delta$ and the Plücker coordinates of D. What happens when $D \subset H$?

2) Prove that V_d is a closed subvariety of $G \times \mathbf{P}(R_d)$, called the *incidence variety* (use 1), for example, and let the plane H vary). Prove that the projections π and p are closed maps (*i.e.*, they send closed sets to closed sets).

3) Determine the fibres $p^{-1}(D)$ for any $D \in G$ (use Riemann-Roch). Deduce that p is surjective and V_d is irreducible. Calculate the dimension of V_d.

4) Assume $d \geqslant 4$. Prove that π is not dominant. Deduce that there is a non-empty open set in $\mathbf{P}(R_d)$ consisting of surfaces which do not contain any line. (We say that the "general" surface of degree $\geqslant 4$ does not contain a line.)

5) Assume $d = 3$. Prove there are only a finite number of lines in the surface $XYZ - T^3 = 0$. (You may either use a direct argument or use 1) and 2) above.) Deduce that π is surjective and that a general cubic surface contains a finite number of lines. Study the surface $XYZ + T(X^2 + Y^2 + Z^2) - T^3 = 0$. (Start by proving that the lines which do not meet the line $X = T = 0$ are defined by equations of the form $Y = aX + bT$, $Z = cX + dT$.)

6) What happens when $d = 1$?

7) We take $d = 2$. Determine the Plücker coordinates of the lines contained in the quadric of equation $XY - ZT = 0$. Prove they form a closed one-dimensional subset of G. Deduce that π is surjective. Are the fibres of π all of the same dimension?

Exercise 1

We work over an algebraically closed base field k.

Let n be an integer $\geqslant 2$. We consider the two polynomials $F = XY - ZT$ and $G = aZ^2 + bZ + c$, where a, b, c are homogeneous polynomials in X, Y, T of degrees $n - 2$, $n - 1$ and n respectively such that a is not a multiple of T. Let C be the subvariety of $\mathbf{P}_k^3$ given by the equations $F = 0$ and $G = 0$. We assume that C is an irreducible curve and $I(C) = (F, G)$.

1) Calculate the degree d and the arithmetic genus p_a of C.

2) Prove that C is birationally equivalent to the plane curve G whose equation is $aX^2Y^2 + bXYT + cT^2 = 0$. Deduce that the geometric genus g of C is such that $g < n(n-1)/2$. Prove that, if $n \geqslant 5$, C is not smooth.

3) Determine the geometric genus of the plane curve of equation $X^2Y^2 + X^2YT + T^4$.

Exercise 2

We work over an algebraically closed base field k.

Let n be an integer > 0, R the ring $k[X_0, \ldots, X_n]$ and $\mathbf{P}^n$ the projective space of dimension n over k. Let r be an integer $\geqslant 0$. We consider for all $i = 0, \ldots, r$ a homogeneous polynomial $F_i \in R$ of degree $d_i > 0$. Let

$$\varphi : \bigoplus_{i=0}^{r} R(-d_i) \longrightarrow R$$

be the R-linear map given by the formula

$$\varphi(G_0, \ldots, G_r) = \sum_{i=0}^{r} F_i G_i$$

and let

$$\psi : \bigoplus_{i=0}^{r} \mathcal{O}_{\mathbf{P}^n}(-d_i) \longrightarrow \mathcal{O}_{\mathbf{P}^n}$$

be the associated map of sheaves.

1) Prove that ψ is surjective if and only if the subvariety $V(F_0, \ldots, F_r)$ in $\mathbf{P}^n$ is empty. For what values of r is this possible?

We now assume $r = n$ and that ψ is surjective. Let $\mathcal{N}$ be the kernel of ψ.

2) Determine the dimension of $V(F_0, \ldots, F_i)$ for $i = 0, \ldots, n$.

3) Prove that for all i such that $2 \leqslant i \leqslant n - 1$ and all $d \in \mathbf{Z}$, $H^i(\mathbf{P}^n, \mathcal{N}(d)) = 0$. Calculate the dimension of $H^n(\mathbf{P}^n, \mathcal{N}(d))$.

4) Prove that the groups $H^1(\mathbf{P}^n, \mathcal{N}(d))$ are not all trivial and give the smallest value of d such that $H^1(\mathbf{P}^n, \mathcal{N}(d))$ is non-trivial.

5) Assume $F_i = X_i^2$ for all i. Calculate the dimension of $H^1(\mathbf{P}^n, \mathcal{N}(d))$ for all $d \in \mathbf{Z}$.

Exam, February 1994

Question 5 is not part of the exam. The symbol ¶ indicates a difficult question.

0 Revision and notations

Throughout the following problem we will work over an algebraically closed base field k. We denote the ring of polynomials $k[X, Y, Z, T]$ by R.

If p is an integer $\geqslant 1$ we denote the binomial coefficients by $\binom{n}{p}$: by convention, this coefficient vanishes if $n < p$.

We denote the vector space of $p \times q$ matrices with coefficients in k by $\mathbf{M}_{p,q}$. This space is naturally an affine variety isomorphic to affine space k^{pq}. We denote by C_s the subset of $\mathbf{M}_{p,q}$ consisting of matrices of rank $\leqslant s$. We recall that C_s is an irreducible closed set of codimension $(p - s)(q - s)$ in $\mathbf{M}_{p,q}$ (cf. Exercise IV, 3).

We recall the following result.

Intersection Theorem 1. *If X and Y are two irreducible affine algebraic varieties in k^n of dimensions r and s, then every irreducible component of $X \cap Y$ is of dimension $\geqslant r + s - n$ (cf. Chapter IV, 2.5).*

(NB: if $X \cap Y$ is empty this theorem gives us no information on their respective dimensions.)

Let r be an integer $\geqslant 2$ and let E and F be two free graded R-modules

$$E = \bigoplus_{j=1}^{r-1} R(-m_j), \qquad F = \bigoplus_{i=1}^{r} R(-n_i),$$

where the (possibly negative) integers m_j and n_i are such that $n_1 \leqslant \cdots \leqslant n_r$ and $m_1 \leqslant \cdots \leqslant m_{r-1}$. Moreover, we assume $m_j \neq n_i$ for all pairs i, j and

$$\sum_{j=1}^{r-1} m_j = \sum_{i=1}^{r} n_i.$$

Let $u : E \to F$ be a graded degree zero (*i.e.*, sending an element of degree n to an element of degree n) R-linear homomorphism. We denote the vectors of the canonical bases of E and F by e_j ($j = 1, \ldots, r-1$) and ε_i ($i = 1, \ldots, r$) respectively. The homomorphism u is given in these bases by an $r \times (r-1)$ matrix $A = (a_{ij})$ whose coefficient of index i, j is a homogeneous polynomial of degree $m_j - n_i$. In particular, this coefficient vanishes if $m_j < n_i$.

The aim of this problem is to give conditions on the integers m_j and n_i under which there exists an injective homomorphism $u : E \to F$ as above such that its cokernel is the ideal of a curve (resp. of a smooth curve) in $\mathbf{P}^3$.

We recall (*cf.* Chapter X, 2.7) that the following results hold:

1) u is injective if and only if all the $r - 1$ minors $\varphi_1, \ldots, \varphi_r$ of A are all non-zero.
2) If we assume the minors φ_i are all non-zero, the following are equivalent:
   i) The polynomials φ_i have no common factor.
   ii) Coker u is the saturated ideal of an ACM curve whose ideal is generated by the polynomials φ_i.

1 Preliminary results

1) Let n and p be integers $\geqslant 1$ and M an $n \times p$ matrix with coefficients in an arbitrary commutative ring S. We assume that M is written in block form

$$M = \begin{pmatrix} M_1 & M_2 \\ 0 & M_3 \end{pmatrix}$$

where M_1 is a $k \times l$ matrix, $1 \leqslant k \leqslant n$, $1 \leqslant l \leqslant p$. Let I (resp. I_1) be the ideal generated by the p-minors of M (resp. by the l-minors of M_1). If $p > n$ (resp. $l > k$), then by convention these minors are zero. Prove that $I \subset I_1$. (Reduce to the case $p = n$ and then argue by induction on n.)

2) Let D be the subspace of $\mathbf{M}_{r,r-1}$ formed of matrices $M = (\mu_{ij})$, $\mu_{ij} \in k$, such that $\mu_{ij} = 0$ for $i \geqslant j+2$ and $i \leqslant j-1$ (*i.e.*, matrices with only two non-zero diagonals: we will call them bidiagonal matrices).

   a) Prove that $D \cap C_{r-2}$ is non-empty and all its irreducible components are of codimension 2 in D.

   b) Let V be a subspace of $\mathbf{M}_{r,r-1}$ containing D. Prove that all the irreducible components of $V \cap C_{r-2}$ are of codimension 2 in V. (Use the intersection theorem applied to D and an irreducible component Z of $V \cap C_{r-2}$, noting that the zero matrix is in Z.)

2 Necessary conditions

Let $u : E \to F$ be a graded homomorphism.

1) Assume that u is injective. Prove that $m_j > n_j$ for all $j = 1, \ldots, r-1$. (Study the restriction of u to the submodule $\bigoplus_{k \leqslant j} R(-m_k)$.)

2) Assume that u is injective and the cokernel of u is the ideal of a curve in $\mathbf{P}^3$. This curve is connected (*cf.* Chapter X, 2.4) but not necessarily reduced. Prove that $\forall j = 1, \ldots, r-1$, $m_j > n_{j+1}$. (Use 1.1.) Deduce that $n_1 > 0$.

3) We assume in addition that C is smooth (and is hence an irreducible variety *cf.* Chapter V, 2.2). Prove that $\forall j = 1, \ldots, r-2$, $m_j > n_{j+2}$. (Use 1.1 to show that otherwise C would contain a curve other than C.)

3 Sufficient conditions, 1

We assume that $\forall j = 1, \ldots, r-1, m_j > n_{j+1}$.

Let H be the set of homogeneous degree 0 homomorphisms $u : E \to F$. We identify H with the set of corresponding matrices $A = (a_{ij})$.

1) Prove that H is a k-vector space and determine its dimension, which we will denote by N.

   We consider the map $\psi : H \times (k^4 - \{0\}) \to \mathbf{M}_{r,r-1}$ defined by $\psi(A, P) = A(P) = (a_{ij}(P))$ and we set $\psi_P(A) = \psi(A, P)$.

2) a) Prove that ψ is a morphism. Prove that for all $P \in k^4 - \{0\}$, ψ_P is a linear map from H to $\mathbf{M}_{r,r-1}$ whose image is the subspace

$$V = \{M = (\mu_{ij}) \in \mathbf{M}_{r,r-1} \mid m_j < n_i \implies \mu_{ij} = 0\}.$$

   We set $v = \dim V$. Determine the fibres of ψ_P and give their dimension.

   b) Prove that V is also the image of ψ and that V contains the space D of bidiagonal matrices.

3) Consider $M \in \mathbf{M}_{r,r-1}$. The fibre $\psi^{-1}(M)$ is a closed subset of $H \times (k^4 - \{0\})$, which we equip with its variety structure. Calculate the dimension of $\psi^{-1}(M)$.

   We set $W = \psi^{-1}(C_{r-2})$. This is a closed subset of $H \times (k^4 - \{0\})$, which we equip with its variety structure.

4) Prove that

$$W = \{(A, P) \in H \times (k^4 - \{0\}) \mid \varphi_1(P) = \cdots = \varphi_r(P) = 0\},$$

where the polynomials φ_i are the $(r-1)$-minors of the matrix A. Prove that W is a variety of dimension $N + 2$.

5) Prove there is a non-empty open set Ω in H such that for any $u \in \Omega$, u is injective and the cokernel of u is the ideal of an ACM curve in $\mathbf{P}^3$. (Study the fibres of the projection π_1 from W to H.)

Such a property, which holds for u in some non-empty open (and hence dense) set in H, will be said to hold for a "general" u.

4 The genus of ACM curves

The aim of this section is to prove that for any integer $g \geqslant 0$ there is a (not necessarily smooth) ACM curve of arithmetic genus g. Let C be an ACM curve with a resolution $0 \to E \to F \to I_C \to 0$, where E and F are as above. It follows that, for $j = 1, \ldots, r-1$, $m_j > n_{j+1}$. We recall that the arithmetic genus g of C is given by the formula

$$g = \sum_{j=1}^{r-1} \binom{m_j - 1}{3} - \sum_{i=1}^{r} \binom{n_i - 1}{3}.$$

1) Prove that

$$g = \Big(\sum_{j=1}^{r-1} \sum_{n_{j+1} \leqslant n < m_j} \binom{n-1}{2}\Big) - \binom{n_1 - 1}{3}.$$

(You may use the formula $\binom{n+3}{3} = \sum_{0 \leqslant k \leqslant n} \binom{k+2}{2}$.)

The integers of the form $\binom{n}{2} = n(n-1)/2$ for $n \geqslant 2$ are called triangular numbers. In the following problem, you may use the following result: any positive integer is the sum of at most three triangular numbers.[1]

2) Prove that $r \leqslant n_1 + 1$. Give the possible values of the integers n_i and the m_j when $n_1 = 1$ (resp. 2). Prove that the genus of C is then either zero or a triangular number (resp. the sum of at most two triangular numbers). Prove that by this method we can construct curves of all genuses which are sums of at most two triangular numbers.

3) Assume $n_1 = 3$. Prove that g is the sum of three triangular numbers. Study the converse and complete the proof of the theorem.

4) We set $n_1 = s$. Prove that

$$g \geqslant s \binom{s-1}{2} - \binom{s-1}{3}.$$

¶ Prove there is no smooth ACM curve of genus 8.

---

[1] This result is due to Gauss (*cf.* Serre, *Cours d'arithmétique*, Chapter IV) but had been previously mentioned by Fermat in the famous (and famously too small) margin of his copy of the works of Diophante.

5 Sufficient conditions, 2

The aim of this section is to prove the following theorem.

Theorem (Gruson-Peskine, 1976). *Assume that k is of characteristic 0. With the notations of Section 0 above the following conditions are equivalent.*

  i) $\forall j = 1, \ldots, r-2,\ m_j > n_{j+2},$
 ii) *If $u : E \to F$ is general enough, u is injective and its cokernel is the ideal of a smooth connected ACM curve.*

1) ¶ We use the above notation, particularly those in Section 3. However, we consider W with its scheme structure. Assume $m_1 > n_r$ (which implies Condition i). We set $S = \psi^{-1}(C_{r-3})$.

  a) Prove that $\dim S < N$ and $W - S$ is a smooth variety.
  b) Consider the restriction of the projection π_1 to the open set $W - S$. Applying the generic smoothness theorem (*cf.* Problem VI and the appendix on schemes), prove there is an open set of H over which the fibres of π_1 are smooth. Complete the proof of the theorem.

2) ¶¶ Prove the Gruson-Peskine theorem in general.

References

[BBM]   E. BALLICO, G. BOLONDI & J. MIGLIORE – The Lazarsfeld-Rao problem for liaison classes of two-codimensional subschemes of $\mathbf{P}^n$, *Amer. J. Math.* **113** (1991), p. 117–128.

[Bbki] N. BOURBAKI – *Elements of Mathematics. Commutative Algebra. Chapters 1–7*, Springer, Berlin, 1998.

[EGA]   A. GROTHENDIECK & J. DIEUDONNÉ – *Éléments de géométrie algébrique*, vol. 4, 8, 11, 17, 20, 24, 28, 32, Publications Mathématiques de l'IHÉS.

[F]   W. FULTON – *Algebraic Curves*, Benjamin, 1969.

[Go]   C. GODBILLON – *Topologie algébrique*, Hermann, Paris, 1971.

[G]   R. GODEMENT – *Topologie algébrique et théorie des faisceaux*, Hermann, Paris, 1958.

[Gr]   A. GRAMAIN – *Topology of surfaces*, BCS Associates, Moscow, ID, 1984, (French edition: 1971).

[GP]   L. GRUSON & C. PESKINE – Genre des courbes de l'espace projectif (II), *Ann. Sci. École Norm. Sup. (4)* **15** (1982), p. 401–418.

[H]   R. HARTSHORNE – *Algebraic Geometry*, Springer, Berlin, 1977.

[L]   S. LANG – *Algebra*, Addison-Wesley, Reading, MA, 1965.

[MDP]   M. MARTIN-DESCHAMPS & D. PERRIN – *Sur la classification des courbes gauches*, Astérisque, vol. 184-185, Société Mathématique de France, Paris, 1990.

[Ma]   H. MATSUMURA – *Commutative Algebra*, Benjamin, 1970.

[M]   D. MUMFORD – *The Red Book of Varieties and Schemes*, Lect. Notes in Math., vol. 1358, Springer, 1988.

[P]   D. PERRIN – *Cours d'algèbre*, Ellipses, Paris, 1996.

[Pes]   C. PESKINE – *An algebraic introduction to complex projective geometry. 1*, Cambridge Studies in Advanced Mathematics, vol. 47, Cambridge University Press, Cambridge, 1996.

[PS]   C. PESKINE & L. SZPIRO – Liaison des variétés algébriques, *Invent. Math.* **26** (1974), p. 271–302.

[Rao]   A.P. RAO – Liaison among curves in $\mathbf{P}^3$, *Invent. Math.* **50** (1979), p. 205–217.

[R]   W. RUDIN – *Real and complex analysis*, McGraw-Hill, New York, NY, 1974.

[S]   P. SAMUEL – *Algebraic theory of numbers*, Houghton Mifflin, Boston, 1970
           (French edition: 1967).
[Sh]   I.R. SHAFAREVICH – *Basic Algebraic Geometry*, Springer, 1977.
[Tohoku]   A. GROTHENDIECK – Sur quelques points d'algèbre homologique, *Tôhoku
           Math. J.* **9** (1957), p. 119–221.

Index of notations

Index

Universitext

Holmgren, R. A.: A First Course in Discrete Dynamical Systems

Howe, R., Tan, E. Ch.: Non-Abelian Harmonic Analysis

Howes, N. R.: Modern Analysis and Topology

Hsieh, P.-F.; Sibuya, Y. (Eds.): Basic Theory of Ordinary Differential Equations

Humi, M., Miller, W.: Second Course in Ordinary Differential Equations for Scientists and Engineers

Hurwitz, A.; Kritikos, N.: Lectures on Number Theory

Huybrechts, D.: Complex Geometry: An Introduction

Isaev, A.: Introduction to Mathematical Methods in Bioinformatics

Istas, J.: Mathematical Modeling for the Life Sciences

Iversen, B.: Cohomology of Sheaves

Jacod, J.; Protter, P.: Probability Essentials

Jennings, G. A.: Modern Geometry with Applications

Jones, A.; Morris, S. A.; Pearson, K. R.: Abstract Algebra and Famous Inpossibilities

Jost, J.: Compact Riemann Surfaces

Jost, J.: Dynamical Systems. Examples of Complex Behaviour

Jost, J.: Postmodern Analysis

Jost, J.: Riemannian Geometry and Geometric Analysis

Kac, V.; Cheung, P.: Quantum Calculus

Kannan, R.; Krueger, C. K.: Advanced Analysis on the Real Line

Kelly, P.; Matthews, G.: The Non-Euclidean Hyperbolic Plane

Kempf, G.: Complex Abelian Varieties and Theta Functions

Kitchens, B. P.: Symbolic Dynamics

Klenke, A.: Probability Theory: A comprehensive course

Kloeden, P.; Ombach, J.; Cyganowski, S.: From Elementary Probability to Stochastic Differential Equations with MAPLE

Kloeden, P. E.; Platen; E.; Schurz, H.: Numerical Solution of SDE Through Computer Experiments

Koralov, L. B.; Sinai, Ya. G.: Theory of Probability and Random Processes. 2nd edition

Kostrikin, A. I.: Introduction to Algebra

Krasnoselskii, M. A.; Pokrovskii, A. V.: Systems with Hysteresis

Kuo, H.-H.: Introduction to Stochastic Integration

Kurzweil, H.; Stellmacher, B.: The Theory of Finite Groups. An Introduction

Kyprianou, A.E.: Introductory Lectures on Fluctuations of Lévy Processes with Applications

Lang, S.: Introduction to Differentiable Manifolds

Lefebvre, M.: Applied Stochastic Processes

Lorenz, F.: Algebra I: Fields and Galois Theory

Luecking, D. H., Rubel, L. A.: Complex Analysis. A Functional Analysis Approach

Ma, Zhi-Ming; Roeckner, M.: Introduction to the Theory of (non-symmetric) Dirichlet Forms

Mac Lane, S.; Moerdijk, I.: Sheaves in Geometry and Logic

Marcus, D. A.: Number Fields

Martinez, A.: An Introduction to Semiclassical and Microlocal Analysis

Matoušek, J.: Using the Borsuk-Ulam Theorem

Matsuki, K.: Introduction to the Mori Program

Mazzola, G.; Milmeister G.; Weissman J.: Comprehensive Mathematics for Computer Scientists 1

Mazzola, G.; Milmeister G.; Weissman J.: Comprehensive Mathematics for Computer Scientists 2

Mc Carthy, P. J.: Introduction to Arithmetical Functions

McCrimmon, K.: A Taste of Jordan Algebras

Meyer, R. M.: Essential Mathematics for Applied Field